AF676225

Surface Displacement Measurement from Remote Sensing Images

SCIENCES

Image, Field Director – Laure Blanc-Feraud

Remote Sensing Imagery, Subject Head – Emmanuel Trouvé and Avik Bhattacharya

Surface Displacement Measurement from Remote Sensing Images

Coordinated by
Olivier Cavalié
Emmanuel Trouvé

ISTE

WILEY

First published 2022 in Great Britain and the United States by ISTE Ltd and John Wiley & Sons, Inc.

Apart from any fair dealing for the purposes of research or private study, or criticism or review, as permitted under the Copyright, Designs and Patents Act 1988, this publication may only be reproduced, stored or transmitted, in any form or by any means, with the prior permission in writing of the publishers, or in the case of reprographic reproduction in accordance with the terms and licenses issued by the CLA. Enquiries concerning reproduction outside these terms should be sent to the publishers at the undermentioned address:

ISTE Ltd
27-37 St George's Road
London SW19 4EU
UK

www.iste.co.uk

John Wiley & Sons, Inc.
111 River Street
Hoboken, NJ 07030
USA

www.wiley.com

© ISTE Ltd 2022
The rights of Olivier Cavalié and Emmanuel Trouvé to be identified as the authors of this work have been asserted by them in accordance with the Copyright, Designs and Patents Act 1988.

Any opinions, findings, and conclusions or recommendations expressed in this material are those of the author(s), contributor(s) or editor(s) and do not necessarily reflect the views of ISTE Group.

Library of Congress Control Number: 2022934057

British Library Cataloguing-in-Publication Data
A CIP record for this book is available from the British Library
ISBN 978-1-78945-083-5

ERC code:
PE6 Computer Science and Informatics
PE6_11 Machine learning, statistical data processing and applications using signal processing (e.g. speech, image, video)
PE10 Earth System Science
PE10_14 Earth observations from space/remote sensing

Contents

Preface

This book is part of ISTE–Wiley's *SCIENCES* series and belongs to the *Image* field of the *Engineering and Systems* department. The *Image* field covers the entire processing chain, from acquisition to interpretation, by analyzing the data provided by various imaging systems. This field is split into seven subjects, including *remote sensing imagery* (RSI). Based on this subject, we proposed a series of books that would portray diverse and comprehensive topics related to advanced remote sensing images and their applications for *Earth observation* (EO). There has been an increasing demand for the monitoring and prediction of our planet's evolution on local, regional and global scales. Hence, airborne, space-borne and ground-based platforms with active and passive sensors have been used to acquire images that measure several features at various spatial and temporal resolutions over the past few decades.

RSI has become a broad multidisciplinary domain, attracting scientists across the diverse fields of science and engineering. The books proposed in this RSI series aim to present state-of-the-art scientific knowledge about the primary sources of images acquired by optical and radar sensors. The books cover the processing methods that have been developed by the signal and image processing community to extract useful information for end-users for an extensive range of EO applications.

Each RSI book focuses on a general topic, such as change detection, surface displacement measurement, target detection, model inversion or data assimilation. We dedicate this second book of the RSI series to the measurement of displacements on the Earth's surface using RSI. It presents the main approaches that are used to derive displacement information from image pairs or image time series using optical and synthetic aperture radar (SAR) data. Following the numerous studies conducted in this domain, we divide this book into methodology and applications.

The first part, Theory, Principles and Methodology, is dedicated to the wide range of methods that have been developed to measure displacement fields from remote sensing images. Chapter 1 provides an overview of the main space missions devoted to EO, which have been used to acquire optical and SAR images for more than 40 years. From the archives of space agencies to future missions, this chapter

Surface Displacement Measurement from Remote Sensing Images,
coordinated by Olivier CAVALIÉ and Emmanuel TROUVÉ.
© ISTE Ltd 2022.

presents the history and the main characteristics of the data processed to measure the Earth's topography and surface displacement. Chapters 2 and 3 present the first group of methods developed, which measure displacement from the radiometry of images acquired at different dates, for optical and SAR images, respectively. These methods based on template matching are known as correlation or offset tracking techniques. These chapters include a brief description of the sensor geometry that is required to transform the measured offsets into velocities and to evaluate the potential and limits of this approach. Chapters 4, 5 and 6 examine the second group of methods, which use the phase of SAR images to derive surface displacement using interferometry (InSAR). The principle of InSAR, the geometric and statistical issues and the first processing steps are described in Chapter 4, including the limitations due to atmospheric artifacts. Chapter 5 shows how time-series InSAR methods, namely persistent scatterer and small baseline approaches, succeed in measuring centimeter- and even millimeter-scale deformations. Finally, Chapter 6 focuses on the methods that have been developed to unwrap the phase and to take the phase closure principle into account, improving the results or discovering hidden phenomena.

The second part, Applications for Surface Displacements, exhaustively covers the use of SAR interferometry and image correlation to retrieve the Earth's surface displacements. Indeed, the first application of InSAR that made the cover of Nature, in April 1993, was for the measurement of earthquake movements. Chapter 7 is thus dedicated to measuring tectonic displacements and showing how InSAR and image correlation have been used over time to monitor more subtle tectonic motions of the Earth. Next, Chapter 8 describes how measuring surface displacement is crucial to understanding the dynamics of volcanoes. In addition to natural phenomena, some of the Earth's surface deformations are induced by human activities. Chapter 9 reviews all kinds of human-induced signals and how InSAR is beneficial to monitoring them. Chapters 10 and 11 focus on surface motions due to gravity, namely landslide and glacier dynamics. Finally, the last chapter opens up the field of image correlation by cleverly taking advantage of the small spatial baseline and the time lag between the instruments imaging the Earth in different spectral bands onboard the very same satellite.

This book offers for the first time a complete overview of the methodological approaches deployed for the deformation measurement of the Earth's surface using RSI and their applications to the monitoring of major geophysical phenomena. Over the last few years, a virtuous circle between space agencies and scientists has developed. Indeed, in the last 20 years, the cost of satellite images has decreased drastically, the quantity of data has increased and the performance of data processing algorithms has greatly improved. As a result, the maturity of the remote sensing community now allows near real-time observation of the Earth's surface deformation, enhancing our capability to detect and mitigate natural hazards substantially.

Olivier CAVALIÉ
Coordinator of the book
Emmanuel TROUVÉ
Coordinator of the book and head of
the "Remote Sensing Imagery" subject
Avik BHATTACHARYA
Head of the "Remote Sensing Imagery" subject

February 2022

Part 1

Theory, Principles and Methodology

1

Relevant Past, On-going and Future Space Missions

Philippe DURAND and Stephane MAY

CNES, Toulouse, France

Earth observation satellites have provided images all around the world for more than half a century. Passive optical imaging systems and synthetic aperture radar (SAR) active sensors are nowadays the main sources of information used to derive surface displacement fields. To provide an overview of the images available through space agencies and their main characteristics, this first chapter describes different space missions with data relevant for ground motion displacement measurements. For radar, it is important to observe the field under the same incidence angle, meaning that this may depend upon the satellite orbit housekeeping and general design. SAR missions where orbit is not maintained are not mentioned (except the Iceye mission): their data seem useless at first sight for SAR interferometry (InSAR, see Chapter 4) and probably for SAR correlation algorithms (offset tracking, see Chapter 3). Data access, particularly when free of charge, is also described for the missions.

1.1. Some key parameters for space missions

1.1.1. *Parameters for both SAR and optical missions*

Remote sensing missions to date have had low-altitude orbits of a few hundred kilometers around the Earth. While optical missions mainly work, except for infrared instruments, along half of the orbit under sun lighting, SAR images can be acquired

Surface Displacement Measurement from Remote Sensing Images,
coordinated by Olivier CAVALIÉ and Emmanuel TROUVÉ.
© ISTE Ltd 2022.

across the whole orbit, day and night, and also through cloud cover for the frequencies examined later in this chapter.

Sun-synchronous missions: Most remote sensing missions are placed on sun-synchronous orbits. This is important for optical imagery to assure the same looking angle with regard to the sunlight (although it slightly changes anyway with the seasons). These orbits are also used for SAR imagery, sometimes because they are on the same platform as optical sensors (e.g. JERS-1, Envisat or ALOS), and also because of power constraints. Radar instruments have high power consumption and thus recent missions are often on a 6:00 and 18:00 (or dawn–dusk) orbit that maximizes the energy received by a solar panel (which can be fixed and always orientated towards the sun). ERS-1 and ERS-2 inherited a platform heritage from the first French optical satellite (SPOT-1), keeping the local hour at 22:30 (10:30 descending): the panel had a fixed direction towards the sun but had to make one rotation per orbit on the spacecraft.

These orbits are also near polar orbits, which assure a large coverage of the Earth, except some areas near the poles, depending on the look angle and swath, and with differences between the south and north poles depending on whether the radar is left- or right-looking. When crossing the equator, these orbits have a fixed local hour, which must be maintained throughout the mission duration (inclination maneuvers): this is a key parameter, and all satellite passes occur at the same hour if observed under the same incidence angle. Note that the reference value of the local hour is when the orbit crosses the equator from south to north (ascending part).

Revisit: This corresponds to the time taken for a satellite to revisit a given geographical location on the Earth. It depends on the satellite orbit (altitude, inclination, etc.) and agility.

Repeat cycle: This key parameter expresses the number of days separating two data takes under the same orbital point of view, which can be equal to or higher than the revisit. At the beginning of the mission, in relation to the sensor field of view or swath, we must establish a total number of orbits N_{tot} and also choose the number of days in the cycle D_{cycle}. Then, we have the following relationship:

$$N_{tot} = [14(\text{or } 15) + p/D_{cycle}] \times D_{cycle}, \qquad [1.1]$$

where $[14(\text{or } 15) + p/D_{cycle}]$ is the number of orbit revolutions in one day. The orbit duration is about 100 min as the orbit altitudes are in the range of 400–900 km: when under 96 min, there will be at least 15 orbits/day. p is an integer chosen with a close relationship to D_{cycle}.

It is possible to change the repeat cycle during a mission, which means a change of altitude for the satellite. For example, ERS-1, with a change of altitude of less than 6 km, had a D_{cycle} that changed from 35 to 3 days. With a constellation where all satellites have the same orbit (but are phased differently), for example, with SPOT,

ERS, CSK, Sentinel and TerraSAR-X/TanDEM-X/PAZ constellations, it is possible to combine images from the different satellites and hence reduce the time interval between data takes while keeping the same orbital point of view, called the "repeat time".

Satellite	Local hour	Entire rev.	p	D_{cycle} (days)	N_{tot}	Altitude (km)
ERS-1, ERS-2 routine	22:30	14	11	35	501	785
ERS-1 phases A, B, D	22:30	14	1	3	43	775
Envisat	22:30	14	11	35	501	785
Envisat end of life	22:30	14	11	30	431	783
Radarsat-1,2	18:00	14	7	24	343	798
JERS-1	22:30	14	43	44	659	568
ALOS	22:30	14	27	46	671	692
TerraSAR-X, TanDEM-X, PAZ	18:00	15	2	11	167	515
CSK, CSG	6:00	14	13	16	237	620
Sentinel-1A/B	18:00	14	7	12	175	693
Radarsat CM	18:00	14	11	12	179	593
ALOS-2	18:00	14	3	14	199	628

Table 1.1. *Main orbital parameters for some sun-synchronous SAR missions*

Satellite	Local hour	Ascending/ descending	Period of orbit (min)	D_{cycle} (days)	Altitude (km)
Cartosat-2A/2B	9:32	D	97.4	5	635
Landsat 8	10:00	D	98.8	28	705
SPOT-7	10:30	D	98.8	26	694
Sentinel-2A/2B	10:30	D	100.6	10	786
Kompsat-3A	10:50	A	95.2	28	528
Co3D	11:00	-	94.6	-	502
WorldView-1	13:30	D	94.6	14	496
Eros B	14:00	D	94.8	4	520

Table 1.2. *Main orbital parameters for some sun-synchronous optical missions*

Common parameters for optical missions: For the optical missions, the same orbital parameters are key elements for the mission. The orbits are in general sun-synchronous. The local hour has a direct impact on the viewed scenes. There is a compromise between viewing a scene early in the morning in order to reduce atmospheric effects (water vapor) and near midday in order to reduce the length of shadows in the images. The local hour is in general chosen between 9:30 and 14:00.

The choice of the orbital altitude has an impact on the spatial resolution, as well as on the viewing angle of the terrestrial scenes if a high revisit rate is required.

High-resolution satellites are, in general, agile and consequently relax the mission constraints. The altitude of low Earth orbit (LEO) satellites is, in general, chosen to be between 420 km, the altitude of the International Space Station (ISS), and 900 km. The final resolution of the disparity map used to derive elevation (see Chapter 2) or offset map used to derive displacement is directly linked to the initial image resolution. So, the spatial resolution of the pixels is one of the key parameters of an optical mission.

Non-sun-synchronous missions: The famous Shuttle Radar Topography Mission (SRTM) was phased, but not sun-synchronous. With an inclination of 57° at 233 km altitude, this mission produced a worldwide digital elevation model (DEM), but it was limited in latitude.

1.1.2. *Parameters specific to SAR missions*

This section addresses some parameters specific to SAR systems. Some important parameters, such as the range and azimuth ambiguity ratios, are not listed because they do not really affect phase or ground motion measurements.

Central frequency and bandwidth: In SAR remote sensing, frequency allocation is a key issue. The International Telecommunication Union Radiocommunication Sector (ITU-R) provides recommendations for remote sensing (ITU-R 2009), and particularly for active microwave sensors with regard to Radiocommunication Requirements (RR) Art. 5 (Table 1.3).

Allocated frequencies by RR Art. 5 (MHz)	Conventional letter	ITU required bandwidth (MHz)
432–438	P	6
1,215–1,300	L	20–85
3,100–3,300	S	20–200
5,250–5,570	C	20–320
8,550–8,650	X	20–100
9,300–9,900	X	20–600

Table 1.3. *Allocated frequencies and bandwidths for active SAR civil sensors in space as mentioned by the ITU-R (2009)*

There are concerns, however, about other ITU attributions that affect these frequencies, in particular in the C-band (wireless LAN, mobile phones) and L-band (radio location and radio navigation), according to the ITU-R (2010). Sometimes military radars are also active in such frequencies. Both can cause real interferences in image data and make them partly useless. Each frequency has its own properties, and all categories have either flown (L, S, C, X) or are being scheduled (P-band with Biomass).

We can note from Table 1.3 that the highest bandwidth is available in the X-band: as a result of this point combined with the more compact antennas needed for such missions and the inferred cost reduction, many new satellites are aiming at the X-band. In the ITU WRC 2015 meeting, an extension to a 1,200-MHz bandwidth (9,200–10,400 MHz) was approved, but some countries have expressed reservations and stick to 9,300–9,900 MHz (ITU-R 2020). In addition, this table is not applicable everywhere: for instance, the P-band is not accepted over the United States for SAR civil sensors. SAR missions of the interest detailed in this chapter and associated frequency bands are listed in Figure 1.1.

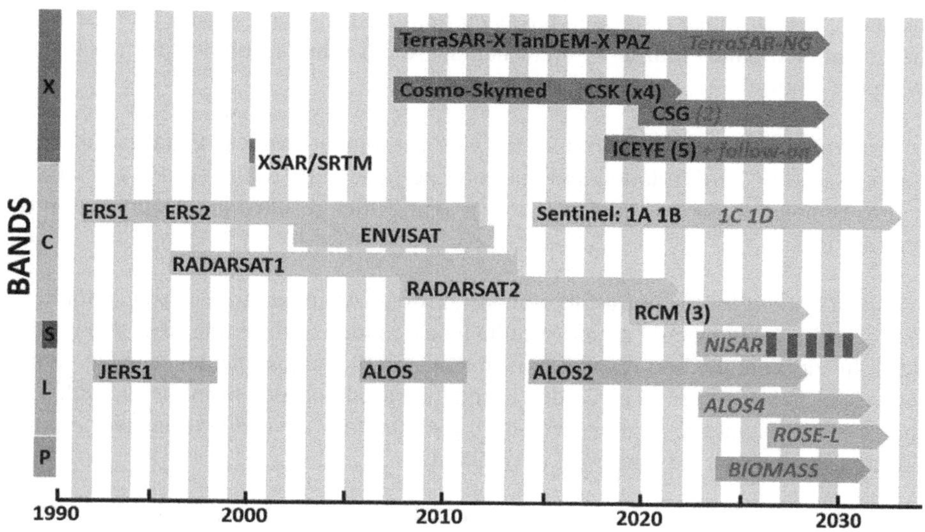

Figure 1.1. *Relevant SAR missions and associated frequency bands (future missions in gray italics). For a color version of this figure, see www.iste.co.uk/cavalie/images.zip*

Radar swath and pulse repetition frequency (PRF): Stripmap is the simplest working mode of a SAR system. More advanced modes can be explained relative to this mode. In stripmap, the resolution after synthesis in the azimuth direction (along-track) is given by half the antenna length $L_a/2$. The Doppler bandwidth in the azimuth is given by $B_{Dop} = v_{s/c}/(L_a/2)$, where $v_{s/c}$ is the satellite velocity. The PRF f_P needs to be higher than B_{Dop} to avoid spectral layover in the azimuth. In contrast, in the range direction (across-track), the PRF must be limited to avoid range ambiguities or, to put it another way, the swath will be maximized if the PRF is minimized: swath $< c/2f_P$. This explains why stripmap swath is lower in X-band compared to C-band missions (a shorter antenna means higher PRFs and narrower swaths).

If higher resolution is needed, it is possible to expand B_{Dop} by electronic steering in the azimuth (the same resolution is kept in range) and thus to increase the length of observation along the orbit: this is called the spotlight mode. The counterpart is that it is no longer possible to obtain a continuous along-track imaging as in stripmap, but only sparse scenes. If a larger swath is needed, then the approach is to use ScanSAR modes: the Doppler bandwidth is shared between different subswaths. By adding an azimuth steering mode like a "reverse" spotlight, it is possible to obtain a Sentinel-1-like mode called the terrain observation by progressive scan (TOPS) mode (Zan and Guarnieri 2006). The range resolution can stay the same as the stripmap mode. In some missions with multi-incidence angles (Envisat or Radarsat-1, for instance), loss of range resolution has been mitigated for some modes by increasing the transmitted bandwidth to assure a better ground-range resolution for the lowest incidence angles.

Yaw steering: Most SAR missions use yaw steering on board, so that the spacecraft compensates for Earth rotation and the point perpendicular to the orbit is effectively at zero Doppler. Radarsat-1 was not yaw steered, which resulted in a Doppler shift of several PRFs.

Incidence angle: In contrast to optical sensors, SAR imaging sensors cannot look to the nadir, as this would create simultaneous echoes in the same pixel from both sides. In addition, the best resolution is obtained at the largest incidence angle, at the expense of lower power received due to the increased distance from the ground. The choice of the incidence angle is critical in mountainous areas due to layover on one side and shadows on the other side of mountains.

Orbit precision: This is an important parameter used to combine SAR images for interferometry. An orbit that is not precisely known will lead to an important phase gradient in the interferogram. There are three kinds of orbit that can be successively delivered:

– the first is a predicted orbit that usually comes along with the product when produced in real time;

– then, a few hours or days after the data acquisition, a restituted orbit is available with more precision;

– finally, after three or four weeks, the most precise orbit can be delivered to users, thanks to dedicated payloads and specific computing.

Nowadays, with precise orbit determination payloads onboard, orbit precision can be known at the order of a few centimeters.

Orbital tube: Recent SAR missions have integrated the use of differential interferometry in their specifications (Sentinel-1, ALOS-2, the Radarsat Constellation mission) and have put constraints on orbit housekeeping. Thus, the orbit must stay in a tube called the "orbital tube", with a radius as small as one hundred or a few hundreds of meters to keep the perpendicular baseline small (see Chapter 4, section 4.2).

Duty cycle, down-link rate and onboard storage: The duty cycle is equivalent to the percentage of time that the radar will be able to work along the orbit. SAR systems need a lot of energy, making it impossible for them to work permanently. Depending on the design of the satellite power unit, a system can deliver between a few seconds of imaging and about 1/4 of the orbit time. There is a huge difference between very small satellites of a few hundreds of kilograms and big platforms of two or three tons. The down-link rate and onboard storage capacity are often two key elements that usually come together. A data relay satellite is sometimes helpful to cope with all the constraints or avoid multiple ground antennas for down-linking, but this usually needs a laser link and data relay commercial contract, which can also be quite expensive.

Instrument noise equivalent σ_0: This parameter is very important when looking at amplitude images. It gives the minimum value that can be reached by the system for an elementary pixel, but entails lower values with multi-looking (see Chapter 3). It varies in range along the swath, and the specification must deliver the worst case: in general, the best values are in the middle of the swath where the antenna gain is maximum. This parameter affects measurements for ground motion: if its value is too poor, then some surfaces that have low backscattering coefficients will not be properly estimated, for instance asphalted roads, tarmacs and even sand in the desert.

Polarization: The design of the antenna sub-system determines what polarization should be implemented. For ground displacement purposes, as the signal-to-noise ratio (SNR) is more favorable, co-polarized data are mainly used, and it is difficult to say whether using HH and VV for offset tracking or InSAR techniques are more advantageous. The use of dual polarization, HH+HV or VV+VH, or quad polarization (HH+HV+VV+VH) is more relevant for other remote sensing applications, such as classification, forest extents and heights, maritime surveillance, pollution at sea or other change detection characterizations. Furthermore, the use of quad polarization generally reduces the swath and azimuth resolution by a factor of two, and thus the size of the archive in this mode is less important.

1.1.3. *Parameters specific to optical missions*

Pushbroom: A pushbroom camera consists of an optical system projecting an image onto a linear array of sensors. Usually, a focal plane is composed of several time delay integration (TDI) image sensors, mounted in a staggered configuration. The image is directly built at the sensor level. Charge-coupled device (CCD) sensors are used where ultra-low noise is preferred, and now complementary metal oxide semiconductor (CMOS) matrix detectors are increasingly used.

Stereo, tri-stereo and more: To build a disparity map (see Chapter 2), at least two stereo images from separate view angles are necessary. It is possible to combine more images (tri-stereo or even more) to build a more precise map. With four images, all faces of buildings may be seen, and it is possible to build a digital elevation model and associated pixel values. The different images over a geographical point are taken from different incidence angles. If they are taken with the same satellite, then they will be asynchronous. If they are taken with a constellation of satellites on the same orbit, then they may be synchronous, such as with the Co3D system. Asynchronous images imply two acquisition instants. If the delay is a few seconds, mobile elements (such as terrestrial vehicles, bots, planes) or clouds will alter the raw displacement measure. If the delay is a few weeks or months, some buildings may appear or disappear. Evolution of agricultural landscape will also alter the match between the images.

Agile satellites: The agility of satellites is a key function. Agile satellites are able to do attitude maneuvers and focus on one scene. This increases the revisit frequency over the same region, and so reduces the time interval between two acquisitions. Such satellites are able to point towards a geographical point thanks to steering mirrors or by changing their whole attitude (yaw, pitch and roll control).

Base-to-height (B/H) ratio: The shift between two images creates a stereoscopic parallax in one direction (see Chapter 2). The stereoscopic angle (also called the base-to-height or B/H ratio) is the ratio between:

– the distance between the two viewing points (base – B);

– and the distance to the observed scene (height – H).

In the older generation of satellites, the B/H ratio was directly fixed with the instrument characteristics. For example, the SPOT-5 satellite had high-resolution HRS instruments to provide large-area along-track stereoscopic images (forward and backward of the satellite). With agile satellites, it is possible to choose the best view conditions and define the choice of B/H ratio. A large B/H ratio favors the observation of different points of view of the scene (e.g. two faces of buildings) but creates constraints on the disparity computation, as some elements of one scene are not seen in the other one. With a lower B/H ratio, the delay between two images is reduced and a scene can be observed under nearly the same conditions, but it adds constraints to the precision of the disparity computation algorithms. In the case of SPOT-5, the shift between panchromatic and XS detectors (19.5 mm in the focal plane) creates a slight stereoscopic angle that allows a stereo reconstruction using the P+XS correlation. In the case of Pléiades, the B/H ratio is chosen by the user.

The key optical parameters of satellite systems are explained in the following:

Spatial resolution: Spatial resolution is a measure of the smallest angular or linear separation between two objects or two pixels on the ground. It is usually expressed in radians or meters. Spatial resolution decreases as the viewing angle increases. Spatial resolutions are, in general, given at the nadir of the satellite. For example, for Ikonos, the spatial resolution is equal to 0.82 m at the nadir and 1 m at 26° off-nadir.

Satellite	Mission characteristics	B/H ratio
SPOT-5	Steering mirror (MCV). Stereoscopic angle up to 54 deg	1.02
SPOT-5	HRS instrument: angle 40 deg, delay 90 s	0.8
Co3D	Synchronous acquisitions	0.20–0.30
SPOT-5	PAN–XS stereoscopic angle: delay 2.25 s	0.018
Pléiades	Satellite agility – stereo or tri-stereo	0.10–0.20
WorldView-3	Satellite agility – stereo or tri-stereo	B/H = 0.20–0.40

Table 1.4. *Examples of satellite missions and B/H ratios*

Spectral resolution or spectral bands: A ground feature or target is defined by its reflectance and spectral emissivity curve. Its response depends on the wavelength ranges. Different classes of objects in an image can thus be distinguished by comparing their responses over distinct wavelength ranges. A large spectral band captures more signal energy and thus enhances the SNR of the image. A panchromatic band defines a wide spectral band, with high SNR and high resolution. In contrast, a multispectral band (XS) captures specific wavelength ranges separated by filters or instruments sensitive to particular wavelengths: the received signal energy decreases as the bandwidth decreases, and therefore the satellite system is designed with a lower resolution to maintain an acceptable SNR. For a given satellite system, regardless of the spectral bands, disparity estimations are performed with the bands that have the highest spatial resolution and the best SNR. Spectral bands may be separated into several families or groups: thermal infrared (TIR) and medium-wavelength infrared (MWIR) depend not only on solar illumination but also on emission from the imaged objects themselves.

Swath: Swath is the geographical width of the image, which is linked to the optical instrument and the number of detectors in the sensor.

Viewing angle/incidence angle: The viewing angle (satellite point of view) is the angle between the nadir point and the angle towards the terrestrial point. The incidence angle (region point of view) is the angle between the vertical to a point and the direction of the satellite. Thanks to steering mirrors or the agility of the whole satellite, it is possible to modify acquisition angle configurations. A low incidence angle (region near the satellite nadir) means that, for example, only roofs can be seen. With a high incidence angle, building facades can be seen, but some occlusion occurs (hidden faces, hidden streets, etc.). Spatial resolution decreases as the viewing angle increases. Hence, spatial resolutions are given at the nadir of the satellite.

Impact of the spatial resolution: The spatial resolution of the satellite directly drives the analysis that can be performed with the images. The definition of optical missions (HR, VHR, etc.) depends on the spatial resolution:

– *1,000–3,000 m*: Analysis of the atmosphere, aerosols and land surface emissivity/temperature. For example: MSG, METOP;

– *250–1,000 m*: Analysis of the atmosphere, aerosols, land surface emissivity/temperature. For example: MODIS, Sentinel-3;

– *30–60 m*: Analysis of land cover (agriculture, forestry, cartography, geology, cryosphere, etc.). For example: Landsat1–7;

– *15–20 m* (medium resolution (MR)): Analysis of large-scale land cover, agriculture, forestry and highway infrastructures and computation of digital terrain models. For example: Sentinel-2 (red edge, shortwave infrared – 20 m);

– *1.5–10 m* (high resolution (HR)): Analysis of land cover, agriculture, regional to city-level coverage, road infrastructures, digital elevation model extraction and medium or large objects, such as boats, and computation of digital terrain models. For example: SPOT-6–7, Sentinel-2 (blue, green, red, near-infrared – 10 m);

– *0.5–1 m* (very high resolution (VHR)): Analysis of many objects in the images. At this resolution, a large number of objects are visible, such as urban elements, houses, vehicles, buildings affected by natural disasters and archeological objects. Extraction of digital surface models can also be undertaken. For example: Pléiades;

– *< 0.5 m* (ultrahigh resolution (UHR)): Detailed analysis of elements of the scene. Many objects are visible in the images with more details, including small vehicles, buildings, archeological objects, etc. Computation of digital surface models can be undertaken. For example: WorldView-3, Cartosat-2.

The main optical satellites that are used for DEM extraction have spatial resolutions better than 10 m. For precise estimation of DEMs, submetric satellites are the best candidates.

Resolution versus swath versus revisit time: The size of telescope mirrors was the limiting factor in reaching high resolutions for the first generations of optical satellites. With the evolution of space technology, the first civil satellites were launched at the beginning of the 2000s with resolutions lower than 1 m. Ikonos was launched in September 1999 and was the first commercial satellite with a resolution of less than 1 m; it was followed by satellites such as Quickbird-2, Eros B, Kompsat-2 and WorldView-1. A satellite with a very high resolution has a reduced swath, in contrast to decametric resolution satellites. A medium-resolution satellite usually has a wider swath and a shorter revisit time. Conversely, meteorological satellites such as the Meteosat Second Generation (MSG) and Meteorological Operational Satellite (MetOp) family usually have a low spatial resolution around 1 or 3 km, with a wide swath between one and several thousands of kilometers, in order to give a view of the meteorological situation. The number of spectral bands, from 500 nm to 15,000 nm, allows a fine analysis of clouds and aerosols. These satellites have a low spatial resolution but a high temporal resolution. Their spatial resolution is not compatible with DEM extraction.

Choice of the spectral bands: The choice of spectral bands combined with the spatial resolution directly drive the applications of a satellite. Many satellite systems exist with many different configurations (PAN+XS configuration, only PAN configuration, only XS configuration, etc.). A significant number of UHR and VHR satellites have four bands (blue, green, red and near-infrared) associated with a panchromatic band. The panchromatic band (P or PAN) usually has a spatial resolution four times higher than the multispectral ones (XS). Pan-sharpening techniques are employed in order to build panchromatic plus multispectral (P+XS) products, which means fused multispectral products at a panchromatic spatial resolution.

Band name	Spectral interval
UV band	300–400 nm
Visible bands (VIS)	400–700 nm
Near-infrared (NIR) bands	700–1,000 nm
Short-wavelength infrared (SWIR)	1,000–2,500 nm
Medium-wavelength infrared (MWIR)	3,000–5,000 nm
Thermal infrared (TIR)/long-wavelength (LWIR)	8,000–15,000 nm

Table 1.5. *Description of spectral bands*

High-resolution (HR) satellites work with lower resolution but include additive infrared bands useful for land cover, water color or atmospheric analysis. To compute digital elevation models and to measure displacements, the spectral band with the highest spatial resolution and the best SNR is generally chosen. For many satellites, this is usually the panchromatic band. For some HR systems with only XS bands, multispectral bands may be used (e.g. Sentinel-2).

1.2. Past and on-going SAR missions

This section presents spaceborne missions that have been used or are used to compute displacement fields by interferometric processing. Some peculiarities or issues are pointed out to help the reader understand them. This is thus not an exhaustive list of SAR space missions. Note that the Magellan mission began to map Venus in September 1990 with an SAR instrument, but its topography was computed using a radargrammetry technique, combining two images from either the same side or from opposite sides: this is rather far from the topic of this book, and thus this very successful mission will not be further detailed here. We must note that most of these missions were organized and designed before the interferometric technique emerged, around the beginning of the 1990s. Thus, certain constraints, such as orbit housekeeping, yaw steering and burst synchronization for ScanSAR modes, were sometimes not taken into account in the designs of the spacecraft or payload.

After a series of C-band and L-band satellites at medium resolution in the 1990s, a new generation of SAR satellites operating in the X-band arrived in Europe at the national level (Germany, Italy, Spain) from mid-2007 onwards: although targeting different goals, they all had significantly improved resolutions and repeat pass capacities, which gave them serious advantages when focusing on localized areas.

With small satellite missions, such as ICEYE, interferometry is usually possible with a short time period: maintaining a small orbital tube is too expensive for the length of the mission and archives on large scales are out of scope; however, such constellations can gather significant data over cities and PS techniques can be considered. NovaSAR is not addressed here, as it is a single satellite on a short-term mission with limited applications for interferometry. Some Chinese and Indian missions are also not addressed due to the limited availability of the datasets. Finally, a significant improvement in interferometric abilities and coverage arrived with missions such as Sentinel-1 and ALOS-2.

1.2.1. *ERS-1, ERS-2 and Envisat*

ERS-1 (1991–2000) and ERS-2 (1995–2011), the European remote sensing satellites, began the era of SAR instruments in Europe and demonstrated the first applications in radar interferometry. The five-year overlap provided the opportunity to develop very nice orbital combinations, such as the so-called tandem phase, allowing acquisitions of the same area within just a one-day interval, which was much more favorable than the native 35-day repeat cycle for InSAR on rapidly changing surfaces.

The Environmental Satellite (Envisat) continued the C-band SAR missions for Europe with the ASAR instrument, as well as numerous added payloads, among which were the DORIS instrument for precise orbit determination and the MERIS spectrometer, giving simultaneous information on the composition of the atmosphere and overlapping most of the SAR swath. Envisat was put on the same 35-day orbit as the ERS series from April 2002 until October 2010, when a drifting orbit had to be set due to the lack of propellant until its end in May 2012.

ERS-2 had an overlap in the years 2007–2008 with Envisat, with a 30 min delay between the two satellites, but the central frequency was shifted from 5.3 GHz (ERS) to 5.33 GHz (Envisat) so that cross-intereferometry was restricted to common bandwidth in limited cases.

The geometry of ERS satellites was quite simple with a 23° fixed incidence angle, swath of about 100 km and one single polarization set to VV. The Envisat ASAR instrument was more complex, with modes divided in ScanSAR and its large swath of 405 km (composed of five sub-swaths), as well as seven stripmap modes (IS1 to IS7). A wave mode was operated over the ocean. Customers could order any kind of mode, so the archive was not homogeneous over the Earth and strongly depended on the area considered: it therefore became more difficult to get two images acquired in the same mode, suitable for interferometry. However, for some closely monitored volcanic islands, it was quite interesting to get different passes in the 35-day cycle with different incidence angles.

Note also that ASAR was not full polarimetric, but authorized different kinds of scheme, such as the alternating polarization (AP) mode, which could emit alternatively in H and V polarizations and receive H and V.

Some limitations of the missions with regard to interferometry: As these missions were decided before differential interferometry emerged, some limitations had to be tackled (Solaas and Coulson 1994). For ERS-1/2, orbital housekeeping looseness led to perpendicular baselines $B_\perp$ (see section 4.2.2 in Chapter 4) higher than the limit of 1,100 m. A preferred value of 600 m was set for $B_\perp$ (see definition in section 4.2), so that a tool to determine which orbits or track frames were compatible for interferometry had to be developed. In addition, the orbit was not known as precisely as the latest SAR missions: orbital fringes had to be removed before interpretation. PRARE, a dedicated instrument for precise orbit determination, failed after several weeks on ERS-1 but lasted several years on ERS-2 and was able to deliver precise ranging accuracy of less than 7 cm.

In contrast to future missions (e.g. ALOS-2 or Sentinel-1), Envisat ScanSAR bursts were not synchronized on board, meaning the Doppler overlaps had random values from one pass to another and were not always compatible. Some results were however published in ScanSAR mode; for instance, for the 2003 Bam earthquake in Iran.

In late October 2010, Envisat got a new orbit cycle of 30 days in a degraded configuration. At this point, interferometry could only work around latitudes of 38° (north and south). As a result, tracks could not be combined with the previous 35-day orbit cycle. However, along-track interferometry in the new geometry could produce interesting results, for example interferograms for the Sendai (Tohoku) earthquake that occurred on March 11, 2011, which led to the Fukushima disaster.

Data access: During the missions, European Space Agency (ESA) products were not free of charge and RAW data were much cheaper. Furthermore, some technical teams preferred to do the radar synthesis themselves to ensure a better signal shape in the range spectrum. Nowadays, as indicated by the ESA (2019), the (A)SAR on-the-fly (OTF) data processing and dissemination service allows end-users to gain direct access to ESA's complete high-resolution (A)SAR archive from the ERS-1, ERS-2 and Envisat missions. The service provides (A)SAR Level 1 high-resolution data, which are processed from Level 0 on user request by the system. Users can register themselves and immediately download Level 1 (A)SAR data products. The service is available via the ESA Online Dissemination Service at https://esar-ds.eo.esa.int/oads/access.

1.2.2. *Canadian C-band satellites: Radarsat-1, Radarsat-2 and RCM*

Radarsat-1 was the first imaging radar developed for the Canadian Space Agency (CSA). It was launched in December 1995 and ended its operations in April 2013.

The satellite had noticeable activity over the Arctic areas for ice monitoring. In the late 1990s, it was the only civilian radar capable of imaging part of the Earth, while ERS could only operate in the visibility of ground receiving stations due to the lack of onboard recording. The satellite offered a variety of imaging modes, including the large swath in ScanSAR (510 km) and narrow ScanSAR (305 km), seven stripmaps (100 km) and three others with 150 km swath, ranging from 20 to 47°, as well as 15 fine modes at incidences above 37° (45 km). The radar operated only in single co-polarization HH. The satellite native design was right-looking imaging, but the CSA conducted a dedicated campaign over Antarctica with a 180° rotation of the spacecraft, making it left-looking, and the first radar mosaic of the continent could thus be produced. Interferometry was demonstrated during the mission, although there were some strong limitations with this spacecraft. Radargrammetry for DEM was also demonstrated, prior to the SRTM mission.

There were several limitations of the mission with regard to interferometry:

– the satellite was not yaw-steered, meaning the Doppler centroid had strong values, exceeding several PRFs, as well as strong variations in the year, and for interferometry, we should take care when combining pairs, the minimum and maximum of Doppler shift being around the solstices;

– the orbit-keeping was quite loose, and baselines reached several kilometers;

– orbital knowledge was a special issue in computing correct interferograms, and the use of fine orbit interpolators, such as Hermite polynomials, was a necessity.

The *Radarsat-2* spacecraft was launched in December 2007 while Radarsat-1 was still operating. The spacecraft added significant improvements in many areas: right- and left-looking, quad polarization, low σ_0, ultrafine modes, ground track maintained within a 500 m wide range, yaw steering and onboard state recorders. Although on the same reference orbit as Radarsat-1, Radarsat-2 images were not pairable with Radarsat-1 for interferometry: the central frequency was shifted from 5.3 to 5.405 GHz and the Doppler centroid was not compatible, due to the difference in yaw steering. At the end of 2020, the spacecraft was still operating, making a transition to the next Canadian SAR mission.

The *Radarsat Constellation Mission (RCM)* aims to replace the Radarsat-2 mission, continuing the traditional Canadian C-band SAR imagery as well as adding automatic identification system (AIS) payloads to improve maritime surveillance. It is composed of three identical satellites that were launched together in June 2019 for an expected lifetime of seven years. The sun-synchronous constellation has a 18:00 ascending node and flies at an altitude range between 586 and 615 km. As the three satellites are equally spaced along the orbit, the repeat time is only four days, and the orbit is maintained within a 120 m radius orbital tube, which gives good advantages for several InSAR applications. Various schemes of polarization are implemented: single polarization, dual co-cross or compact polarimetry are available on all modes; dual

HH–VV is available for specific modes, as well as a quad-polarization mode. Radar modes are numerous, from spotlight (1 × 3 m resolution, one look) to wide ScanSAR (100 m, eight looks)

Data access: Since April 2019, 36,000 images of archive data acquired by the Government of Canada all over the world have been freely available (see CSA 2019). Unfortunately, not all acquired data can be accessed at the time of editing (2021).

1.2.3. *Japanese L-band satellites: JERS-1, ALOS and ALOS-2*

The Japanese Earth Resources Satellite (JERS-1) was the first of a series of Japanese SAR sensors working in the L-band, followed later by the Advanced Land Observing Satellite (ALOS) and ALOS-2. All three projects were conducted under governmental institutions. Launched in February 1992, JERS-1 lost its onboard recording capacity in August 1997, which limited its use over direct receiving ground stations, and the spacecraft finally terminated its mission in October 1998. The SAR sensor operated at 1.275 GHz (wavelength of 23.5 cm), with a bandwidth of 15 MHz. Only co-polarization in HH was implemented. The nominal swath was a stripmap mode of 75 km, at a fixed look angle of 35°. Tracks were acquired regularly, minimizing the time between two consecutive swaths, which was a great advantage for large mosaics and radiometric stability. Interferometry showed great coherence for vegetated areas compared to C-band radar, as in ERS, with comparable resolution. As an optical system (OPS) was also used, the satellite was put on a sun-synchronous orbit with a descending node at 10:30 to 11:00.

ALOS was the second Japanese satellite and was able to deliver L-band SAR images through its PALSAR instrument operating at a 1,270 MHz central frequency. The launch took place in January 2006, and the local hour was set at 10:30 descending node, with an onboard optical sensor (PRISM, AVNIR-2). The bandwidth was increased to 28 MHz in single-polarization HH or VV and 14 MHz in dual-polarization HH+HV VV+VH or quad polarization with different fine beams, with 70 km and 40 km swath, respectively. The ScanSAR mode only worked in co-polarization HH or VV, with a 350 km swath sub-divided into five beams. The spacecraft stopped operating in May 2011.

In April 2014, *ALOS-2* followed the ALOS program with an L-band radar payload, without the optical part later put on ALOS-3. Compared to ALOS, the ALOS-2 PALSAR instrument improved the resolution (tunable between 14 and 84 MHz), the revisit cycle (14 days instead of 46 for ALOS) and right and left imaging capacities and provided new modes, such as spotlight and ScanSAR with burst synchronization for ScanSAR interferometry.

Data access: We can note that the background mission of these satellites was orchestrated by a basic operational scenario, i.e. a long-term homogeneous and consistent programming of the whole Earth in different modes, with regular updates. It is possible to freely access JERS-1 and ALOS data through the ESA EO portal at https://earth.esa.int/web/guest/-/jers-1-sar-level-1-single-look-complex-image and https://earth.esa.int/web/guest/-/alos-palsar-fbs-fbd-and-plr-products, respectively. The datasets contain all ESA acquisitions over the ADEN zone (Europe, Africa and the Middle East), plus some worldwide products received from JAXA.

1.2.4. *SRTM and X-SAR*

The Shuttle Radar Topography Mission was aimed at computing a semi-global digital elevation model from the Earth. This was achieved in February 2000 on board the US space shuttle Endeavour carrying the SIR-C/X-SAR payload, which previously flew in 1994 in L-, C- and X-bands. This time, the payload was reconfigured as a single-pass interferometric mission with a fixed baseline as long as the 60 m mast that separated the dual antennas in two bands: the C-band instrument from the Jet Propulsion Laboratory (JPL) and the X-band instrument from the Italian Space Agency (ASI) and German Space Agency (DLR). For the C-band, swath reached 225 km thanks to a four-beam ScanSAR, and for the X-SAR instrument, there was a stripmap swath of 50 km. The mission duration was limited to 11 days: only the C-band swath could assure the semi-global DEM, while the X-band covered a subpart of it. Due to the 57° inclined orbit of the shuttle at an altitude of 223 km, the DEM covered only latitudes ranging from 56° south to 60° north.

Data access: DEMs in C-band at 3 sec (90 m) spacing and then at 1 sec (30 m) spacing were released worldwide in 2003 and 2015, respectively, and are still commonly used 20 years after the mission. They are freely available after registration at https://earthexplorer.usgs.gov/. The more sparse X-SAR DEM is available through the geoportal https://download.geoservice.dlr.de/SRTM_XSAR/.

1.2.5. *TerraSAR-X, TanDEM-X and PAZ*

TerraSAR-X (TSX), TanDEM-X (TDX; TerraSAR-X add-on for digital elevation measurement) (Germany) and PAZ (Spain) are similar satellites launched in June 2007, June 2010 and February 2018, respectively. They are compact “Toblerone” satellites, including a SAR antenna and a solar panel fixed on the main body. This shape is compliant with an heliosynchronous dawn–dusk orbit and has a very low altitude of 515 km compared to previous SAR missions. The two German satellites produced a worldwide DEM, following the incomplete coverage of the SRTM/X-SAR mission, still with the interferometry technique and simultaneous bistatic acquisitions: flying together, various baselines were chosen along the project to refine a 12 m DEM at a specified 2 m vertical relative accuracy (slopes less than 20°). The SAR mission started with different beam modes, such as four-beam ScanSAR (100 km), stripmap (30 km), spotlight (10 km) and high-resolution spotlight (azimuth extension limited to 5 km instead of 10 km). Then, in 2013, six-beam wide ScanSAR (194–266 km) and Staring Spotlight (7.5 km max and limited azimuth swath of 2.5–2.8 km) modes were added as operational modes. Staring Spotlight has a 0.18 m resolution in the azimuth and can use a 300 MHz bandwidth to afford a 0.6 m slant range resolution. Polarization can be in co-pol HH or VV, and dual polarization is available with different combinations depending on the mode used. Quad polarization is only available as an experimental mode. In 2009, a noticeable experiment assessed the TOPS mode interferometric feasibility, before the launch of Sentinel-1 (see Chapter 2, section 2.8; see also Prats-Iraola *et al.* (2012)).

PAZ data can be combined with TSX with a four- or seven-day time interval.

Data access: TSX and TDX data can be obtained at a low cost for scientific use after submitting a relevant proposal to the DLR. The 3 sec (90 m) DEM has been freely available with registration since October 2018: https://download.geoservice.dlr.de/TDM90/. The most precise DEM is commercially distributed by Airbus Defence and Space: https://www.intelligence-airbusds.com/imagery/reference-layers/.

1.2.6. *COSMO SkyMed constellations*

COSMO SkyMed (CSK) is the first Italian SAR constellation. Four satellites operate in the X-band on the same dawn–dusk orbit and the ascending node is at 6:00, whereas other satellites often use 18:00. The platform is inherited from the Radarsat-2 prima platform, with right- and left-looking abilities. The launches of the four satellites occurred in June and December 2007, October 2008 and November 2010.

CSK is a dual civil and military mission, and the submetric spotlight is for defense use only. Other modes are accessible for commercial use, i.e. two different ScanSAR modes (100 and 200 km), one stripmap (40 km) and one civil spotlight mode (10 km). The satellites operate in a single-polarization scheme with HH, HV, VV or VH selectable.

COSMO SkyMed New Generation (CSG) is composed of two satellites: the first one was launched in December 2019. Orbiting at the same altitude as the first generation, this new system has significant improvements in terms of agility, resolution and polarizations. It is possible to obtain simultaneous swaths in stripmap mode and also different spotlight images on the same orbital span. The system also adds quad-polarization possibilities in stripmap. For civilians, three spotlight modes are possible, in single or dual polarization. Stripmap always has the same 40 km swath in single polarization as well as in dual polarization. As for the first generation, a narrow ScanSAR of 100 km and wide ScanSAR of 200 km are possible, again in single or dual polarization.

Data access: Archived products can be seen and commercially ordered through the eGEOS portal at http://catalog.e-geos.it/.

1.2.7. *SAOCOM1*

Satélite Argentino de Observación con Microondas (SAOCOM) is an L-band SAR program from Argentina's space agency the Comisión Nacional de Actividades Espaciales (CONAE). The two satellites SAOCOM-1A and SAOCOM-1B were launched in October 2018 and late August 2020, respectively. Agreements with the ASI made it possible to put SAOCOM satellites on the same orbit as the CSK constellation, providing the opportunity of obtaining X-band and L-band images at a 10-minute interval. CONAE implemented a TOPSAR mode, as in Sentinel-1, for the narrow ScanSAR of 170 km and a wide ScanSAR of 350 km, as well as a stripmap mode of 40 m at 10 m resolution. Single or dual polarization HH, HH+HV, VV or VH or as well as quad polarization, are possible.

Data access: Some archive images can be accessed through the portal at https://catalog.saocom.conae.gov.ar/catalog after registration.

1.2.8. *Sentinel-1*

Sentinel-1 belongs to the Copernicus European program, but the conception and operations were delegated to the European Space Agency. This is the first long-term program that has really taken InSAR as an operational objective, with a very small orbital tube of 100 m radius and 12-day repeat pass, which comes down to six days with two satellites (S1A and S1B), and possibly four days with three satellites operating, as Sentinel-1C and 1D have already been authorized. Sentinel-1A and 1B were launched in April 2014 and April 2016, respectively.

In contrast to the Envisat mission, a focus on a consistent archive is of prime interest and most emerged land is monitored, with the maximum capacity of any system operating in Europe. There are four basic modes: extra wide (EW) ScanSAR of 400 km in five subswaths, interferometric wide (IW) ScanSAR of 240 km in three subswaths, six stripmap modes about 80 km each, and a wave mode (20 km images every 200 km at incidences of 23° and 36°) for open ocean. IW is the privileged mode for land, avoiding gaps, and sometimes uses single-polarization VV, but most of the time dual polarization has been used since Sentinel-1B was launched.

IW is the main operational mode: the ScanSAR mode includes the TOPS mode successfully tested on TerraSAR-X (see section 1.2.5). Agility for the antenna is introduced along-track (azimuth) from backward to forward, for each burst of about 20 km. Each IW subswath consists of 9–10 bursts delivered with their own auxiliary parameters. Each burst is precisely synchronized between each acquisition so that interferometry is always possible.

The next two Sentinel-1 satellites, S1C and S1D, are quite similar to S1A and S1B and are in a building/storage phase untill launch, which will probably occur in

2022/2023 and 2024/2025, with the same operating modes for SAR images, but with increased capacity for maritime surveillance (AIS payload). This will hopefully insure continuous SAR data from October 2014 to beyond 2035, which has never happened in a space program to date. Hopefully, the system will operate with three satellites and four-day repeat pass interferometry on a large scale, at least over Europe.

Some limitations of the mission with regard to interferometry: In contrast to stripmap, the IW mode introduces additional complexity with regard to interferometry: to avoid phase jumps in the final image, very precise coregistration and phase examination in burst overlaps must be achieved. Products do not always cover the same area on the ground and splitting/gathering bursts in several products may be necessary.

Data access: All Sentinel-1 products are freely accessible via the Copernicus hub (https://scihub.copernicus.eu/dhus/) or on mirror sites such as PEPS (https://peps.cnes.fr/). In contrast to the Copernicus hub, PEPS do not archive RAW data, but only processed single-look complex (SLC), ground range detected (GRD) and wave mode data. The archive began on October 3, 2014. Other products directly derived from the Sentinel-1 data and showing ground motion for European land are freely available through a Copernicus service (https://www.copernicus.eu/en/events/events/online-european-ground-motion-service-copernicus).

1.2.9. *ICEYE*

ICEYE-X1, launched in January 2018, is the first SAR microsatellite of Finland's operational commercial constellation Iceye Ltd, a startup company from Espoo. The ICEYE-X1 cube (70 cm side) weighs under 100 kg, which is less than a twentieth of Sentinel-1. This size reduction produces a strong reduction in the cost too. The global imaging constellation will consist of 18 small satellites, allowing less than a few hours for revisit. ICEYE's satellite constellation in the X-band is growing, with additional spacecraft being produced and launched each year: in December 2018, ICEYE-X2 was launched. The ICEYE-X3 payload, launched on May 5, 2019, was a demonstration mission, dedicated to the US Army. ICEYE-X4 and -X5 were launched on July 5, 2019, with a Soyuz-2-1b rocket, and ICEYE-X6 and -X7 were launched on September 28, 2020. ICEYE-X8 and -X9 followed in January 2021. The Finnish company launched four additional SAR satellites at the end of 2020 and planned to launch an additional eight in 2021. The company offers X-band SAR data in several imaging modes, including very high resolutions in single-look complex and ground range detected image formats. ICEYE's satellites are capable of imaging in stripmap, spotlight and ScanSAR modes in VV polarization.

Data access: See https://www.iceye.com. The Finnish New Space Economy company ICEYE is now providing radar imaging data from its commercial SAR satellite constellation to the International Charter "Space and Major Disasters" at no cost for use in monitoring and response activities.

1.3. Future SAR missions

This section presents the main SAR missions that have been decided on and that will use interferometric measurements. We must note that many SAR systems are planned to arrive on the market, more than ever before, with very cheap and light payloads. In most of the cases, the orbit is not maintained, and if interferometry is theoretically possible between two passes, it must be seen more like an opportunity rather than a dedicated application. For these reasons, these missions are not detailed here.

1.3.1. *TerraSAR-NG*

TerraSAR-NG is an improved system continuing the X-band TSX/TDX/PAZ missions. Initially, one satellite is forecast, and one of the aims is to implement a 1,200 MHz bandwidth and reach a 0.25 m resolution on the ground. ITU regulations incorporated the demand in 2015 (see section 1.1.2 on central frequency and bandwidth). Another improvement is the implementation of a TOPSAR system for different ScanSAR modes.

1.3.2. *ALOS-4*

ALOS-4, planned to be launched in 2022, is a successor of the Japanese SAR L-band missions ALOS and ALOS-2. Its PALSAR3 instrument will be primarily activated with right-side looking and in a stripmap mode with an increased 200 km swath (compared to 50 km with the ALOS-2 PALSAR2), while maintaining the high temporal resolution with a revisit time of 14 days and covering an incidence angle from 30 to 44°; the other beams (ScanSAR and spotlight modes) and left-side observation are used for quick disaster monitoring. The ALOS-4 will be equipped with an AIS receiver, as with the ALOS-2, for maritime surveillance purposes. To deal with the deforestation issue, ALOS-4 will conduct observations more frequently, with five times more precision, in order to detect smaller deforestation areas than ALOS-2. ALOS-4 will acquire data more frequently than ALOS-2; thus, interferometric data will lead to more efficient infrastructure maintenance.

1.3.3. *NISAR*

The NASA and Indian Space Research Organisation (ISRO) NISAR mission with both L-band and S-band wavelengths is dedicated to and optimized for studying hazards and global environmental change. NISAR, planned to be launched in 2022, will observe Earth's land and ice-covered surfaces at 3–10 m resolution globally, with 12-day regularity on ascending and descending passes, sampling Earth on average every six days for a baseline three-year mission. As NISAR will be only left-looking, it will be a major asset for mapping Antarctica, while Sentinel-1, with a similar large mission coverage, is already undertaking right-looking operations.

NASA contributions include the L-band SAR instrument, with a 12 m diameter deployable mesh reflector, 9 m deployable boom and octagonal instrument structure. In addition, NASA will deploy a high-capacity solid-state recorder (approximately 9 Tbits at end of life), GPS, a 3.5 Gbps Ka-band telecom system and an engineering payload to coordinate command and data handling with the ISRO spacecraft control systems.

The ISRO will provide the spacecraft and launch vehicle, as well as the S-band SAR electronics to be mounted on the instrument structure. The coordination of technical interfaces among subsystems is a major focus area in the partnership. Orbit control within 350 m and a pointing control shorter than 273 arcseconds will enable good accuracy to follow deformation by InSAR.

Data access: As mentioned by NASA (2019), the science teams and algorithm development teams at NASA and the ISRO will work jointly to create a common set of product types and software. The project will deliver NISAR data to NASA and the ISRO for archiving and distribution. NASA and the ISRO have agreed to a free and open data policy for these data. All NISAR science data (L- and S-band) will be freely available and open to the public, consistent with the long-standing NASA Earth Science open data policy.

The target strategy assigns a single radar mode to a given area on Earth. Where target areas overlap, the modes are compatible so that no science discipline loses information. This set of global target types and associated radar modes will provide each of the individual disciplines with the data they need for their science. The observation plan calls for nearly continuous global coverage over land and ice. India has planned specific radar modes to fulfill ISRO's science requirements for the mission. For the rest of the globe, the most inclusive radar mode was chosen where conflicting science discipline needs were identified.

1.3.4. *Biomass*

This mission has been selected as part of the ESA's Earth Observation Program (Earth Explorer 7) and will be able to operate, for the first time in space, a P-band imaging radar at 70 cm wavelength. At this frequency (435 MHz), the ITU allocation is limited to a 6 MHz bandwidth: the program will address key information issues over forested areas, at low-resolution cells. A large circular antenna of 12 m will be used.

Different applications are foreseen to exploit penetration through tropical forests, with SAR polarimetry and PolInsar interferometry as well as tomography. Different phases and orbit changes will be used along the mission to achieve different purposes and fulfill these applications, with different baselines from different orbits. The first option is a coverage with a double baseline using three interleaved swaths, a low repeat cycle and then a change of altitude to ensure a new coverage; complete coverage would need five months before coming back to the same tracks. The second option is a sub-cycle of three or four days, and a total cycle of 27 or 36 days, then a repetition every five months (Hélière *et al.* 2016). The coverage is determined by three adjacent stripmap modes with swaths of 60, 50 and 40 km, implying a roll maneuver to change swath (Arcioni *et al.* 2013).

Launch is foreseen in 2023 in Kourou on a Vega launcher to put Biomass on a dawn–dusk sun-synchronous orbit about 600 km high.

1.3.5. *ROSE-L*

The Copernicus L-band ESA SAR Sentinel-1–2 radar mission, named ROSE-L (Radar Observing System for Europe), provides enhanced continuity for a number of Copernicus services and downstream commercial and institutional users. It should be launched in 2028 and will work in synergy with other Copernicus elements to specially address emergency management service requirements. It should offer the same coverage and same frame as Sentinel-1 (swath widths around 250 km) to exploit synergy with a repeat coverage of less than one day in Arctic areas, three days in Europe and six days everywhere using two satellites (Pierdicca *et al.* 2019). Spatial resolution is expected to be better than 5 m x 10 m.

Data access: The main characteristics of the ROSE-L mission, dedicated to operational services, include reliability, data timeliness, a European free and open data policy, systematic acquisitions, a reduction in the number of payload modes and coordinated operations with Sentinel, and these distinguish ROSE-L from the other SAR mission initiatives.

1.4. Optical imaging missions

1.4.1. *Past optical missions*

In this section, we present some optical satellites that have taken images that are considered as references today. These satellites were launched between 1984 and 2004. Most of them are inactive today (but not all of them). Such satellites were heavy, and many of them had considerable longevity. At the beginning of the 2000s, new satellites arrived, such as Ikonos and Quickbird-2: these were agile satellites, with submetric resolution and increased dynamic range. For various satellites, several parameters, such as altitude, local time, weight, year of launch, spatial resolution, panchromatic spectral range and swath, are given in Tables 1.6 and 1.7. As we focus on ground motion displacement measurements and DEM extraction, the highest spatial resolution (panchromatic band in general), with its corresponding swath, is given.

Satellite	Nation	Altitude (km)	Local time	Period of orbit (min)	Inclination (deg)	Weight (kg)
Landsat 5	USA	705	9:45	99	98.2	2,200
SPOT-1	France	822	10:30	-	-	1,907
SPOT-2	France	822	10:30	-	-	1,907
SPOT-3	France	822	10:30	-	-	1,907
SPOT-4	France	832	10:30	-	-	2,755
Landsat 7	USA	705	10:00	99	98.2	1,973
Terra ASTER	USA–Japan	705	10:30	98.88	98.3	4,864
Ikonos	USA	681	10:30	-	-	817
Quickbird-2	USA	460	10:30	93.5	97.2	1,100
SPOT-5	France	822	10:30	101.4	98.7	3,030
CBERS-2/2B	China–Brazil	778	10:30	100.26	98.5	1,450
Formosat-2	Taiwan	888	9:30	-	99.14	764

Table 1.6. *Past mission characteristics and parameters*

Satellite	Year of launch	End of life	Highest spatial resolution (m)	Panchroma spectral range (nm)	Number of XS bands	Swath width for highest resolution (km)	Dynamic range (bits)
Landsat 5	1984	2013	30	-	11	185	-
SPOT-1	1986	2003	10	500–730	3	60	8
SPOT-2	1990	2009	10	500–730	3	60	8
SPOT-3	1993	1996	10	500–730	3	60	8
SPOT-4	1998	2013	10	510–730	4	60	8
Landsat 7	1999	-	15	520–900	7	185	8
Terra ASTER	1999	-	15	-	15	60	8
Ikonos	1999	2015	0.82	450–900	4	11.3	11
Quickbird-2	2001	2015	0.65	450–900	4	18	11
SPOT-5	2002	2015	2.5	480–710	4	60	8
CBERS-2/2B	2003,2007	2007,2010	20	510–730	4	113	-
Formosat-2	2004	2016	2	450–900	4	24	8

Table 1.7. *Past mission instrument characteristics*

In the following, some optical satellites are presented to illustrate their interest for surface displacement measurement using archive satellite imagery:

– *Landsat 5:* Landsat 5 delivered Earth imaging data for nearly 29 years and set the Guinness World Record for the "longest operating Earth observation satellite". Its longevity provides a reference for time-series analysis of the Earth.

– *SPOT-1:* A commercial high-resolution optical satellite initiated by CNES in the 1970s. Since 1986, the SPOT family has taken more than 10 million high-quality images, which constitute reference data.

– *SPOT-4:* The second generation of the SPOT satellites. The panchromatic band is associated with four multispectral (XS) bands at 20 m resolution. The platform is three-axis stabilized, and the viewing direction is selected with a mirror.

– *Landsat 7:* This satellite, launched in 1999, is considered a high-accuracy calibrated Earth-observing satellite. Its radiometric measurements are accurate compared to the same measures made on the ground. In October 2008, the USGS made all Landsat 7 data free to the public. It led to a 60-fold increase of data downloads. The satellite is still active in 2020.

– *ASTER:* The ASTER instrument on the Terra satellite creates a detailed map of land surface, and scientists have made an elevation model with this data. The first global digital elevation model (GDEM) was released to the public in 2009. It covers the polar regions and complements NASA's SRTM. The spatial resolution of pixels of 15 m is conducive to the creation of a 30 m resolution DEM. The last version of the ASTER DEM was released in 2019. Since 2016, all satellite images have been available at no charge for all users. The satellite was still active in 2020.

– *Ikonos:* Ikonos was a breakthrough among the commercial satellites. The whole platform is used to point towards a geographical region. The whole satellite is agile and, consequently, the satellite weight is reduced. Commercial images with submetric resolution are available. The dynamic range of images is increased.

– *Quickbird-2:* Quickbird-2 features enhanced commercial spatial resolution. It is the highest-resolution commercial satellite to date.

– *SPOT-5:* Third-generation satellite of the SPOT family operated by Spot Image. It takes 2.5 m panchromatic images and 10 m multispectral ones. Pointing is done with steering mirrors. The HRS instrument is composed of two telescopes pointing around 20° forward and backward. The HRS takes native stereoscopic images with a resolution of 5 m (along track) $\times$ 10 m (across track).

– *Formosat-2:* This satellite has a daily revisit for event/disaster monitoring, thanks to a body-pointing capability of $\pm 45°$ in roll and pitch.

1.4.2. *On-going optical missions*

This section presents optical satellites that are still operational and whose data may be used for DEM extraction and ground motion displacement measurement. Their spatial resolution is below 15 m. The list is not exhaustive. The key parameters of the missions are summarized in Tables 1.8 and 1.9, and the main characteristics of the instrument are listed in Tables 1.10 and 1.11. Some figures are different depending on the web sources and may differ from the real ones in some cases. Swath width and dynamic ranges are given for the highest resolution (mainly panchromatic bands).

During the last 15 years, the main drivers were increases in the spatial resolution and the revisit time. This is conducive to the conception of constellations of agile satellites (simpler, if possible, and smaller). It is possible to observe several evolution patterns for optical satellite missions:

– agility is often the reference: it simplifies the instrument and reduces use of steering mirrors;

– operators reduce the satellite altitude in order to find a compromise between smaller aperture and better spatial resolution. Examples: Cartosat-2C and Cartosat-3A;

– a large aperture is chosen to optimize the resolution. Examples: WorldView-3 and Gaofen-8;

– satellite constellations are increasingly proposed: this decreases the weight of the satellite, the payload and platform complexity and the satellite unitary cost, while increasing the number of satellites (constellation) and improving the revisit time. Constellations of high-resolution satellites are used to retrieve change detection information, geo-statistics, etc. Examples: SkySat, SuperView and RapidEye;

– usually, a focal plane is composed of several time delay integration (TDI) image sensors, mounted in staggered configuration. CCD sensors are used where ultralow noise is preferred. However, CMOS detectors with low power, high frame rate and low cost are also used. The use of matrix detectors (CCD or CMOS) adds video mode functions into the satellites. Examples: Zhuhai and BlackSky constellations.

In the following, some missions with a specific interest for DEM computation or displacement measurement are presented:

– *Asnaro:* The overall aim of this project is to develop a new generation of mini-satellite buses with high-performance characteristics. It is based on open-architecture techniques and manufacturing methodologies to reduce the cost and development period.

– *Carbonite-2:* This satellite is a technology demonstration of low-cost video-from-orbit with commercial off-the-shelf (COTS) technologies. At only 100 kg, it delivers 1 m ground sample distance (GSD) images. The swath width is very limited.

Satellite	Nation	No. of satellites	Altitude (km)	Local time	Period of orbit (min)	Inclination (deg)	Weight (kg)
Alos	Japan	1	692	10:30	-	98.2	4,000
Alsat-1B	Algeria	1	700	10:30	98.5	98	103
Alsat-2A/2B	Algeria	2	637	10:30	-	-	130
Asnaro	Japan	1	504	11:00	-	97.42	495
BKA (Belka-2)	Belarus	1	505	-	-	97.49	473
BlackSky constellation	USA	20	500	10:30	-	53	56
Carbonite-2	UK	1	505	10:30	94.6	97.5	100
Cartosat-1 (IRS-P5)	India	1	618	10:15	-	97.87	680
Cartosat-2A/2B	India	2	635	9:32	75	90	690
Cartosat-2C–2F	India	4	505	9:30	94.72	97.4	727
Cartosat-3/3A/3B	India	1	505	9:30	-	97.5	1,625
CBERS-4/4A	China–Brazil	2	779	10:30	100.32	98.54	1,980
CESAT-1	Japan	1	500	-	-	97.45	50
Deimos-1	Spain	1	686	10:30	-	-	91
Deimos-2	Spain	1	600	10:30	-	97.9	310
Dove PlanetScope	USA	55	420	9:30–11:30	-	51.6	-
Dove PlanetScope 2 (PS2)	USA	100–150	475	9:30–11:30	-	98	-
DubaiSat-1	UAE	1	685	10:30	98.2	98.13	200
DubaiSat-2	UAE	1	600	10:30	-	97.1	300
DubaiSat-3 (KhalifaSat)	UAE	1	613	10:30	-	-	330
Eros B	Israel	1	520	14:00	94.79	97.45	350
Eros C	Israel	1	510	-	-	-	400
Gaofen-1	China	1	635	10:30	-	98.36	1,080
Gaofen-2	China	1	608	10:30	98.02	-	2,100
Gaofen-6	China	1	634	11:00	-	98.05	1,080
Gaofen-7	China	1	502	-	-	-	2,400
GeoEye-1	USA	1	770	10:30	98	-	1,955
Jilin-1/7/8	China	3	658	10:30	-	-	450
Kanopus-V-IK	Russia	1	510	10:30	-	-	473
Kanopus-V1–V6	Russia	5	510	10:30	-	-	450
KazeoSat 1	Kazakhstan	1	759	10:15	100	98.54	830
KazeoSat 2	Kazakhstan	1	630	10:30	-	-	185
Kompsat-2	South Korea	1	685	10:50	-	-	800
Kompsat-3	South Korea	1	685	10:50	-	98.14	980
Kompsat-3A	South Korea	1	528	10:50	-	97.5	1,100

Table 1.8. *Mission characteristics and parameters (1/2)*

Satellite	Nation	No. of satellites	Altitude (km)	Local time	Period of orbit (min)	Inclination (deg)	Weight (kg)
Landsat 8	USA	1	705	10:00	98.8	98.22	2,780
Nemo-HD	Slovenia–Canada	1	500	-	-	-	70
Pléiades-1A	France	1	695	10:30	-	-	980
Pléiades-1B	France	1	695	10:30	-	-	980
Prisma	Italy	1	614	10:30	99	98.19	830
RapideEye	Germany–USA	5	630	11:00	-	-	156
ResourceSat-2	India	1	817	10:20	-	-	1,206
Resurs-P 1, 2, 3/GEOTON	Russia	3	473	10:30	-	-	6,570
Resurs-P 1, 2, 3/SHMSA-VR	Russia	3	473	10:30	-	-	6,570
Sentinel-2-A/B	Europe (ESA)	2	786	10:30	-	98.5	1,200
SkySat-1	USA	1	450	13:00	-	-	110
SkySat-2-21	USA	6	450	10:30–13:00	-	-	110
SPOT-6	France	1	694	10:30	98.8	98.2	712
SPOT-7	France–Azerbaijan	1	694	10:30	98.8	98.2	712
SuperView-1–4	China	4	530	10:30	97	-	560
TeLEOS-1	Singapore	1	550	-	-	15	400
TH-1	China	3	500	13:30	-	97.35	1,000
TripleSat/DMC3/ SSTL-S1-2-3	UK–China	3	651	10:30	97.7	-	350
TripleSat/DMC3/ SSTL-S4	UK–China	1	636	10:30	97.7	-	458
UK-DMC2	UK	1	665	10:00	97.9	97.8	120
Vision-01	UK	1	583	10:30	96.2	97.5	-
VRSS-1	Venezuela	1	635	10:30	97.48	97.7	880
VRSS-2	Venezuela	1	645	10:30	97.6	98.05	1,000
WorldView-1	USA	1	496	13:30	94.6	97.5	2,500
WorldView-2	USA	1	770	10:30	100	98.58	2,800
WorldView-3	USA	1	617	10:30	97	98	2,800
WorldView-4	USA	1	617	10:30	97	97.97	2,485
Zhuhai-1 OHS-2A-2D/ 3A-3D	China	8	500	10:15	-	97.3	50
Zhuhai-1 OVS-1A/1B/2	China	3	500	10:15	-	97.3	50–90
Ziyuan-3/ZY-3 01-03	China	3	506	10:30	94.7	97.4	2,630

Table 1.9. *Mission characteristics and parameters (2/2)*

Satellite	Year of launch	End of life	Highest spatial resolution (m)	Panchromatic spectral range (nm)	No. of XS bands	Swath width for highest resolution (km)	Dynamic range (bits)
Alos	2006	2011	2.5	520–770	4	70	-
Alsat-1B	2016	-	12	-	3	-	-
Alsat-2A/2B	2010, 2016	-	2.5	-	4	17.5	-
Asnaro	2014	-	0.5	-	-	10	12
BKA (Belka-2)	2012	-	2.1	-	4	23	-
BlackSky constellation	2018 (2), 2019 (2), 2020 (4), 2021	-	0.8–1	-	-	-	-
Carbonite-2	2018	-	1	-	4	5(x5)	-
Cartosat-1 (IRS-P5)	2005	-	2.5	500–850	0	30	-
Cartosat-2A/2B	2008, 2010	-	1/0.8	500–850	0	9.8	-
Cartosat-2C–2F	2016–2018	-	0.65	500–850	4	10	-
Cartosat-3/3A/3B	2019	-	0.28	500–850	4	17	-
CBERS-4/4A	2014, 2019	-	5	510–730	16	60	8
CESAT-1	2017	-	1	-	-	6x4	-
Deimos-1	2009	-	22	-	3	600	-
Deimos-2	2014	-	0.75	450–900	4	12/24	10
Dove PlanetScope	2018	-	2.7–3.2	-	4	24x8	12
Dove PlanetScope 2 (PS2)	2020	-	3.7–4.9	-	4/8	24x16, 32x19.6	12
DubaiSat-1	2009	2017	2.5	420–720	4	20	8
DubaiSat-2	2013	-	1	420–890	4	12/24	-
DubaiSat-3 (KhalifaSat)	2017	-	0.70	420–890	4	16/12	-
Eros B	2006	-	0.7	500–900	0	7	10
Eros C	2019	-	0.38	450–900	4	11.5	12
Gaofen-1	2013	-	2	450–900	8	70	-
Gaofen-2	2014	-	0.8	450–890	4	45	-
Gaofen-6	2018	-	2	450–900	-	90	-
Gaofen-7	2019	-	<1 m	-	-	-	-
GeoEye-1	2008	-	0.46	450–800	4	15.2	-
Jilin-1/7/8	2015	-	0.72–0.8	450–900	4	48	-
Kanopus-V-IK	2018	-	2.5	540-860	6	23	12
Kanopus-V1–V6	2012–2018	-	2.1	540–860	4	23	12
KazeoSat 1	2014	-	1	450–750	4	20	12
KazeoSat 2	2014	-	6.5	-	5	77	12
Kompsat-2	2006	-	1	500–900	4	15	10
Kompsat-3	2012	-	0.72	450–900	4	16	14
Kompsat-3A	2015	-	0.55	450–900	5	12	14

Table 1.10. *Instrument characteristics (1/2)*

Satellite	Year of launch	End of life	Highest spatial resolution (m)	Panchromatic spectral range (nm)	No. of XS bands	Swath width for highest resolution (km)	Dynamic range (bits)
Landsat 8	2013	-	15	500–680	10	185	12
Nemo-HD	2020	-	2.8	400–900	4	10	-
Pléiades-1A	2011	-	0.7/0.5	457–848	4	20	12
Pléiades-1B	2012	-	0.5	457–848	4	20	12
Prisma	2019	-	5	400–700	66 + 171	30	12
RapideEye	2008	2020	6.5/5	-	5	77	12
ResourceSat-2	2016	-	5.8	-	11	23/70	-
Resurs-P 1, 2, 3/GEOTON	2013, 2014 2016	-	1	580–800	5	38	-
Resurs-P 1, 2, 3/SHMSA-VR	2013, 2014 2016	-	12	430–700	6	97	-
Sentinel-2-A/B	2015, 2017	-	10	-	13	290	-
SkySat-1	2013	-	0.8	450–900	4	2/8	-
SkySat-2-21	2014	-	0.65–0.8	450–900	4	2/8	-
SPOT-6	2012	-	1.5	455–745	4	60	12
SPOT-7	2014	-	1.5	455–745	4	60	12
SuperView-1–4	2016, 2018	-	0.5	450–890	4	12	11
TeLEOS-1	2015	-	1	-	0	12	10
TH-1	2010, 2012, 2015	2018, 2020	2–5	510–690	4	60	-
TripleSat/DMC3/ SSTL-S1-2-3	2015	-	1	450–650	3	23.8	10
TripleSat/DMC3/ SSTL-S4	2018	-	1	450–650	3	23.8	10
UK-DMC2	2009	-	22	-	-	600	-
Vision-01	2018	-	0.9	450–650	4	20.8	10
VRSS-1	2012	-	2.5	450–900	4 + 4	57	10
VRSS-2	2017	-	1	-	4 + 5/6	-	-
WorldView-1	2007	-	0.46	448–895	0	17.6	11
WorldView-2	2009	-	0.46	464–801	4 + 4	16.4	11
WorldView-3	2014	-	0.31	450–800	8 + 8 + 12	13.1	11
WorldView-4	2016	2019	0.31	450–800	4	13.1	11
Zhuhai-1 OHS-2A-2D/3A-3D	2018, 2019	-	10	-	-	150	-
Zhuhai-1 OVS-1A/1B/2	2017, 2018	-	0.9–2.0	-	-	22.5	-
Ziyuan-3/ZY-3 01-03	2012, 2016, 2020	-	2.5	500–800	4	51	10

Table 1.11. *Instrument characteristics (2/2)*

– *Pléiades:* Pléiades represents the new generation of French satellites with submetric resolution. The main characteristics of Pléiades are the high image quality, the high agility of the satellites, high image geolocation, the ability to provide images with a short delay (exemplified with the International Charter "Space and Major Disasters") and the ability to take stereo or tri-stereo images on a single track.

– *RapidEye:* The first constellation of Earth observation satellites, with five identical satellites launched simultaneously with five spectral bands and 5 m resolution. The aim of the constellation was to provide Earth observation imagery for decision-making and geospatial services. In April 2020, Planet Labs decided to end the operation of the RapidEye constellation.

– *Sentinel-2:* An Earth observation mission from the Copernicus Program to systematically acquire imagery at a high resolution (10–60 m). It was developed and is being operated by the ESA. Its main characteristics are high swath (290 km), systematic acquisition of land surfaces and a large revisit (five days with two satellites) and a large number of multispectral bands with high SNR. However, the main breakthrough is the free and open data policy. After many years of acquisition of decametric images by several Earth observation satellites, Sentinel-2 has boosted the general use of satellite images at 10 m resolution. The easy access to satellite images opens the door to many new applications: agriculture (crop monitoring), land cover change, detailed vegetation monitoring, ice extent mapping, flood mapping, etc.

– *SPOT-6:* SPOT-6 is the proposition by Airbus Defence and Space to ensure the continuity of SPOT-4 and SPOT-5 satellites. The swath is similar, with a resolution of 1.5 m. It is thus able to cover large areas for agriculture, deforestation and environmental monitoring, as well as for industries, coastal surveillance and defense.

– *WorldView-1:* This satellite offers a high capacity and high resolution (0.46 m). It is based only on panchromatic acquisitions. WorldView-1 has an excellent geolocation capability and rapid targeting of in-track stereo images.

– *WorldView-2:* This very-high-resolution satellite adds multispectral bands to the panchromatic one for enhanced spectral analysis. WorldView-2 has a higher agility and works at a higher altitude. Consequently, the revisit time of any place on Earth is reduced to 1.1 days. Despite its higher altitude, the spatial resolution remains the same as WorldView-1. WorldView-2 has the capacity for direct tasking: some customers can load imaging profiles directly to the satellite and it delivers the data directly to their own ground stations.

– *WorldView-3:* While WorldView-3 is similar to WorldView-2 in terms of conception, several multispectral instruments are added: it contains eight multispectral bands (visible to near-infrared), eight short-wave infrared (SWIR) bands and 12 CAVIS bands (at 30 m resolution). It is the first super-spectral high-resolution commercial satellite.

– *WorldView-4:* In 2019, WorldView-4 suffered a failure in one of its control moment gyroscopes. It is therefore considered no longer usable.

– *Dove PlanetScope:* PlanetScope is one of the satellite constellations operated by the company Planet. It has almost daily coverage worldwide with more than a hundred nanosatellites. Each PlanetScope is a CubeSat 3U with dimensions 10 cm × 10 cm × 30 cm, producing images at 3.7 m resolution at nadir, with four spectral bands (red, green, blue and near-infrared).

1.4.3. *Future optical missions*

Regarding the new missions, it is possible to outline the main strategic drivers as follows:

1) satellite operators replace the existing missions that provide intensively used products, across the whole range of resolutions (<10 m spatial resolution in general). Examples: WorldView Scout 1-6, ResourceSat-3/3A;

2) new commercial systems try to reach higher-spatial-resolution markets. Examples: PleiadeNeo, WorldView Legion 1-6;

3) the value of the data is linked to the acquisition time after an event. Operators want to decrease the satellite revisit time by increasing the number of agile satellites. Thanks to the evolution of technologies, it is possible to reduce satellite dimensions and satellite weights compared to the earlier generations. It is also possible to reduce the cost of a single satellite (and the launch cost). In order to reach a high spatial resolution, a lower altitude is chosen. Examples: new constellations such as Co3D, Vivid-i and PlanetScope. However, many new space constellation projects have also been canceled due to technical and financial constraints;

4) new technologies open the door to new in-orbit functionalities; for example, new sensors provide new imagery and video capabilities;

5) new hyperspectral satellites are coming, such as Enmap and Prisma, with 30 m resolution. The hyperspectral images may be used to obtain information about soil characteristics such as humidity and may be very useful for the analysis of soil displacements.

The technical specifications of some future optical missions, where available, are given in Tables 1.12 and 1.13. Some of them may be of specific interest for future 3D and surface displacement measurements:

– *Co3D:* These extremely agile satellites with all-electric platforms will provide new ways of acquiring, processing and transferring images. They will provide very-high-resolution 3D data;

– *PléiadesNeo:* This system corresponds to the new generation of Pléiades satellites. With a 1.5 m diameter telescope, the satellites will provide 30 cm images. The constellation will produce 40 TB of data every day;

– *EarthDaily:* EarthDaily provides an alternative choice for building global world coverage for change detection activities with this satellite constellation working at a 5 m resolution. This system complements a very-high-resolution service;

– *Vivid-i:* This constellation is planned with 15 satellites, with a weight lower than 100 kg, providing imagery and video capabilities at less than 1 m resolution;

– *WorldView Legion:* These satellites correspond to the new generation of WorldView satellites. Launches are planned for 2021;

– *WorldView Scout:* This constellation will have six or more small satellites, which will be able to revisit some key areas 40 times a day at a 0.80 m resolution.

Satellite	Nation	No. of satellites	Altitude (km)	Local time	Weight (kg)
Co3D	France	4–2	502	11:00	300
Enmap	Germany	1	653	11:00	871
PléiadesNeo	France	4	620	10:30	750
ResourceSat-3/3A	India	1	795	10:30	1,200
EarthDaily	Canada	8	-	-	340
Vivid-i constellation (Earth-i)	UK	15	500	-	100
WorldView Legion 1–6	USA	6	-	-	-
WorldView Scout 1–6	USA–Saudi Arabia	6	-	-	-

Table 1.12. *Future mission characteristics and parameters*

Satellite	Year of launch	Highest spatial resolution (m)	No. of XS bands	Swath width for highest resolution (km)
Co3D	2023	0.5	4	7 x 5
Enmap	2021	30	8 + 154	30
PléiadesNeo	2021	0.3	≥4	14
ResourceSat-3/3A	2021	10	4	280
EarthDaily	2022	5	9	360
Vivid-i constellation (Earth-i)	-	0.6–1	3	5.2 x 5.2
WorldView Legion 1–6	2021	0.30	-	-
WorldView Scout 1–6	2021	0.8	-	-

Table 1.13. *Future instrument characteristics*

1.5. Conclusion

With SAR imaging missions, free data availability, as with, for example, Sentinel-1 and future NISAR or ROSE-L missions, is of real interest for scientists in the field of displacement observations, in particular from missions with high repetitiveness and controlled orbits. For other missions, data of interest will hopefully be accessible through scientific announcements as they used to be. As for new low-cost constellations delivering X-band SAR data, the usefulness for remote sensing displacement applications is still unknown. Even if high resolution is often the highlighted improvement for these missions, the lack of systematic acquisitions over large areas reduces the potential interest of these data for ground displacements all over the Earth.

For the optical imaging missions, nowadays, many Earth observation images are available across many spectral bands and many spatial resolutions. Thanks to the Sentinel and Landsat programs, a large number of images are freely available, allowing many new applications for scientists and commercial companies. Temporal resolution becomes as important as spatial resolution. Many satellite optical missions are commercial ones, with high competition in terms of the prices of provided images, the freshness of the data and the value-added services. With some applications, the satellite image as a product becomes less important. New space actors directly provide geo-statistics services based on the aggregated satellite data. The latest generations of satellites with very high resolutions (30 cm) open the door to more precise measurements of ground displacements.

1.6. Acknowledgments

We thank Roger Fjørtoft and Jean-Marie Nicolas for their careful reviews and comments that helped to improve this chapter.

1.7. References

Arcioni, M., Bensi, P., Fois, F., Gabriele, A., Hélière, F., Lin, C., Massotti, L., Scipal, K. (2013). ESA's Biomass mission system and payload overview. *Proceedings of the ESA Living Planet Symposium*, Edinburgh, UK.

CSA (2019). Open data: Over 36,000 historical RADARSAT-1 satellite images of the earth now available to the public. Canadian Space Agency, April 1, Saint-Hubert, Quebec [Online]. Available at: https://www.canada.ca/en/space-agency/news/2019/03/open-data-over-36000-historical-radarsat-1-satellite-images-of-the-earth-now-available-to-the-public.html.

ESA (2019). (A)SAR on-the-fly: Data dissemination and processing service user manual. ASAR-OTF-UM, version 2.1 [Online]. Available at: https://earth.esa.int/eogateway/documents/20142/37627/ASAR-OTF-user-manual.pdf.

Hélière, F., Carbone, A., Fonseca, N., Ayllon, N., Barnes, A., Fehringer, M. (2016). Biomass P-band SAR. *EUSAR Proceedings*, Hamburg, Germany.

ITU-R (2009). Frequency bands and required bandwidths used for spaceborne active sensors operating in the earth exploration-satellite (active) and space research (active) services. ITU-R recommendation RS.577-7 [Online]. Available at: https://www.itu.int/rec/R-REC-RS.577-7-200902-I/en.

ITU-R (2010). The essential role and global importance of radio spectrum use for Earth observations and for related applications [Online]. Available at: https://www.itu.int/dms_pub/itu-r/opb/rep/R-REP-RS.2178-2010-PDF-E.pdf.

ITU-R (2020). Eco frequency information system table ITU-region1 [Online]. Available at: https://efis.cept.org/view/search-general.do/.

NASA (2019). *NASA-ISRO SAR (NISAR) Mission Science Users' Handbook*. NASA Jet Propulsion Laboratory, CL# 18-1893, California, USA.

Pierdicca, N., Davidson, M., Chini, M., Dierking, W., Djavidnia, S., Haarpaintner, J., Hajduch, G., Laurin, G., Lavalle, M., López-Martínez, C., Nagler, T., Su, B. (2019). The Copernicus L-band SAR mission ROSE-L (Radar Observing System for Europe). *Proceedings of the SPIE Remote Sensing 2019*, Strasbourg, France.

Prats-Iraola, P., Scheiber, R., Marotti, L., Wollstadt, S., Reigber, A. (2012). Tops interferometry with TerraSAR-X. *IEEE TGRS*, 50(8).

Solaas, G. and Coulson, S. (1994). ERS-1: Interferometric baseline algorithm verification. ERS-1 SAR Interferometry Orbit Listing User Note [Online]. Available at: https://earth.esa.int/ers/eeo4.10075/0010d.html.

Zan, F.D. and Guarnieri, A.M. (2006). TOPSAR: Terrain observation by progressive scans. *IEEE TGRS*, 44(9), 2352–2360.

2

Image Matching and Optical Sensors

Marc PIERROT DESEILLIGNY and Ewelina RUPNIK

LaSTIG, ENSG-IGN, Gustave Eiffel University, Saint-Mande, France

2.1. Introduction, definition and applications

Image matching is a well-known problem in the Earth sciences and features in two main applications: 1) the computation of 3D geometry from images acquired from different viewpoints and 2) the computation of surface deformations from images acquired from similar points of view. Many scientific communities (photogrammetry, robotics, computer vision, Earth science, etc.) have investigated image matching problems, which has led to many different naming conventions. In this chapter, we will interchangeably use the terms *image matching* and *image correlation* to describe the same concept, which is introduced in the following section. The image matching concepts and methods that are described in this chapter are presented in the context of optical data. Most of them can also be applied to SAR amplitude images. However, since the radiometry and geometry of SAR sensors are different from optical sensors, methods specific to SAR imagery will be presented in Chapter 3.

2.1.1. *Problem definition*

Given a pair of images, let us assume that for some physical reasons, one image is a deformation of the other. The goal of image matching is to recover this deformation,

Surface Displacement Measurement from Remote Sensing Images,
coordinated by Olivier CAVALIÉ and Emmanuel TROUVÉ.
© ISTE Ltd 2022.

given the images. In mathematical terms, let I_1 and I_2 be the images we want to *match*, and let I_2 be the deformation of I_1 by a certain geometric function G. For each pixel p, we can then write:

$$I_1(p) = I_2(p') \quad \text{with} \quad p' = G(p). \qquad [2.1]$$

Equation [2.1] assumes that radiometry is directly comparable between the two images. This hypothesis is true in optical flow approaches (Horn and Schunck 1981) because the images are acquired from nearly the same viewpoints and at small time intervals. In Earth sciences, when images are collected (a) from different orbits over non-Lambertian scenes, (b) at different dates and (c) often using different spectral bands, this hypothesis does not hold true. To account for the different phenomena, we assume that the radiometry of I_2 is altered by some *unknown* factor, and that the induced change is only locally constant (i.e. it depends on the surface material, the scene geometry, texture and atmospheric conditions). We then introduce R, the local radiometric deformation, and rewrite equation [2.1] as follows:

$$I_1(p) = R(p, I_2(G(p))) \qquad [2.2]$$

We can now reformulate the matching problem: given two images I_1 and I_2, where one image is a deformation of the other following equation [2.2], recover G, and optionally R. Note that, formulated in this way, image matching is an ill-posed problem because there are an infinite number of solutions satisfying the equation. As evidence of this, by taking the simpler equation [2.1] and setting $I_1 = I_2$, it can be seen that the solution is known up to a geometric transformation G that preserves the radiometry of I_1.

2.1.2. *Geometry of matching*

Matching can be carried out in several geometries, but for the purpose of this chapter, we only distinguish between two simple cases:

– One-dimensional (1D) matching: for each pixel, we compute only one deformation component; this is applied to epipolar image matching (see section 2.5.2.3) where we know that corresponding pixels are found on the same lines, i.e. $G(x, y) = (x + p_x, y)$; this case occurs when the scene is static, the images are taken from different points of view and the matching is used to compute a 3D surface.

– Two-dimensional (2D) matching: here, no *a priori* about the matching direction exists, and the deformation function G can be any arbitrary function; this case occurs when matching is used to compute the movement of a deformable scene (e.g. due to an earthquake).

In this chapter, we treat the two cases identically. The reader should keep in mind that in the 1D case, the search region is a 1D interval, whereas in the 2D case, it is a 2D region (e.g. square, rectangle, circle, ellipse).

2.1.3. *Radiometric and geometric hypothesis*

The most common image-matching approach is the so-called *vignette* or template matching. It allows us to narrow down the solutions to equation [2.2], which is justified by the following hypotheses about the R and G deformations:

– G is globally small, hence it is sufficient to match it only in the neighborhood of a pixel (equation [2.3]);

– G is locally small, so its gradients are close to identities (equation [2.4]), justifying comparisons of windows of pixels instead of single pixels;

– the gradient of R relative to p is small, hence it can be initially ignored (equation [2.5]).

$$G(p) \approxeq p \qquad [2.3]$$

$$\frac{\partial G}{\partial x} \approxeq 1 \;\; and \;\; \frac{\partial G}{\partial y} \approxeq 1 \qquad [2.4]$$

$$\frac{\partial R}{\partial x} \approxeq 0 \;\; and \;\; \frac{\partial R}{\partial y} \approxeq 0 \qquad [2.5]$$

Equations [2.3] and [2.4] may seem restrictive; however, in practice, it is sufficient that G is a smooth function and that we know an approximation $\tilde{G}$ of G (e.g. by assuming the mean altitude of the scene). When this is the case, we can compute a rectified image $\tilde{I_2}$ of I_2 that is defined by $\tilde{I_2}(p) = I_2(\tilde{G}(p))$, and compute the image deformation Δ_G between I_1 and $\tilde{I_2}$. Finally, we set the output deformation to $G = \tilde{G} \circ \Delta_G$. This said, for the sake of simplicity, we assume in the following that equations [2.3] and [2.4] apply.

2.2. Template matching

2.2.1. *Matching as a similarity measure*

Assuming that equation [2.3] is true, we can state that the correspondence of a pixel p is a pixel $p' = p + \delta(p)$ with $|\delta(p)| < d_{max}$, where d_{max} is a given threshold. When d_{max} is sufficiently small, thus avoiding ambiguity and long computation, the matching problem boils down to computing, for each pixel p, its most similar pixel in the search region given by $D \in \langle -d_{max}, -d_{max} \rangle$ (see Figure 2.1).

For a given pair I_1, I_2, let $DS(p, p')$ be the dissimilarity between two pixels p and p'. If N_d is the neighborhood of size d_{max}, the matching algorithm is simply:

$$M(p) = argmin_{p' \in N_D}(DS(p, p')) \qquad [2.6]$$

We now define a similarity measure that is both non-ambiguous and efficient to compute.

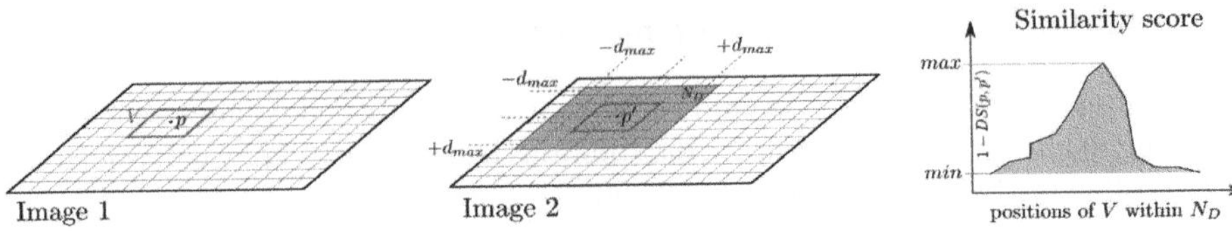

Figure 2.1. *A toy example of the basic 2D template matching. A window V centered at p in Image 1 is moved in the search area N_D of Image 2. At each position, a similarity score is computed for the pair of overlapping windows. The maximum score points to the best match of pixel p. For a color version of this figure, see www.iste.co.uk/cavalie/images.zip*

2.2.2. *Basic template approach*

Ignoring the radiometric deformation R, let us have a closer look at equation [2.1]. Assuming hypothesis [2.4], we define a dissimilarity measure using the following formula, where V is the neighborhood that defines a window:

$$DS(p, p') = \sum_{v \in V} |I_1(p + v) - I_2(p' + v)| \qquad [2.7]$$

Note that the formula of equation [2.7] has a strong implication – it means that if p' is the corresponding pixel of p, then for any v in the neighborhood V, $p' + v$ is the corresponding pixel of $p + v$. In other words, it is the hypothesis expressed in equation [2.4]; however, it is given in a discrete form. Of course, the key issue here is the size of the neighborhood V: if V is too small, then matching will be ambiguous; if V is too big, the above discrete hypothesis will not be true.

2.2.3. *Normalized template approach*

Let us now consider equation [2.2]. Due to radiometric deformation, for a pair of corresponding pixels p and p', $I_1(p) \approxeq I_2(p')$ no longer holds true. Therefore, the formula from equation [2.7] cannot be used. The general guideline to handle this inconsistency is as follows:

– using equation [2.5], assume that, in a local neighborhood, the radiometric deformations do not depend on p; consequently, we have $R(p, I_2) = R(I_2)$;

– select a parametric model for R;

– when testing the hypothesis that p is the correspondence of p', use all pairs of radiometries $(I_1(p + v), I_2(p' + v), v \in V)$ to estimate the parametric model of R.

2.2.3.1. *Rank normalization*

Rank normalization is one of the most general parametric models, as it only imposes that R be a growing function. In the continuous case, i.e. with an infinite number of measurements, the normalization consists of histogram normalization. In the discrete case, as in the matching of a *window* of a finite size, this comes down to a rank normalization. That is, for each image I_k, and each pixel q in the neighborhood of p, compute the number of pixels with a value lower than q in the neighborhood V:

$$Rank_k^p(q) = Card(\{r \in V(p)/I_k(r) < I_k(q)\}) \quad [2.8]$$

$$DS_{Rank}(p, p') = \sum_{v \in V} |Rank_1^p(p+v) - Rank_2^{p'}(p'+v)| \quad [2.9]$$

One of the drawbacks of this normalization is that it is relatively slow to compute. If N is the cardinal of V, a naive implementation can lead to a cost of N^2 complexity. Using a quick sort algorithm to globally sort all values in $V(p)$, the computation complexity can be reduced to $N\log(N)$.

2.2.3.2. *Normalized cross-correlation*

Most often, rank normalization is too general; therefore, more restrictive similarity measures are sought. A common model presumes that the radiometric deformation between I_1 and I_2 can be represented locally by an affine transform, with a being the translation and b the scaling factor:

$$I_1(p) = a + bI_2(p) \quad [2.10]$$

Using this assumption, we can compute normalized images $\overline{\overline{I_k}}$, independent of any affine transform within the window. This can be done in two steps:

1) render the radiometry invariant to translation by subtracting the average of the radiometries (equation [2.11]);

2) render the radiometry invariant to scaling by dividing it by its standard deviation (equation [2.12]).

$$\overline{I_k^p}(q) = I_k(q) - \mu_k(p) \quad [2.11]$$

$$\overline{\overline{I_k^p}}(p) = \frac{\overline{I_k^p}(p)}{\sigma_k(p)} \quad [2.12]$$

where N is the number of pixels in the template (i.e. $N = (2n+1)^2$ for a square window of size n). For instance, in Figure 2.1, $n = 1$ and $N = 9$.

The average and standard deviations are computed from:

$$\mu_k(p) = \frac{\sum_{v \in V} I_k(p+v)}{N} \quad [2.13]$$

$$\sigma_k(p) = \frac{\sum_{v \in V} \overline{I_k^p}^2 (p+v)}{N} = \frac{\sum_{v \in V} I_k^2(p+v)}{N} - \mu_k(p)^2 \qquad [2.14]$$

Immediately we can see that $\overline{\overline{I_k}}$ is invariant to any translation scaling in the radiometry:

$$\overline{\overline{I_k}} = \overline{\overline{a + bI_k}} \qquad [2.15]$$

The normalized cross-correlation is then defined as a dissimilarity measure DS:

$$DS_{NCC}(p,p') = \sum_{v \in V} (\overline{\overline{I_1^p}}(p+v) - \overline{\overline{I_2^p}}(p'+v))^2 \qquad [2.16]$$

Alternatively, and as is often encountered in the literature, it is known as:

$$NCC(p,p') = \sum_{v \in V} \overline{\overline{I_1^p}}(p+v)\overline{\overline{I_2^p}}(p'+v) \qquad [2.17]$$

The above two expressions are closely related by:

$$DS_{NCC}(p,p') = 2(1 - NCC(p,p')) \qquad [2.18]$$

The DS_{NCC} dissimilarity measure remains the most popular similarity. Its wide use is due to numerous advantages:

– it can be easily interpreted as a distance between normalized images;

– the radiometric normalization to translation and scaling is justified by physical phenomena (e.g. changing illumination conditions);

– DS_{NCC} is always in the interval of $[0 - 4]$, whereas NCC is always in the interval of $[-1, 1]$. They can be interpreted as distance and dot products between unit vectors (see equation [2.12]), which facilitates their interpretation or the fixing of certain parameters (e.g. thresholds, regularization parameters, etc.);

– they are fast to compute. For instance, using the integral images, the computation time can be made to be independent of the size of the window.

NOTE.– The above discussion is limited to image cross-correlation in the spatial domain. Alternatively, cross-correlation can be defined in the frequency domain, where it is equivalent to finding the phase difference between Fourier transforms of a pair of windows. For an extensive example of ground displacement computation in the Fourier domain, see Leprince *et al.* (2007).

2.2.3.3. *Census coefficient*

Rank distance and normalized cross-correlation use all the elements of the window to normalize the radiometry. As a result, in the presence of outliers, the normalization will be contaminated by false measurements. Outliers are not rare in image matching; for instance, when performing image matching for 3D geometry computation, outliers can arise systematically due to, for example, occlusions.

Census is a similarity measure that is robust to outliers and thus addresses the above shortcoming. To introduce census, let us define a binary operator $>_k^p$:

$$>_k^p (p+v) = I_k(p+v) > I_k(p) \quad [2.19]$$

The operator applied to a window, $\sum_{v \in V} I_k(p+v)$, converts it to a binary vector, where all elements larger than the value of the central pixel $I_k(p)$ are true. By comparing such binary vectors computed for a pair of windows (centered at p and p' in images I_1 and I_2, respectively), we obtain the census coefficient:

$$Census(p,p') = \sum_{v \in V} >_1^p (p+v) \neq >_2^{p'} (p'+v) \quad [2.20]$$

What makes census robust is that the normalization is undertaken using only the value of the central pixel. Consequently, in the sum in equation [2.20], if there is an outlier, it will not influence the output score. Conversely, because of the strict normalization, the census coefficient is not very discriminant. With this in mind, we should apply census: (i) with large window sizes in order to bring "more" information or context into the measurement vector; or (ii) in combination with the regularization techniques described in section 2.3. A use case where census performs well is the computation of 3D scene geometry with severe occlusions.

2.2.3.4. *Other similarity measures*

There are many similarity measures and listing them all is beyond the scope of this chapter. Here, we mention several common alternatives:

– Non-centered normalization: the only difference with the measure described in section 2.2.3.2 is that the average is not subtracted; as a result, it is less discriminant, as well as less sensitive to noise in textureless areas.

– Mutual information (Viola and Wells 1997): this measure of similarity calculates the joint entropy between two random signals (e.g. two overlapping images). In general, the entropy is high if the information content is high (i.e. many variations of intensities in a textured region), and vice versa, the entropy is small for low information content (i.e. a flat region). For a pair of perfectly matched images, the joint entropy is low because one image predicts the content of the other, and therefore can be used as the matching cost. The similarity measure is particularly suitable to modeling important radiometric distortions; for example, image pairs acquired in different illumination conditions or even with sensors of different radiometric sensitivity.

– Learning-based similarity measures (Žbontar and LeCun 2016): with the advances in machine learning and in particular deep learning, many state-of-the-art methods learn to predict the similarity of a pair of patches. The advantage of learning approaches is that they are able to learn an arbitrary function and hence model arbitrary geometric and radiometric deformations. On the downside, the parameters of the function are learnt by heart from specific training data and generalize poorly to other datasets.

2.3. Handling "large" deformations

2.3.1. *Introduction*

Limiting the matching problem to the definition of a similarity measure is possible when the search space is limited. When the search interval is increased, two limitations arise. First, the likelihood of erroneous matching grows with the growing number of tested hypotheses and leads to noisy results. Second, the computation time increases proportionally to the number of tested hypotheses.

In this section, we describe the most typical ways to solve the above-mentioned problems. The methods presented in section 2.3.2 address the robustness aspect, while in section 2.3.3, both the robustness and time issues are dealt with.

2.3.2. A posteriori *filtering and regularization approach*

As described in section 2.1.3, to solve the matching problem certain hypotheses about the regularity of the unknown geometric deformation G must be developed. However, by simply computing the most similar correspondence of a given pixel, there is no guarantee that the outcome deformation will comply with the regularity hypothesis. To impose it, the common practice is to force the geometric deformation to follow the hypothesis in equation [2.4].

A straightforward way to implement this behavior is to use an *a posteriori* filtering technique on the computed deformation values, for example:

– a local mean filter – suitable for scenarios without outliers and when the expected deformation is a smooth function;

– a local median filter – suitable for scenarios with outliers and when the expected deformation contains discontinuities;

– a combination of the median and mean filters to eliminate the outliers and obtain a smooth deformation.

However, *a posteriori* filtering is only a post-processing technique and does not use all the available information to recover the "correct" deformation. Let us look at the scenario illustrated in Figure 2.2. Over a relatively large area, two alternative geometric deformations are presented as a blue line and a dashed line, respectively.

The blue line is wrong, although it results from the most similar correspondences, and the dashed line is the real deformation but the second best result. In this scenario, filtering techniques are not capable of recovering the real deformation.

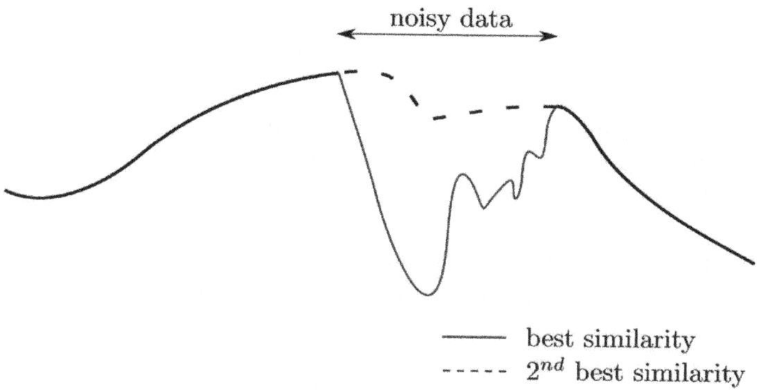

Figure 2.2. *In the noisy data region, two geometric deformations are shown as a blue line and a dashed line resulting from the best and second best similarity scores, respectively. The second best result is the true deformation and can be recovered with regularization techniques. For a color version of this figure, see www.iste.co.uk/cavalie/images.zip*

Ideally, we would like to guide the algorithm to perform the following: compute a deformation that maximizes the similarity measure *but* if there is a deformation with good enough similarity that is smoother (therefore more likely) than the first, choose the latter. In other words, the algorithm searches for a trade-off between similarity and regularity.

To formalize this, for a given deformation G, let us define the dissimilarity DS and the regularity Reg terms:

$$DS(G) = \sum_{p \in I_1} DS(I_1, I_2(G)); \qquad [2.21]$$

$$Reg(G) = \sum_{p \in I_1} |\partial(G - Id)|^{\alpha}; \qquad [2.22]$$

where the terms are defined over all pixels p of image I_1, $\partial(G - Id)$ is the gradient of G minus identity and α can take practically any value, but it should be chosen based on the characteristics of the deformation G (see Figure 2.3). Of course, the remarks at the end of section 2.1.3 also apply here, and Id can be replaced by any smooth *a priori* approximation of G.

Our aim is to find a solution that simultaneously minimizes DS and Reg. With the concepts presented up to now it is impossible, because the solution minimizing DS is the argument of the minimum for each pixel independently (defined in equation [2.6]), while the solution minimizing the regularity is simply the identity. An energetic approach allows us to handle this trade-off in a neat way by defining a function E:

$$E(G) = DS(G) + \beta Reg_{\alpha}(G) \qquad [2.23]$$

Here, the deformation that we look for is defined as the deformation G that minimizes E. In equation [2.23], the term DS, known as the data term, depends only on the images, while Reg_{α}, known as the regularization, represents the *a priori* knowledge of the smoothness of G. The β factor allows us to control the relative importance of the data and regularization terms.

Currently, the energetic approach is the most common way to solve the matching problem, both in academic and commercial software solutions. It is an elegant and quite general way to combine the information brought by the data with the *a priori* information. When embarking on this approach, two non-trivial issues must be investigated: (1) the relation between the regularization term and the solution that will be favored; (2) once the function E is defined, the choice of the algorithm that can efficiently compute the deformation, minimizing equation [2.23]. An in-depth answer to these two questions is far beyond the scope of this chapter; however, in the following, we provide a brief commentary on the current methods.

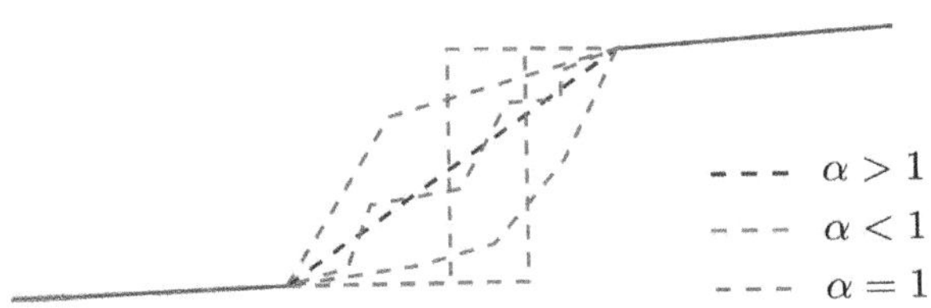

Figure 2.3. *The impact of the regularization term in equation [2.22] on the solution. The red lines represent the deformation G (displacement or the surface). In the dashed line zones, the image information is not discriminant enough to compute the deformation. The way the gap is filled with information depends on the value of the α parameter, which is represented by the blue, green and gray dashed lines. For a color version of this figure, see www.iste.co.uk/cavalie/images.zip*

Regarding the impact of the regularization term, Figure 2.3 illustrates how α given in equation [2.22] can be used to favor diverse geometries. The dashed interval represents no image information and the red line corresponds to a region for which the image information is good enough to find the solution (i.e. deformation). Let us suppose that the solution in the information gap is steered only by the regularization term. Depending on the value of α, we can obtain different solutions:

– $\alpha > 1$: there is a unique solution, and it is the straight blue line that joins the red line on either end; this choice favors a smooth solution and is adequate for computing the 3D geometry of natural scenes without mountains;

– $\alpha < 1$: any path making the jump in a single step is the solution; this choice favors the solution with discontinuities and is adequate for computing the 3D geometry in urban areas;

– $\alpha = 1$: all the monotonous paths are the solutions; this is a neutral choice and adequate if the geometry is composed in the majority of smooth surfaces with some discontinuities, as in the case of coseismic deformations.

Concerning minimization algorithms, the most common methods are mincut–maxflow, semi-global matching (SGM) and variational methods. The mincut–maxflow computes the exact minimum of equation [2.23]; however, it imposes that $\alpha \geq 1$ and is relatively costly in time and space (i.e. it has high memory consumption). The SGM is a version of the dynamic programming method. It is not limited by the value of α and is efficient in both time and space. On the downside, it does not compute the exact minimum, which sometimes leads to unpleasant artifacts. Finally, the variational approach uses nonlinear least squares to solve equation [2.23]. As the deformation G is computed globally, there is no need to use the template method; as a result, the solution does not suffer from the artifacts specific to the window-based method. A drawback is that the variational approach requires good initialization of G.

2.3.3. *Multi-resolution approach*

When dealing with large deformations, the regularization approach partially solves the robustness issue but not the CPU issue. In fact, it increases the total computation time by having to include the regularization. Multi-resolution processing, a particular case of the "coarse to fine" strategy, can further improve the two aspects (Pierrot-Deseilligny and Paparoditis 2006; Rothermel *et al.* 2012). The strategy is to first obtain an approximate and fast (i.e. coarse) solution and then to use it to compute a finer solution. A basic multi-resolution approach can be described as follows:

– compute down-sampled images at half the resolution: $\underline{I_k}(p) = I_k(2 \cdot p)$;

– compute the geometric deformation $\underline{G}$ between $\underline{I_1}$ and $\underline{I_2}$; note that the images have four times fewer pixels, and the research intervals are two or four times smaller (in 1D or 2D matching, respectively), hence the computation is theoretically 8–16 times faster;

– use $\underline{G}$ as a predictor to limit the research interval in full-resolution images.

This procedure is typically used recursively by computing a pyramid of images and starting the computation with the lowest resolution images. For example, in MicMac software (Rosu *et al.* 2015), when processing large satellite images, the downsampling currently goes to $1/32$ or $1/64$ at the lowest resolution.

2.4. Discrete nature of the image and sub-pixel matching

Until now we have made the implicit assumption that for a given pixel p with integer coordinates, its corresponding pixel is also a pixel q with integer coordinates. The drawback of this assumption is that we can only compute integer deformations and this thus reduces the accuracy of the measurement. Sub-pixel deformations cannot be determined because they are invisible to the matching algorithm. Typically, if we work with satellite images of 10 m resolution, this "integer" approach would limit the accuracy to 5 m.

There are two ways to overcome this shortcoming, both based on interpolation, which we refer to as the *interpolation of similarity* and the *similarity of interpolation*. In the *interpolation of similarity*, for a given pixel, we (1) compute and memorize the similarity calculated in the search region, using integer pixels, (2) use some quadratic function to model the similarity function and finally (3) extract its maximum. In the *similarity of interpolation*, we select an interpolating algorithm and use it to extract the value of I_2 using non-integer pixels (i.e. at sub-pixel positions). This approach is generally slower but more general. To limit the signal aliasing, precautions must be taken in selecting the interpolating kernel.

Using such techniques, it is possible to achieve the accuracy of $\frac{1}{10}$ of a pixel. Nevertheless, the accuracy remains largely dependent on the images (e.g. the sensor, the scene type, etc.) and may vary from $\frac{1}{2}$ in an unfavorable condition to $\frac{1}{100}$ in very favorable conditions.

2.5. Optical imaging sensors

In the previous sections, we used an implicit assumption that in both forms of matching (1D and 2D), we could approximately predict the position of corresponding pixels. This prediction is possible thanks to the known geometry of the sensors with which the images are acquired. The geometries describe position and rotation in 3D space and enable relating images taken from different viewpoints.

2.5.1. *Sensor geometries*

The two common classes of sensor geometries are frame and pushbroom geometries. Frame cameras dispose of an area array sensor in which all the pixels share the same optical center and are exposed to light simultaneously[1]. Pushbroom cameras, in contrast, are composed of a linear sensor array and scan the scene along

1. With the exception of frame sensors that use rolling shutters and scan the image line by line, thus violating this assumption.

the trajectory of the sensor. As a result, each image line is acquired at a different time and from a unique position in space, following the perspective projection along the linescan and the orthogonal projection in the perpendicular direction. In fact, the real pushbroom sensors used in spaceborne imaging have a complex layout that is often composed of several detectors situated in the focal plane (Lier *et al.* 2008). To facilitate the use of the images, the images undergo resampling to create an *ideal* sensor. The ideal sensor simulates a single-line pushbroom sensor, free from optical distortions and from the jitter artifacts caused by high-frequency attitude changes (De Lussy *et al.* 2005).

Until recently, frame cameras were more common in terrestrial and airborne applications, whereas pushbroom scanners were a standard in spaceborne imaging. Thanks to technological advances and lower costs, we can observe more and more of the frame arrays used in satellites (e.g. Planet's CubeSat constellations).

2.5.2. *Sensor orientation in 3D space*

2.5.2.1. *Physical model*

Sensor orientation in 3D space is carried out in two steps by applying the so-called *intrinsic* and *extrinsic* calibration parameters. The intrinsic parameters allow us to associate a pixel measurement in the 2D image coordinates with a bundle in 3D space. The extrinsic parameters orientate the camera in the world coordinate frame; in other words, they allow us to georeference a camera. These steps are realized by three consecutive operations: $\mathbb{R}^3 \mapsto \mathbb{R}^3$ transformation from the world to the camera coordinates; $\mathbb{R}^3 \mapsto \mathbb{R}^2$ perspective projection from 3D to 2D camera coordinates; and finally $\mathbb{R}^2 \mapsto \mathbb{R}^2$ from a 2D camera to image coordinates. This mapping can be represented in a compact equation form, referred to as the *collinearity equation* (see Figure 2.4):

$$\mathbf{p}_k = \mathcal{I}\left(\pi\left(\mathbf{R}_k\left(\mathbf{P} - \mathbf{C}_k\right)\right)\right) \qquad [2.24]$$

where $\mathbf{R}_k$ and $\mathbf{C}_k$ are the *extrinsic* orientation parameters of some image k; in particular, $\mathbf{R}_k$ is the rotation matrix that moves from world coordinates to camera coordinates, and $\mathbf{C}_k$ is the image perspective center in the world coordinates; $\pi\left[x, y, z\right]^T = \frac{[x,y]^T}{z}$ represents the perspective projection; $\mathcal{I}$ indicates the intrinsic parameters (i.e. the focal length, the principal point and the distortions) and $\mathbf{p}_k$ the image coordinates $(i, j)_k$ of the point $\mathbf{P}$ in some image k.

Equation [2.24] holds true for frame cameras, i.e. when all pixels are exposed at the same instant. To take into account pushbroom cameras, we should generalize the above equation by including the time component t:

$$\mathbf{p}_k = \mathcal{I}^t\left(\pi^t\left(\mathbf{R}_k^t\left(\mathbf{P} - \mathbf{C}_k^t\right)\right)\right). \qquad [2.25]$$

The projection equation of this form is rigorous because it describes the passage from the 3D ground scene to the 2D image space using physical parameters. The physical nature of the parameters is considered advantageous because they (i) are seldom correlated, (ii) can be easily interpreted, and (iii) can be optimally adjusted. Nonetheless, for several reasons, the rigorous model is rarely used in practice. One of its main drawbacks is the lack of standardization among the satellite image providers, hence the difficulty in maintaining those models in various software solutions. Another drawback is due to its complexity and, in particular, the fact that the number of unknowns grows with the number of image lines. In response to this, several sensor replacement models have been proposed over the last 40 years. Among them are the polynomial image geometry model; the grid interpolation model; the *rational function model* (RFM), also known as the *rational polynomial coefficient* (RPC) model; and the *replacement sensor model* (RSM). Nowadays, the RPC model is the prevalent choice to describe the geolocation of a satellite sensor; therefore, we discuss it further in the next section.

Note that equation [2.25] does not explicate the ray's refraction effects but assumes rays to be straight lines. In practice, image rays are bent as they pass through the different layers of the atmosphere, and to compensate for this phenomena, image corrections are applied. The atmospheric refraction correction is inversely proportional to the square of the focal length; therefore, it is less pronounced for narrow-angle satellite sensors. For example, for near-vertical acquisitions by Pleiades satellites with an incidence angle of 15°, the geolocalization error reaches approximately 2 m (see the Pléiades Imagery User Guide). For more details on sensor models, see OpenGIS Consortium (1999) and McGlone (2013).

2.5.2.2. *Empirical model – rational polynomial coefficients*

The RPC model relates the ground coordinates and the image coordinate using two rational polynomial functions:

$$\bar{j} = g(\lambda, \phi, h) = \frac{p_1(L, P, H)}{p_2(L, P, H)},$$
$$\bar{i} = f(\lambda, \phi, h) = \frac{p_3(L, P, H)}{p_4(L, P, H)}, \quad [2.26]$$

where $(\bar{j}, \bar{i})$ are the normalized sample (column) and line (row) indices; (λ, ϕ) are the latitude and longitude coordinates in the WGS-84 geodetic system and (L, P) are their normalized equivalents; h and H are the ellipsoidal altitude and its normalized equivalent, respectively; and p_i indicates polynomials of the third order; the normalization is defined in the range $\langle -1, 1 \rangle$.

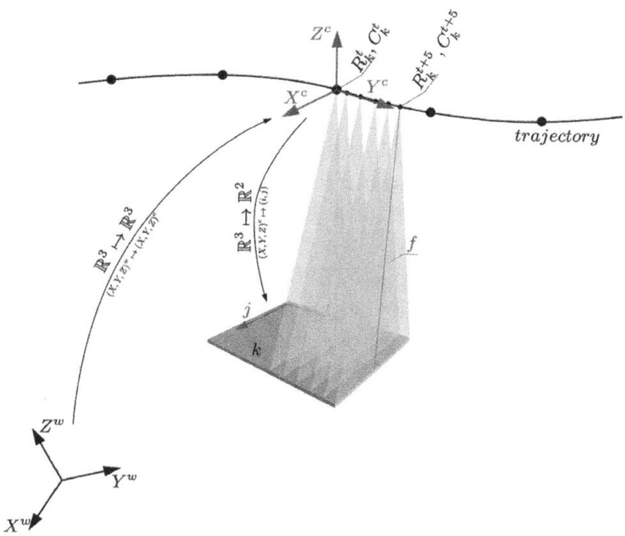

Figure 2.4. *Sensor orientation in 3D space.* $(X, Y, Z)^w$ *and* $(X, Y, Z)^c$ *are the respective world and camera frames.* R_k^t, C_k^t *are the extrinsic parameters of some image* k *at time* t, (i, j) *are the image coordinates and* f *is the focal length. For a color version of this figure, see www.iste.co.uk/cavalie/images.zip*

NOTE.– Equation [2.26] is a generic formulation from which special cases can be derived. For example, if the denominator $p_2 = p_4 = 1$, the equation becomes the conventional polynomial model. If the polynomial order is limited to 1, and $p_2 = p_4$, the equation becomes a 2D affine transformation, also known as the *direct linear transform* (DLT) (Tao and Hu 2001b). As a result, we may decompose the different terms according to their meaning: the ratio of the first-order polynomial describes the ground to image projection; the second-order terms account for the Earth curvature, refraction and lens distortions; finally, the third-order terms compensate for high-frequency errors (e.g. sensor vibrations) (Tao and Hu 2001a).

The accuracy of the RPC model is conditioned by the estimation method. We can distinguish between terrain-dependent (from ground control points (GCPs)) and terrain-independent (from the physical sensor model) estimation methods. If the RPCs are generated from the physical sensor model, the accuracy is generally as good as that of the model that served to estimate the coefficients. If the physical sensor is not available (e.g. due to poor performance of the on-satellite direct measurement devices), the quality of the RPCs is highly dependent on the number as well as the distribution of GCPs and deteriorates quickly with insufficient number points (Tao and Hu 2001b). It should be noted that the rational polynomial functions are generally not capable of modeling high-frequency distortions caused, for example, by a satellite's vibrations. We owe their accuracy to the image pre-processing step that transforms the real sensor to an *ideal* one (De Lussy *et al.* 2005).

Even though RPC-encoded geolocation is precise with regard to the physical model, the accuracy – i.e. the absolute precision with respect to some Earth-fixed coordinate frame – is not guaranteed. The RPCs delivered customarily with Pléiades, SPOT-6/7, QuickBird, WorldView, PlanetScope, GeoEye, Ikonos, Cartosat, etc., can be misaligned by up to 10 pix (CE90), equivalent to several meters on the ground (Oh and Lee 2015). Such deviations are due to errors in direct measurements on board the satellite (i.e. GNSS, gyros, star trackers), especially errors in attitude and position (Fraser and Hanley 2005). These may hinder further processing; for example, the digital surface model generation with matching algorithms (d'Angelo 2013). Note that due to the short-term correlation of onboard measurements, errors in RPC geolocation from in-orbit acquisitions are less pronounced than errors from multi-orbit and multi-epoch acquisitions. The RPC model is therefore usually subject to refinement. For narrow field-of-view sensors, as in the satellite pushbroom camera, the attitude errors induce the same net effect on the ground as do the in- and across-track positioning errors. As a result of this correlation, sensor errors can be modeled with two correction functions that are defined in the image coordinates: (1) δf in the line direction, to absorb the in-track orbit and pitch errors; and (2) δg in the sample direction, to absorb the equivalent across-track errors (Grodecki and Dial 2003):

$$\begin{aligned} j &= g(\lambda, \phi, h) \cdot scale_j + offset_j + \delta g \\ i &= f(\lambda, \phi, h) \cdot scale_i + offset_i + \delta f \end{aligned} \qquad [2.27]$$

where δg and δf are polynomials. The parameters of the correction functions are estimated in a least squares adjustment with image correspondences and GCPs used as observations. To strengthen the solution, the equations can include additional constraints imposing that the $(\delta g, \delta f)$ displacement fields follow a global sensor rotation (Rupnik *et al.* 2016). The GCPs are commonly collected using existing databases of orthophoto images and global digital elevation models; for instance, the Sentinel-2 and Shuttle Radar Topography Mission (SRTM) DEMs (Kääb *et al.* 2017).

2.5.2.3. *Epipolar model*

Given a pair of overlapping images and their known geometries, in order to predict positions of their corresponding points, we often resort to the so-called epipolar geometry. Epipolar geometry is mainly useful in two tasks: relative orientation of an image and image matching for digital surface model (DSM) generation. During the relative orientation of frame cameras, the epipolar constraint is written into the essential matrix from which the rotation and translation of two images can be decomposed. DSM computation reduces the correspondence search from a 2D problem (i.e. a pixel in one image matched against every pixel in the other image) to a line or a curve of candidates.

To illustrate the epipolar geometry of the original image pair, let us look at Figure 2.5. By applying the intrinsic and extrinsic parameters in equations [2.24]–[2.25] to some pixel p in image I_1, we obtain a 3D bundle B_p associated with that position. This bundle, projected to an overlapping image I_2 via a plane π anchored at the projection centers O_1 and O_2, will intersect the focal plane of the image I_2, giving rise to the epipolar line e_p (see Figure 2.5). In other words, for some pixel p, the above relationship predicts the position of its candidate corresponding points in another image.

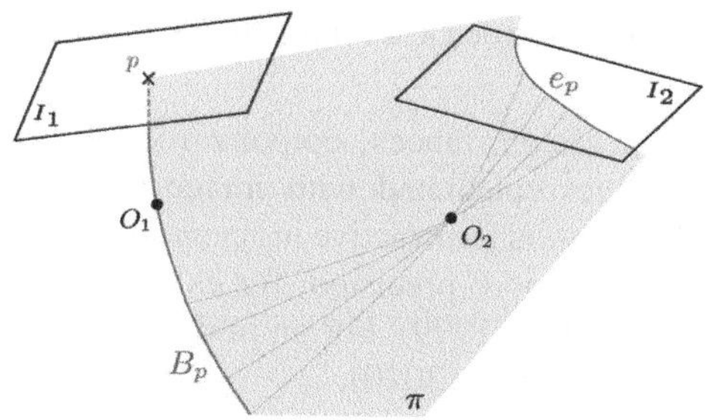

Figure 2.5. *Epipolar line e_p as a projection of a bundle B_p via the epipolar plane π. For a color version of this figure, see www.iste.co.uk/cavalie/images.zip*

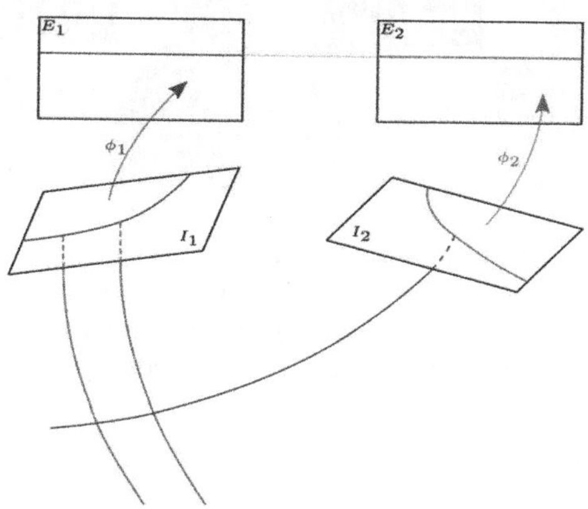

Figure 2.6. *An image pair in the original geometry (bottom) and rectified to the epipolar geometry with functions ϕ_1, ϕ_2 (top). For a color version of this figure, see www.iste.co.uk/cavalie/images.zip*

In image matching, it is useful to separate the correspondence search from orientation parameters. To undertake this, an image pair can be rectified to an equivalent geometry, so that the corresponding epipolar lines are aligned and parallel to image rows (see Figure 2.6). From then on, the image matching comes down to finding conjugate points across lines with the same y-coordinate, with no regard to the image geometry. For frame cameras, the rectification to the equivalent geometry involves applying two 2D projective transformations to both images of a stereo pair. Unlike the frame cameras, pushbroom sensors acquire each image row from a different perspective center, and as a result, the epipolar lines are neither straight lines nor are they conjugate across the image (Gupta and Hartley 1997). *State-of-the-art* approaches leverage this by either rectifying the image per patch, assuming that locally the epipolar curves are straight lines (de Franchis *et al.* 2014), or mapping the epipolar curves across the images, followed by a rectification transformation that straightens the curves to lines (Oh 2011; Pierrot-Deseilligny and Rupnik 2020).

NOTE.– Under certain circumstances, epipolar rectification is possible using only image correspondences calculated with image processing techniques (e.g. SIFT (Lowe 2004)). This method is attractive in scenarios when the image geometry is unavailable or known with poor precision, for example in processing historical images such as Corona (Madden 1999). For an in-depth discussion on the method, see Pierrot-Deseilligny and Rupnik (2020).

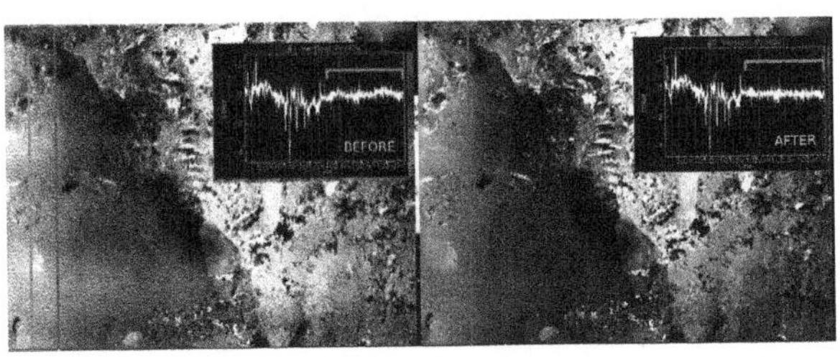

Figure 2.7. *The effect of unmodeled sensor errors on image matching. The left image corresponds to a displacement map before correction and the right image corresponds to a displacement map after correction (Delorme* et al. *2020). For a color version of this figure, see www.iste.co.uk/cavalie/images.zip*

2.5.2.4. *Unmodeled sensor errors*

Independent of the orientation model, refined sensor parameters still reveal errors related to the unmodeled satellite's attitude and the CCD chip's misalignment (Ayoub *et al.* 2008). In Figure 2.7 we can observe the effect such discrepancies have on image matching. The left image shows a displacement map obtained from matching of two ortho-rectified images. The lower left part of the image is corrupted with an oscillating signal along the vertical axis, caused by the satellite's jitter. The induced artifacts can be up to several pixels and, therefore, for high accuracy measurements, must be eliminated. A common technique consists of modeling and subtracting the undulating signal in post-processing (Delorme *et al.* 2020; Michel *et al.* 2020).

2.6. Acknowledgments

We thank Renaud Binet and Amaury Dehecq for their careful reviews and comments that helped to improve this chapter.

2.7. References

d'Angelo, P. (2013). Automatic orientation of large multitemporal satellite image blocks. *Proceedings of the International Symposium on Satellite Mapping Technology and Application*, 1–6.

Ayoub, F., Leprince, S., Binet, R., Lewis, K.W., Aharonson, O., Avouac, J.-P. (2008). Influence of camera distortions on satellite image registration and change detection applications. *IEEE International Geoscience and Remote Sensing Symposium*, Boston, USA, 6–11 July.

De Lussy, F., Kubik, P., Greslou, D., Pascal, V., Gigord, P., Cantou, J.P. (2005). Pleiades–HR image system products and geometric accuracy. *Proceedings of the International Society for Photogrammetry and Remote Sensing Workshop*, Hannover, Germany.

Delorme, A., Grandin, R., Klinger, Y., Pierrot-Deseilligny, M., Feuillet, N., Jacques, E., Rupnik, E., Morishita, Y. (2020). Complex deformation at shallow depth during the 30 October 2016 Mw 6.5 Norcia earthquake: Interference between tectonic and gravity processes? *Tectonics*, 39(2), e2019TC005596.

de Franchis, C., Meinhardt-Llopis, E., Michel, J., Morel, J.-M., Facciolo, G. (2014). On stereo-rectification of pushbroom images. *IEEE International Conference on Image Processing (ICIP)*, Paris, France, 27–30 October.

Fraser, C.S. and Hanley, H.B. (2005). Bias-compensated RPCS for sensor orientation of high-resolution satellite imagery. *Photogrammetric Engineering and Remote Sensing*, 71(8), 909–915.

Grodecki, J. and Dial, G. (2003). Block adjustment of high-resolution satellite images described by rational polynomials. *Photogrammetric Engineering and Remote Sensing*, 69(1), 59–68.

Gupta, R. and Hartley, R.I. (1997). Linear pushbroom cameras. *IEEE Transactions on Pattern Analysis and Machine Intelligence*, 19(9), 963–975.

Horn, B.K. and Schunck, B.G. (1981). Determining optical flow. *Artificial Intelligence*, 17(1–3), 185–203.

Kääb, A., Altena, B., Mascaro, J. (2017). Coseismic displacements of the 14 November 2016 Mw 7.8 Kaikoura, New Zealand, earthquake using the Planet optical Cubesat constellation. *Natural Hazards and Earth System Sciences*, 17, 627–639.

Leprince, S., Barbot, S., Ayoub, F., Avouac, J.-P. (2007). Automatic and precise orthorectification, coregistration, and subpixel correlation of satellite images, application to ground deformation measurements. *IEEE Transactions on Geoscience and Remote Sensing*, 45(6), 1529–1558.

Lier, P., Valorge, C., Briottet, X. (2008). *Imagerie spatiale : des principes d'acquisition au traitement des images optiques pour l'observation de la terre.* Editions Cépadues, Toulouse, France.

Lowe, D.G. (2004). Distinctive image features from scale-invariant keypoints. *International Journal of Computer Vision*, 60(2), 91–110.

Madden, F. (1999). The Corona camera system – Iteks contribution to world security. *JBIS*, 52, 379–396.

McGlone, C. (ed.) (2013). *Manual of Photogrammetry*. American Society of Photogrammetry and Remote Sensing, Bethesda, MD, USA.

Michel, J., Sarrazin, E., Youssefi, D., Cournet, M., Buffe, F., Delvit, J., Emilien, A., Bosman, J., Melet, O., L'Helguen, C. (2020). A new satellite imagery stereo pipeline designed for scalability, robustness and performance. *ISPRS Annals of the Photogrammetry, Remote Sensing and Spatial Information Sciences*, 2, 171–178.

Oh, J. (2011). Novel approach to epipolar resampling of HRSI and satellite stereo imagery-based georeferencing of aerial images. PhD thesis, Ohio State University, Ohio, USA.

Oh, J. and Lee, C. (2015). Automated bias-compensation of rational polynomial coefficients of high resolution satellite imagery based on topographic maps. *ISPRS Journal of Photogrammetry and Remote Sensing*, 100, 14–22.

OpenGIS Consortium (1999). The openGIS Abstract Specification – Topic 7: Earth Imagery Case.

Pierrot-Deseilligny, M. and Paparoditis, N. (2006). A multiresolution and optimization-based image matching approach: An application to surface reconstruction from SPOT5-HRS stereo imagery. *ISPRS Workshop on Topographic Mapping from Space*, Ankara, Turkey.

Pierrot-Deseilligny, M. and Rupnik, E. (2020). Epipolar rectification of a generic camera. Preprint [Online]. Available at: https://hal.archives-ouvertes.fr/hal-02968078.

Rosu, A., Pierrot-Deseilligny, M., Delorme, A., Binet, R., Klinger, Y. (2015). Measurement of ground displacement from optical satellite image correlation using the free open-source software MicMac. *ISPRS Journal of Photogrammetry and Remote Sensing*, 100, 48–59.

Rothermel, M., Wenzel, K., Fritsch, D., Haala, N. (2012). Sure: Photogrammetric surface reconstruction from imagery. *LC3D Workshop*, Berlin, Germany.

Rupnik, E., Deseilligny, M.P., Delorme, A., Klinger, Y. (2016). Refined satellite image orientation in the free open-source photogrammetric tools Apero/MicMac. *ISPRS Annals of the Photogrammetry, Remote Sensing and Spatial Information Sciences*, 3, 83.

Tao, C.V. and Hu, Y. (2001a). A comprehensive study of the rational function model for photogrammetric processing. *Photogrammetric Engineering and Remote Sensing*, 67(12), 1347–1358.

Tao, C.V. and Hu, Y. (2001b). Use of the rational function model for image rectification. *Canadian Journal of Remote Sensing*, 27(6), 593–602.

Viola, P. and Wells, W.M. (1997). Alignment by maximization of mutual information. *International Journal of Computer Vision*, 24(2), 137–154.

Žbontar, J. and LeCun, Y. (2016). Stereo matching by training a convolutional neural network to compare image patches. *The Journal of Machine Learning Research*, 17(1), 2287–2318.

3

SAR Offset Tracking

Silvan LEINSS[1,2], Shiyi LI[1], Yajing YAN[2] and Bas ALTENA[3]

[1] *Institute of Environmental Engineering, ETH Zurich, Switzerland*
[2] *LISTIC, Savoie Mont Blanc University, Annecy, France*
[3] *IMAU, Utrecht University, The Netherlands*

In order to provide the basics for the different image correlation methods used for SAR offset tracking, we start with a brief introduction to the SAR imaging geometry and the imaging method and provide a description of radar speckle. Then, we explain various methods of SAR image correlation, followed by two advanced methods using SAR image time series to make offset tracking more robust against decorrelation. We conclude this chapter with the fusion of displacement observations from different orbits into three-dimensional displacement maps.

3.1. Basics of SAR imaging

3.1.1. *Imaging geometry and resolution of SAR systems*

Synthetic aperture radar (SAR) systems illuminate the scene sideways with a broad antenna beam and sample the backscattered signal at different slant-range distances R (Figure 3.1). The measurement is repeated at different positions along the flight direction (satellite orbit), defining the second image dimension, named the azimuth. Except for interferometric systems, radar systems cannot measure the exact direction (look angle θ) at which the scene is observed. Complementary to radar systems, optical systems can only measure the direction in which radiation is emitted and, except for

Surface Displacement Measurement from Remote Sensing Images,
coordinated by Olivier CAVALIÉ and Emmanuel TROUVÉ.
© ISTE Ltd 2022.

lidar systems, distances cannot be measured. With this perspective, the slant-range imaging of radar systems can be considered as orthogonal to the imaging geometry of an optical system.

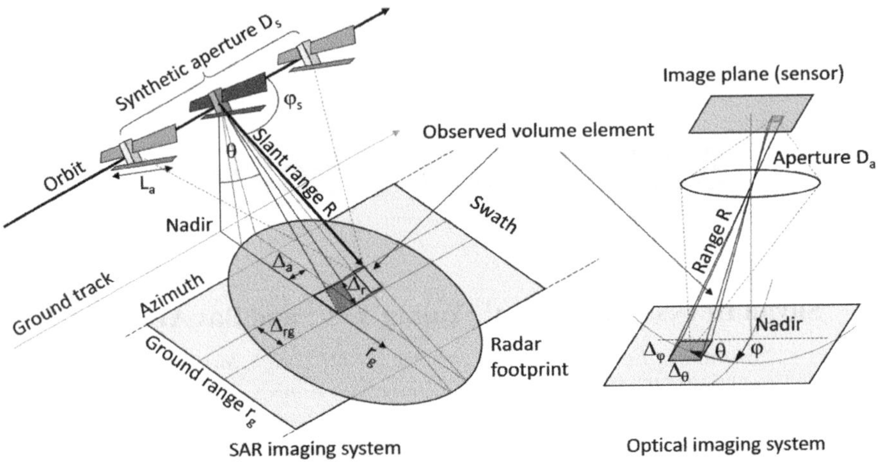

Figure 3.1. *SAR imaging geometry compared to the imaging geometry of a simple optical system. The radar system samples the scene in azimuth and slant-range coordinates (R) and cannot measure the illumination angle θ. The optical system samples the scene in 2D angular coordinates (θ, φ) but cannot measure the (slant) range R. For a color version of this figure, see www.iste.co.uk/cavalie/images.zip*

The orientation of the slant-range and azimuth direction is given by the satellite's orbit. To image nearly every part of the Earth as they rotate underneath satellites, and because SAR systems require several kW of electrical power, all spaceborne radar sensors fly in a near-polar, sun-synchronous dusk–dawn orbit (see Chapter 1, section 1.1.1), where solar panels can always be illuminated by the sun. This makes the azimuth direction roughly aligned with the north/south direction (inclination $\approx 98°$) and results in overpass times at early morning and early evening. The orbit of a satellite moving from the south to the north is called the ascending orbit, whereas it is called the descending orbit when the satellite moves from the north to the south (Figure 3.2). Considering that most SAR systems use a right-looking imaging geometry with respect to the flight direction (azimuth), SAR sensors look approximately towards the east from ascending orbits and approximately towards the west from descending orbits.

SAR systems can provide high-resolution imagery through range and azimuth compression of the received, unfocused, raw backscatter signal (Olmsted 1993; Oliver and Quegan 2004). After compression (focusing), a radar pixel (resolution cell) represents all power scattered back from objects located at a certain range and azimuth, defined by the volume element shown in Figure 3.1. Similar to optical systems, where the diameter D_a of the aperture lens limits the imaging resolution, the length of the synthetic aperture D_s of the radar system limits the azimuth resolution (Figure 3.1). Within the synthetic aperture, an object is illuminated by the antenna beam. The width of the antenna beam is diffraction-limited and depends on the length L_a of the radar antenna (the real aperture). As the opening angle of the illuminating antenna beam does not depend on distance, the azimuth resolution Δ_a is independent of the distance. In theory, the azimuth resolution for the stripmap imaging mode is only given by the antenna length: $\Delta_a \approx L_a/2$. In practice, signal power, pulse repetition frequency, azimuth ambiguities, etc., limit the azimuth resolution to values somewhat larger than $L_a/2$. In the slant-range direction, the resolution $\Delta_r \approx B_R/(2c)$ is given by the inverse of the radar bandwidth B_R and the speed of light c. After focusing, the spatial resolution, or more precisely, the size of the radar resolution cell, is given by the slant-range resolution Δ_r and the azimuth resolution Δ_a (see also Chapter 1, section 1.1.2).

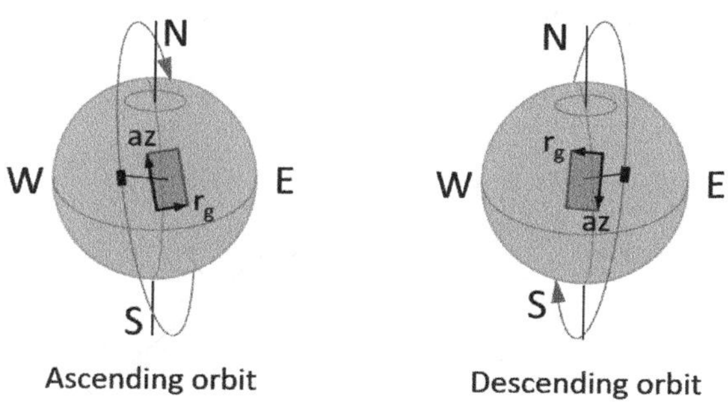

Figure 3.2. *Look-and-flight directions (black and red arrows) for SAR acquisitions from ascending and descending orbits. For a color version of this figure, see www.iste.co.uk/cavalie/images.zip*

Most modern SAR sensors have polarimetric imaging capabilities, either as a default mode (e.g. Sentinel-1) or as optional modes (e.g. TerraSAR-X). The geometric and radiometric resolution as well as the noise floor of acquisitions depends on the system design, but they can be lower for dual- or quad-polarimetric modes. Therefore, single-polarization acquisitions are often preferred for offset tracking.

As SAR images are generated by measuring distances, an oblique imaging geometry is required to obtain a good spatial resolution of the observed scene. However, the resolution depends on the terrain slope. For a horizontal terrain, the ground-range resolution $\Delta_{rg} = \Delta_r/\sin\theta_{\text{inc}}$ is given by projecting the slant-range resolution Δ_r to the ground (Figure 3.3). The local incidence angle θ_{inc} is defined with respect to the vertical. For mountainous terrain, the slope of the local topography can be defined by the angle ζ, measured in the slant-range plane, perpendicular to the azimuth direction. In this plane, a positive slope, $\zeta > 0$, faces the radar and a negative slope, $\zeta < 0$, faces away from the radar. The spatial resolution, parallel to the slope, is then given by

$$\Delta_{\text{slope}} = \frac{\Delta_r}{\sin(\theta_{\text{inc}} - \zeta)}. \qquad [3.1]$$

The best resolution is obtained for a slope-parallel incident beam ($\theta_{\text{inc}} - \zeta = 90°$). Steeper slopes, which face away from the radar and cannot be illuminated, are located in the radar shadow (dark gray in Figure 3.3). For slopes facing the radar, the denominator in equation [3.1] approaches zero at $\zeta = \theta_{\text{inc}}$, where all resolution is lost. For steeper slopes, multiple areas on the terrain exist that have the same distance to the radar. The backscatter signals of these areas overlap; hence, they are imaged into the same radar pixel. This situation is called the radar layover or foldover. With increasing incidence angles, the extent of shadow increases while the extent of layover decreases. For mountainous terrain, incidence angles minimizing the extent of shadow and layover are in the range of 45–55 degrees.

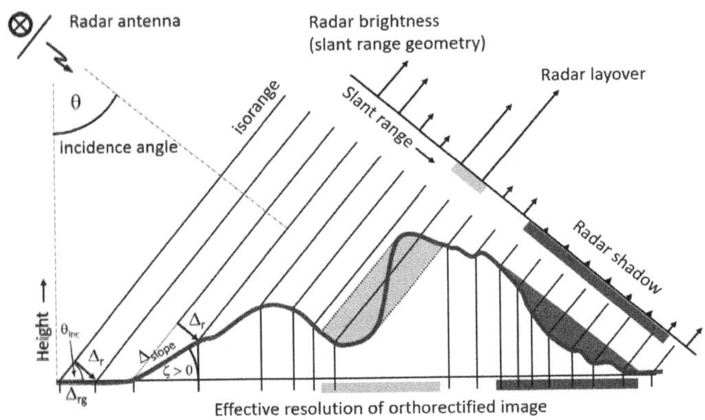

Figure 3.3. *Sampling of the topography in slant-range coordinates. The x-axis indicates the effective resolution on the ground. The areas with the same range fall into layover (light gray). The areas in the radar shadow (dark gray) cannot be imaged. The image was redrawn from Rosen* et al. *(2000)*

Figure 3.4(a) shows an example of a real SAR image with radar coordinates, acquired over the Great Aletsch Glacier (Switzerland) from a descending orbit. In this figure, strong geometric distortions appear because of the oblique, slant-range imaging geometry. Mountains appear distorted or even folded over lower terrain (layover) and point towards the sensor on the right side of the image. Layover appears as bright but narrow ridges, and radar shadow appears on the west-facing mountain slopes in black. Orthorectification alters the pixel spacing (see also Figure 3.3) and projects the image into a planar map geometry (Figure 3.4b); slopes facing away from the sensor appear compressed, while slopes facing the radar get stretched over larger areas. The areas affected by shadow or layover are indicated in blue and red in the shaded relief (Figure 3.4(c)). Orthorectification is performed using a lookup table for interpolation, which is based on the orbit information and an external digital elevation model (DEM).

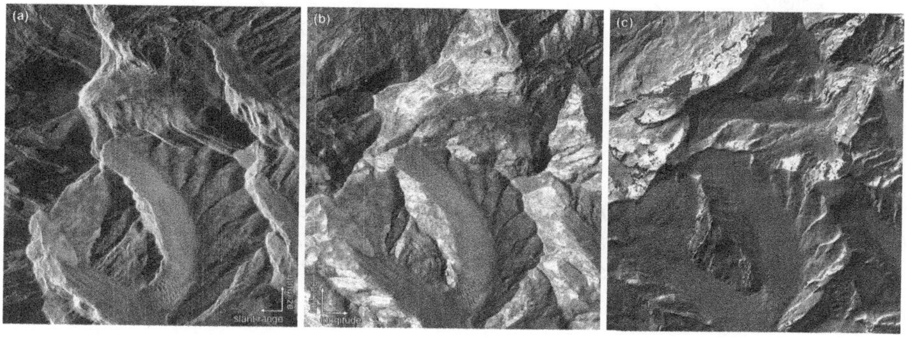

Figure 3.4. *(a) Great Aletsch Glacier (Switzerland) imaged in the slant-range geometry from a descending orbit (TerraSAR-X, ©DLR 2021; incidence angle $\theta = 32°$). (b) Orthorectified radar image. (c) Shaded relief with masks for layover (red) and radar shadow (blue). For a color version of this figure, see www.iste.co.uk/cavalie/images.zip*

3.1.2. *Radar speckle and speckle reduction with multi-looking*

Almost all radar imagery is affected by radar speckle, which can make imagery appear very noisy. The speckle pattern is temporally variable and depends on the surface characteristics. It can either hide or provide information required for displacement measurements. Therefore, we provide here a brief introduction to radar speckle.

Speckle originates from interference. Both ways of image formation, the focusing of light with a lens and the focusing of microwaves with a synthetic aperture, are based on the fact that radiation scattered at a point of the observed scene is collected by an aperture and is then superposed at a location on the image sensor (by a lens for the optical system) or at a slant-range azimuth location in the SAR image (focused numerically). For the two methods, the superposition of radiation generates an interference pattern at the detector. However, during the exposure time of optical systems, the interference pattern is averaged out in time because natural light has a very short temporal coherence. The radar microwave pulses, however, are identical copies of each other and hence temporally coherent. When all the radiation received during the synthetic aperture is superimposed by focusing, the full interference pattern of scattered amplitudes appears in the radar pixels. Notably, SAR systems measure complex-valued backscatter amplitudes containing phase information, whereas optical systems can only measure intensity.

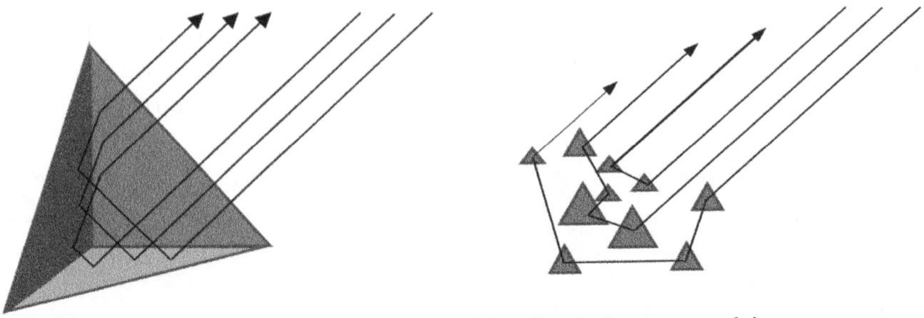

Figure 3.5. *The scattering amplitude and phase of a coherent target is independent of the exact location where scattering occurs; they can appear very bright. The scattering from distributed targets is described by random sum-scattered amplitudes resulting in radar speckle*

The interference pattern measured by the SAR system depends on the observed target. For coherent scatterers, such as corner reflectors of "sheet metal" (left side in Figure 3.5), all scattered microwaves interfere constructively because they have exactly the same travel time. This makes such targets appear very bright with a well-defined scattering amplitude and phase. In contrast, natural surfaces can be described by an ensemble of many small distributed scatterers. The backscatter amplitudes of all these scatterers sum up coherently and form the measured signal of the distributed objects. The amplitude A and phase ϕ of the corresponding pixel of a single-look complex (SLC) radar image can then be written as a sum of random complex numbers (Oliver and Quegan 2004). For the expectation value of the backscatter intensity $I = |A|^2$, it can be shown that the standard deviation of every speckle-affected backscatter estimate is equivalent to its mean, or, in other words, every measurement has an uncertainty of 100% (Oliver and Quegan 2004). This is why SAR images of natural surfaces are characterized by granular or "salt and pepper" textures, as is visible in Figure 3.6(a). The effect is known as radar speckle, commonly also referred to as speckle noise due to the noise-like appearance.

Speckle makes the interpretation of radar images difficult and can hide trackable features. To improve interpretation, incoherent averaging of multiple intensity measurements can be applied to obtain a better estimate of the local intensity I. While the expectation value $\langle I \rangle$ of the intensity remains the same, the standard deviation is given by $\langle I \rangle / \sqrt{L}$. Here, L represents the number of independent measurements or "looks" at the scene, and therefore this process is called multi-looking. There are various methods to generate a multi-looked SAR image, but the most common method is spatially averaging the backscatter intensity, assuming that adjacent pixels represent similar types of scattering (looks at adjacent pixels, ensemble average). This, however, results in a loss of spatial details (Figure 3.6(b)). In contrast, averaging of a number of radar images acquired at different times (temporal multi-looking) can greatly enhance the visibility of stable, non-moving details in the scene. Unfortunately, temporal information is lost and moving surfaces show motion blur (Figure 3.6(c)). Despite the reduction of speckle, the loss of spatial or temporal resolution in multi-looked radar images can deteriorate the precision of tracking.

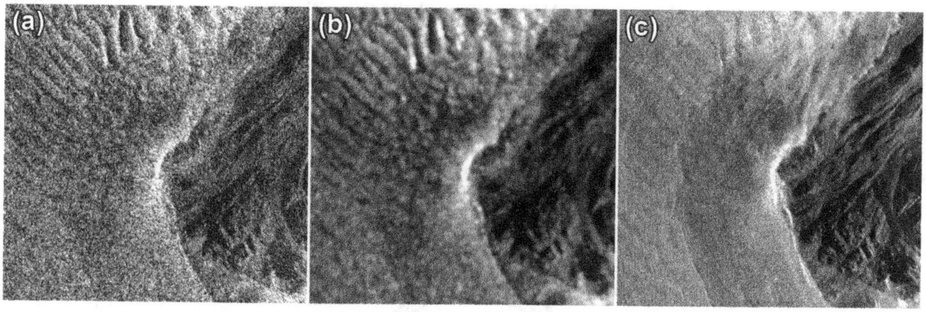

Figure 3.6. *Radar image of a glacier (left side) flowing around a rock ridge (right side). (a) The single-look backscatter intensity is strongly affected by speckle. (b) Speckle reduction using spatial multi-looking with a 5x5 pixel Gaussian smoothing kernel blurs the image. (c) Speckle reduction using temporal multi-looking from 10 consecutive images (repeat interval: 11 days) reveals details on the rock, but causes motion blur on the glacier. Images by TerraSAR-X (©DLR 2021)*

For SAR offset tracking, the intensity distribution of measured backscatter values is relevant. It can be shown that the measured intensity I can be written as σn, where n is a multiplicative term, carrying all random speckle noise, and σ is the true backscatter value (Oliver and Quegan 2004). It follows that speckle is exponentially distributed with $P_n(n) = e^{-n}$ with $n > 0$, and that every measurement can be regarded as its true backscatter value σ multiplied by the speckle noise n, which has a mean value $\langle n \rangle = 1$. The multiplicative nature of speckle makes it possible to separate the noise from the expectation value σ by considering the log of the measured intensity, commonly done by transforming the intensity I to decibels (dB):

$$I_{\text{dB}} = 10 \log_{10} I = 10 \log_{10}(\sigma n) = \sigma_{\text{dB}} + n_{\text{dB}}. \qquad [3.2]$$

Applying the log on the intensity compresses the (exponential) dynamic range of the image and standardizes the noise, which makes it independent of the mean value σ (Oliver and Quegan 2004). This can improve the cross-correlation because features with a relatively low backscatter intensity gain more weight in the correlation.

Finally, for very temporally stable surfaces, the speckle pattern can show temporal coherence such that either the complex radar image or its magnitude can be used as input for offset tracking. Furthermore, strong persistent scatterers that outshine the speckle pattern (radar reflector, Figure 3.5) can be used as trackable point-like targets. For very high resolution radar sensors, speckle can be reduced or even made almost absent when only a very limited number of scatterers are present in each resolution cell.

3.1.3. *Spectral support of the backscatter intensity*

When calculating the intensity from the SLC amplitude, it should be kept in mind that radar images are bandwidth-limited images and that the spectrum of the intensity requires double spectral support (Werner *et al.* 2005). To illustrate this, Figure 3.7 shows the typical spectrum of an SLC radar image (black line). The shape of the spectrum is defined by a hamming window, which is used to reduce side lobes of very bright scatterers. The spectrum shown requires a slightly smaller spectral support than twice the Nyquist frequency f_N because the SLC data are slightly oversampled. The spatial sampling frequency of the data shown is given by $1/(2f_N)$.

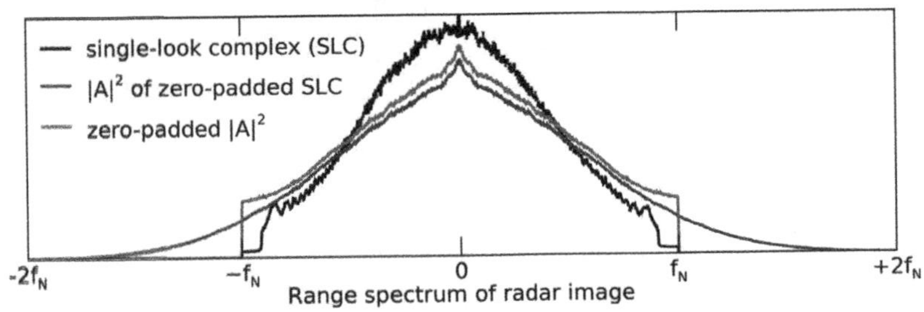

Figure 3.7. *The slant-range spectrum of the single-look-complex amplitude image (black) requires half of the spectral support of the intensity (blue) computed by zero-padding the SLC. In the spectrum of the intensity computed without zero-padding the SLC (red), the wings of the blue curve are folded into the spectral range between the Nyquist frequencies of the SLC image $\pm f_N$, which deteriorates the image quality. For a color version of this figure, see www.iste.co.uk/cavalie/images.zip*

When calculating the intensity of image components, for which the amplitude shows a spatial variation faster than half the Nyquist frequency, $I = |A(t)|^2 = |A\cos(\omega t)|^2| = \frac{A^2}{2}[1 + \cos(2\omega t)]$, it becomes clear that the frequency doubles; this is why the intensity requires double spectral support to avoid aliasing (blue line in Figure 3.7). For SAR data, oversampling is performed by zero-padding the Fourier transform of the image to double the spectral bandwidth, and then applying an inverse Fourier transform before calculating the backscatter intensity (De Zan 2014). This oversampling of the (complex-valued) amplitude can enhance the precision of offset tracking. In practice, and as shown in Figure. 3.7, SAR data are often already slightly oversampled and the spectral weighting with a hamming window reduces high-frequency components of the images, which can make oversampling superfluous. Nevertheless, to attain good sub-pixel precision, oversampling of the complex amplitude before correlating the intensity is more reasonable than oversampling the intensity (red line in Figure 3.7).

3.1.4. *Offsets due to radar penetration*

Radar images are generated by measuring the distance to an observed surface. However, this distance is not necessarily the geometric distance to a metallic target, and atmospheric delays and penetration into the observed medium can result in an apparent shift of the observed surface. These may be (seasonal) height changes due to a varying radar penetration into vegetation or snow or sand appearing as a slant-range displacement. Furthermore, the surface accumulation of snow or ablation (e.g. snow fall, ice melt) can result in slant-range shifts, which might be misinterpreted as displacement (Leinss and Bernhard 2021). This needs to be kept in mind when interpreting generated displacement maps.

3.2. SAR offset tracking

Offset tracking originates from image registration methods. Lucchitta and Ferguson (1986) visually inspected time series of Landsat imagery and found persistent features on outlet glaciers in Antarctica which could be manually tracked. Bindschadler and Scambos (1991) used the high sensitivity of the human eye (optical flow) and flickered consecutive Landsat imagery on a computer screen to identify shifts in the low-frequency undulation field of an ice stream in Antarctica. They used these shifts for coregistration of the images and then applied two methods on the high-pass filtered images: a manual point-picking method to track features and a numerical cross-correlation method to match patterns in a reference area with patterns from a search area. Scambos *et al.* (1992) provide the details of the method and use image-to-image cross-correlation software applied on small templates of an image pair. For SAR imagery, offset tracking originates from SAR interferometry, which requires the precise coregistration of two SAR images to obtain interferometric coherence and fringe visibility (see section 4.4.1.1.). However, while with SAR interferometry only slant-range displacement can be measured, Gray *et al.* (1998) showed that radar speckle can be tracked in the slant range and azimuth direction with cross-correlation. In the following sections, we first describe the method of offset tracking using cross-correlation in detail and then describe various variants of cross-correlation with SAR data.

3.2.1. *Cross-correlation*

3.2.1.1. *Cross-correlation in the spatial domain*

Cross-correlation is a commonly used similarity measure in signal processing. Given a reference signal $r(t)$ and a search signal $s(t)$, the cross-correlation between the two signals can be defined as a function of the shift τ of s relative to r:

$$\gamma(\tau) = \int_{-\infty}^{\infty} r^*(t)s(t+\tau)\mathrm{d}t \quad [3.3]$$

with $r^*(t)$ being the complex conjugate of r. Equation [3.3] expresses the cross-correlation for a 1D continuous signal. Considering images as discrete signals sampled in 2D space, equation [3.3] can be rewritten as:

$$\gamma(\Delta x, \Delta y) = \sum_{x,y}^{\Omega} \left(r(x,y)s(x+\Delta x, y+\Delta y)\right), \tag{3.4}$$

with (x, y) being the image coordinates, $(\Delta x, \Delta y)$ the image shifts and Ω the domain of the image templates, i.e. the size of the correlation window. Equation [3.4] would result in strongly varying values for γ for areas of different brightness in the images. To make the correlation values of different image areas comparable, the image templates need to be adjusted by their mean values μ_r, μ_s and normalized by their standard deviation, which results in the normalized cross-correlation (NCC):

$$\overline{\gamma}(\Delta x, \Delta y) = \frac{\sum_{x,y}^{\Omega} \left(r(x,y) - \mu_r\right)\left(s(x+\Delta x, y+\Delta y) - \mu_s\right)}{\sqrt{\sum_{x,y}^{\Omega} \left(r(x,y) - \mu_r\right)^2 \sum_{x,y}^{\Omega} \left(s(x+\Delta x, y+\Delta y) - \mu_s\right)^2}}. \tag{3.5}$$

The numerator of equation [3.5] is scaled by $1/N$ (N defines the total number of pixels in the template domain Ω) and thus cancels out the $1/N$ in the standard deviation of the templates. The values of the NCC range between $\overline{\gamma} = \pm 1$, and the maximum c_p of the cross-correlation peak is also called the normalized cross-correlation coefficient.

The location of the maximum of $\overline{\gamma}$ provides an estimate of the offset $(\Delta\hat{x}, \Delta\hat{y})$ between the image templates. Oversampling of the input data r, s and interpolation or curve-fitting of the cross-correlation peak (a brief overview is given by Tong *et al.* (2019)) allows for tracking with sub-pixel accuracy with different levels of precision (Debella-Gilo and Kääb 2011). However, depending on the method used, "pixel-locking" or "peak-locking" can occur, i.e. the probability that integer offsets are determined by peak-fitting is higher than non-zero sub-pixel offsets (Chen and Katz 2005).

3.2.1.2. *Confidence measure of offset estimation*

An offset can only be reliably estimated by unambiguously identifying the cross-correlation peak. To quantify the confidence of offset estimation, two parameters are often used: the normalized cross-correlation value (height of the correlation peak) c_p and the signal-to-noise ratio (SNR) of the cross-correlation. Higher confidence is normally characterized by a higher correlation peak c_p, but multiple peaks can exist when periodic patterns are present in the images. The SNR is usually defined as the ratio between the height of the NCC peak and the average of the ambient correlation field $\overline{c_f^2}$:

$$\mathrm{SNR} = \log_{10}\left(c_p^2 / \overline{c_f^2}\right) \tag{3.6}$$

A high SNR indicates a high likelihood that a unique peak is present in the cross-correlation, whereas a low SNR can either be related to strong noise in the cross-correlation or to periodic patterns showing multiple peaks of high correlation c_p. Further methods exist to improve the confidence in the estimated offset (see Chapter 11, section 11.3.4.1).

3.2.1.3. *Cross-correlation in the frequency domain*

The direct calculation of the cross-correlation in the spatial domain is computationally expensive, especially for the relatively large templates required for SAR data as the computational time scales with the number of pixels per template. To reduce the computational complexity, we can calculate the cross-correlation in the frequency domain. For equation [3.3], flipping $r^*(t)$ as $r^*(-t)$ and letting $u = -t$, we have

$$\gamma(\tau) = \int_{-\infty}^{\infty} r^*(u)s(\tau - u)\mathrm{d}u$$

which is a convolution integral. Using the convolution theorem, which states that the Fourier transform of a convolution is identical to the product of the two spectra

$$\mathcal{F}\{f * g\} = \mathcal{F}\{f\} \cdot \mathcal{F}\{g\}$$

with $\cdot$ being element-wise multiplication, we can calculate the normalized cross-correlation in the frequency domain as

$$\hat{\gamma} = \frac{\mathcal{F}\{r\}^* \cdot \mathcal{F}\{s\}}{\sqrt{|\mathcal{F}\{r\}|^2 |\mathcal{F}\{s\}|^2}} \qquad [3.7]$$

with $\mathcal{F}\{r\}$ and $\mathcal{F}\{s\}$ being the (fast) Fourier transformation of the two templates, and $|\mathcal{F}\{\dots\}|^2$ the total of the power spectrum of r and s. Here, we used $\mathcal{F}\{r^*\} = \mathcal{F}\{r\}^*$. The frequency-domain correlation $\hat{\gamma}$ can be transformed back through the inverse Fourier transformation $\overline{\gamma} = \mathrm{real}[\mathcal{F}^{-1}\{\hat{\gamma}\}]$ to track the cross-correlation peak in the spatial domain.

3.2.1.4. *Sub-pixel estimation in the frequency domain*

Displacements can not only be determined by sub-pixel localization of the cross-correlation peak but also directly from the cross-correlation in the frequency domain. When an integer shift is present and the phase-correlation matrix (see section 3.2.2.4) is transformed back to the spatial domain, the resulting correlation surface will be a unit impulse. However, to ensure the discrete nature of the Fourier transform, the impulse is diluted (or convoluted) by a 2D sinc, which smears most of the energy to immediate adjacent neighbors, but surrounded by steep dips (Foroosh *et al.* 2002). Thus, peak-locking effects appear (Chen and Katz 2005).

In order to estimate displacements through frequency methods, it is beneficial to directly estimate them in the phase plane. The phase values in the cross-spectrum have a saw-tooth pattern, where the ridges align perpendicular to the direction of the displacement, and the number of cycles corresponds to the magnitude of the displacement. Hence, many algorithms use unwrapping or work out a solution along one axis (see Balci and Foroosh (2006)). Along two axes, the Radon transform, random sample consensus (RANSAC) or single value decomposition (SVD) can also be used (see a review in Tong *et al.* (2019)).

Alternatively, a two-step procedure can be used, where the integer displacement is first estimated. This integer step is used to realign the template. Then, the displacement is below one pixel displacement and the saw tooth can be approximated by a plane, making direct estimation possible (Stone *et al.* 2001; Hoge 2003) or easing the convergence of iteration (Leprince *et al.* 2007). Phase-plane fitting is well-established using optical imagery. However, its application on SAR data is a current topic of research because the high-frequency noise from SAR speckle makes the application of phase correlation difficult.

3.2.2. *SAR offset tracking using image pairs*

Equations [3.5] and [3.7] are widely used for offset tracking with optical and SAR imagery. Depending on the speckle statistics of the SAR images, different types of input information are used for the templates r and s, and thus different sub-types of SAR offset tracking are described in the following, including coherence tracking, speckle tracking and intensity tracking.

3.2.2.1. *Coherence tracking*

For coherence tracking, the single-look-complex values $Ae^{j\phi}$ of a SAR image are used in the NCC (equation [3.5]). Coherence tracking is successful when the phase and the speckle pattern of two SAR images are correlated (top row in Figure 3.8). Finding the peak of the NCC then corresponds to a coherence maximization process. Derauw (1999) proposed this method to derive azimuth shifts in addition to interferometric estimates of line-of-sight displacements (see Chapter 4). For coherence tracking, relatively small correlation windows of, for example, 8×8 single-look pixels can be sufficient (Strozzi *et al.* 2002), so that very high spatial resolution in the offset field can be obtained. A disadvantage of coherence tracking is that phase gradients within the template (due to deformation gradients such as shear motions at the glacier edge) can reduce or even eliminate the correlation peak (Joughin 2002). These phase gradients can be corrected for by removing the topographic phase or locally estimating the phase gradient in the template's interferogram (Trouvé *et al.* 1998; Werner *et al.* 2005). Coherence tracking can be used to complement InSAR displacement and obtain displacements in the azimuth direction. For example, coherence tracking was used by Funning *et al.* (2005) to derive 3D earthquake displacements.

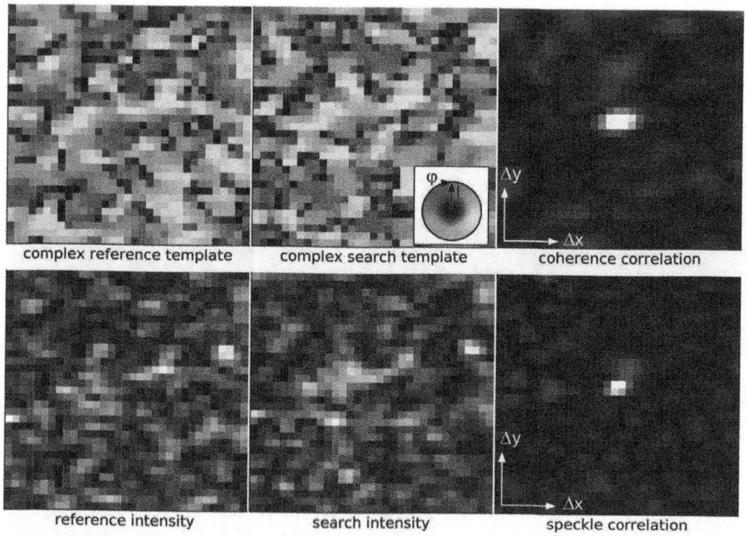

Figure 3.8. *Top row: phase ϕ and intensity I (color; brightness) of reference and search template (32 × 32 pixels); right: the resulting coherence NCC. Bottom row: intensity of the same reference and search template; right: the resulting speckle NCC. For a color version of this figure, see www.iste.co.uk/cavalie/images.zip*

3.2.2.2. *Speckle tracking*

For speckle tracking, the radar amplitude A or intensity I is used for equation [3.5]. Speckle tracking is successful when the speckle pattern is at least partially preserved between an image pair. Speckle tracking was suggested by Gray *et al.* (1998), who used the cross-correlation of detected radar images (intensity) to derive glacier velocity maps with both the range and azimuth direction. Similar to coherence tracking, speckle tracking allows for a very high spatial resolution in the offset field. For correlated speckle patterns, the speckle can be considered as correlated high-frequency noise, which generates a very sharp and narrow peak in the speckle cross-correlation function (Figure 3.8). A good description of speckle tracking combined with coherence tracking is given by Joughin (2002). Compared to coherence tracking, a slightly sharper peak can be obtained, as the intensity $I = |A|^2$ doubles spatial frequencies in the image (see section 3.1.3 and Figure 3.8).

3.2.2.3. *Intensity tracking*

When the speckle pattern is completely uncorrelated, only larger features extending over multiple intensity pixels can be used for cross-correlation (Figure 3.9). In this case, the speckle pattern can be considered as uncorrelated (or incoherent) noise. To distinguish between intensity tracking of speckle and intensity tracking of image features containing uncorrelated speckle, the latter is also called *incoherent* intensity tracking, whereas speckle tracking can be considered as *coherent* intensity tracking.

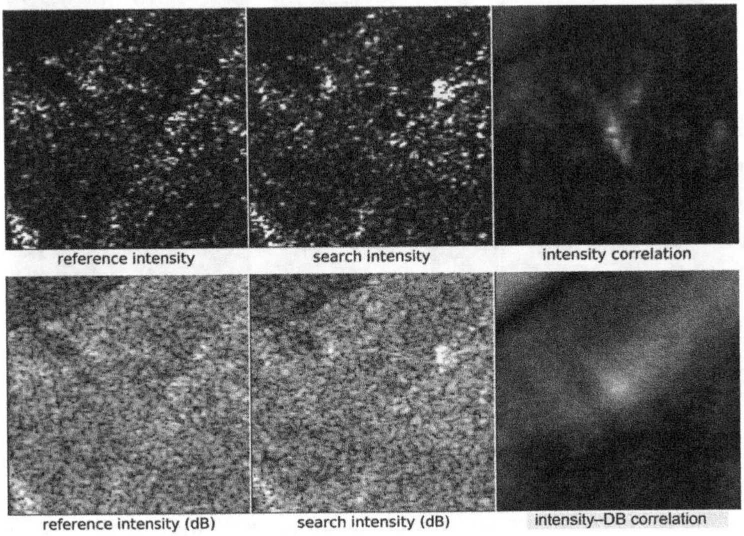

Figure 3.9. *Top row: linearly scaled radar intensity of reference and search area (128 × 128 pixels) and the resulting incoherent intensity cross-correlation. Bottom row: intensity scaled in dB (same reference and search area as above) and the resulting intensity cross-correlation*

To average out the incoherent speckle noise in the cross-correlation, considerably larger templates are required for intensity tracking. Typical template sizes vary between 64 × 64 and 256 × 256 pixels (Strozzi *et al.* 2002; Werner *et al.* 2005; Paul *et al.* 2015). Incoherent intensity tracking can be compared to offset tracking in optical imagery, as in both cases physical features are tracked. However, in optical imagery, speckle are absent and much smaller template sizes (e.g. 16 × 16 or 32×32 pixels) can be used.

As mainly large-scale features contribute to intensity tracking, the correlation function shows a much broader peak (Figure 3.9). As larger templates are required, computation of the cross-correlation can become computationally expensive. In the early days of computers, Vesecky *et al.* (1988) applied a pyramidal approach for calculating the correlation field to estimate the motion of sea ice. Today, with the availability of the fast Fourier transform (FFT), calculation of the cross-correlation can be sped up significantly when calculated in the frequency domain (see section 3.2.1.3).

When considering that speckle represents multiplicative noise for incoherent intensity tracking, it can be of advantage to convert the intensity into dB before cross-correlation. Thus, speckle can be considered as additive noise that reduces outliers and makes the correlation more robust (lower row in Figure 3.9).

3.2.2.4. *Filters for image pre- and post-processing*

To better extract the relevant feature patterns from the input images, various filters can be applied to the images before the cross-correlation. Spatial high-pass filters were shown to be effective in improving the coverage of successful tracking and the SNR of the cross-correlation (equation [3.6]) by de Lange *et al.* (2007).

The "phase correlation" method can be considered as a special case of a (high-pass) frequency filter, in which the frequency spectrum of the input data is completely flattened (normalized). Due to the flattened spectrum, only the phase terms $e^{i\phi}$ of each frequency component are correlated, and hence this method is called "phase correlation" even though only real-valued intensity data are used as input. The normalization reduces the influence of long-wavelength structures and emphasizes high-frequency components in the image. For large enough templates and low enough noise, an extremely sharp correlation peak can be obtained (Schaum and McHugh 1991; Michel *et al.* 1999; Stone *et al.* 2001; Foroosh *et al.* 2002; Zitová and Flusser 2003). Nevertheless, emphasizing high-frequency image components makes the phase correlation method especially sensitive to speckle noise. A similar variant is "orientation correlation" where the orientation of intensity gradients is used as input for cross-correlation (Fitch *et al.* 2002). This method is often preferred for optical images with low noise, as it is less sensitive to illumination changes and shadow (Heid and Kääb 2012; Dehecq *et al.* 2015). However, due to its high-pass character, it is only of limited advantage for SAR images due to speckle.

Feature-based filters provide another method for pre-processing of SAR imagery. For example, stable point-like targets on mountain slopes can be discriminated from temporally variable speckle and their displacement can be tracked (Hu *et al.* 2013; Li *et al.* 2019). Edge-detection methods have been applied on sea ice followed by phase correlation on the speckle-free product (Karvonen *et al.* 2012).

Finally, before tracking results are used as input for geophysical models, post-processing can constrain the tracking results to the most probable or the most physical solution. An overview focused on glaciers is given in Chapter 11, sections 11.3.4 and 11.3.5.

3.2.3. *SAR offset tracking using image series*

With the increasing availability of SAR image series, offset tracking can be performed using many image pairs of the same area (data stack). The information contained in image series provides multiple independent measurements, which can make offset tracking more robust and/or precise. To exploit image series for offset tracking, stacking methods can be used by creating a data cube in which the stacked images represent the same area on the ground but are acquired under different conditions (e.g. different acquisition times). The image stack can be further processed to obtain a stack of cross-correlations or offset fields. Stacking methods have been suggested in various fields; for example, Meinhart *et al.* (2000) suggest them to improve tracking in particle image velocimetry (PIV) using image series. They subdivide the tracking process into three steps: image acquisition, cross-correlation and peak detection. In each of the three steps, stack-averaging can be applied for: 1) image acquisition by averaging each set of two image stacks before cross-correlation of the averaged stacks; 2) image correlation by averaging the stack of cross-correlations of consecutive image pairs before peak detection; and 3) peak detection by averaging the stack of offset vectors obtained from a series of cross-correlations. Each of the three variants has different advantages and disadvantages: for imagery with a low particle density (PIV), the density of the trackable particles is increased by image-averaging; satellite imagery, however, containing dense image content generally suffers from motion blur in the case of steadily moving features, but for features that can be considered stable during the averaging period, image noise can be reduced. Cross-correlation averaging can improve the signal-to-noise ratio, but for variable displacements, the correlation peak becomes blurred. Furthermore, only a single displacement estimate is undertaken after correlation averaging, whereas the averaging of offset vectors provides better statistics on the displacement dynamics and the estimation accuracy. The following two sections describe how to adapt the first two methods of cross-correlation averaging and image averaging to improve SAR offset tracking. Variants of the third method for post-processing of offset fields are probably most widely used and are described in Chapter 11, section 11.3.5.

3.2.3.1. *Cross-correlation stacking*

For a pair of non-identical image templates, their cross-correlation (or NCC) can be regarded as a noisy measurement of the offset between them. To enhance tracking, multiple NCCs can be stacked and averaged to reduce the noise and to improve the measurement accuracy. This principle, called the "average correlation method" by Meinhart *et al.* (2000), has been applied to estimate the velocity of quickly decorrelating ice falls using optical imagery, where it is referred to as "ensemble matching" (Altena and Kääb 2020). Similarly, Li *et al.* (2021) used this principle to estimate the velocity of alpine glaciers from SAR imagery strongly affected by speckle (incoherent intensity correlation, see section 3.2.2.3).

Let us assume that image templates are composed of a signal $\hat{s}$ and noise $\hat{n}$ as $\hat{I} = \hat{s} + \hat{n}$, with ^ indicating the Fourier transform. The NCC in the frequency domain [3.7] can be decomposed as:

$$\hat{\gamma} = \frac{\hat{s}_1^* \hat{s}_2 + \hat{s}_1^* \hat{n}_2 + \hat{n}_1^* \hat{s}_2 + \hat{n}_1^* \hat{n}_2}{\sqrt{E\{|\hat{I}_1|^2\} \cdot E\{|\hat{I}_2|^2\}}}. \quad [3.8]$$

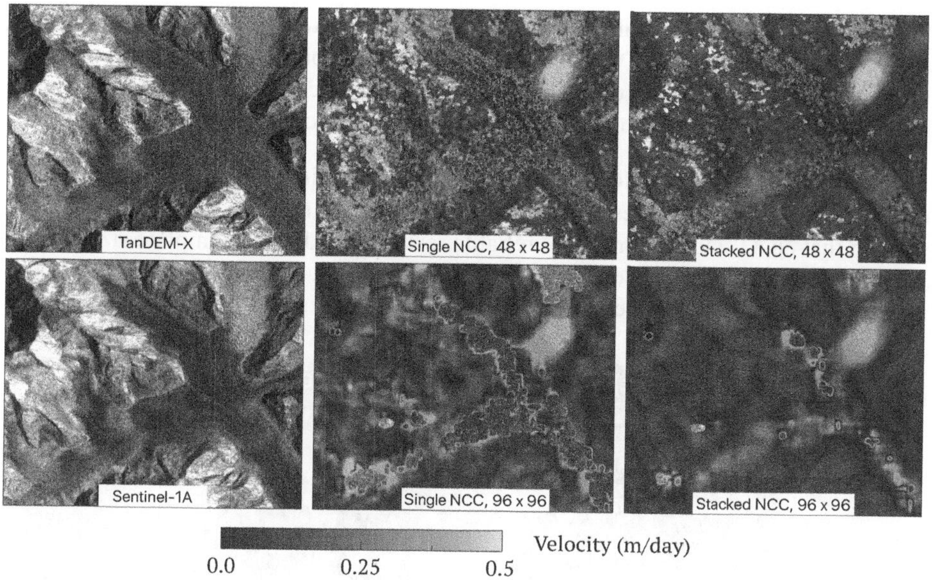

Figure 3.10. *Top row: Great Aletsch Glacier intensity image (left) and velocity images obtained from pairwise NCC (middle) and by stacking seven NCC pairs of TerraSAR-X imagery and templates of 48 × 48 pixels (©DLR 2021). Bottom row: Sentinel-1A image (left) and velocity images obtained from pairwise NCC (middle) and by stacking 11 NCCs (right) using templates of 96 × 96 pixels (©ESA 2021). For a color version of this figure, see www.iste.co.uk/cavalie/images.zip*

Here, the signal $\hat{s}$ is assumed to contain all the correlated content that contributes to the correct template matching, and the noise $\hat{n}$ to represent all the uncorrelated content. In equation [3.8], the shifted auto-correlation $\hat{s}_1^*\hat{s}_2$ generates the signal to track, and the remaining components ($\hat{s}_1^*\hat{n}_2$, $\hat{n}_1^*\hat{s}_2$ and $\hat{n}_1^*\hat{n}_2$) add noise to the NCC and thus bring uncertainty to the measurement. In order to attain reliable offset estimates, a high SNR for the NCC must be achieved, and thus the term $\hat{s}_1^*\hat{s}_2$ must generate a dominant peak that greatly surpasses the noise.

For NCC stacking, a time series of N coregistered images is collected with a revisit time of Δt, and an NCC is calculated from each pair of image templates with a temporal spacing of $T = a\Delta t$ ($a = 1, 2, 3...$ indicating integer multiples of the revisit time Δt). This results in a stack of $N - a$ consecutive NCCs, in which each NCC captures a snapshot of the shift between the template pair. Assuming that image features move with constant velocities during acquisition of the image series, all NCCs in the stack represent identical offsets. Hence, the peaks of $\hat{s}_1^*\hat{s}_2$ in all the NCCs should be located at the same position, whereas the noise term $\hat{s}_1^*\hat{n}_2 + \hat{n}_1^*\hat{s}_2 + \hat{n}_1^*\hat{n}_2$ remains independent and can be reduced by averaging the NCC stack. The principle of

NCC stacking can be generalized to multiple image time series, such as image series acquired in different years, with different spectral or polarimetric channels or even from different sensors (e.g. optical and SAR). NCC stacking is relatively simple to implement but is currently limited to NCC pairs of identical temporal spacing T.

Figure 3.10 shows that NCC stacking can effectively increase the spatial coverage of reliable velocity estimates for SAR offset tracking. In addition, Li *et al.* (2021) showed that NCC stacking produces reliable results already for a template size half as large as required for pairwise NCC. They also achieved a good performance gain for small stack sizes and thus the compromise on temporal resolution is limited.

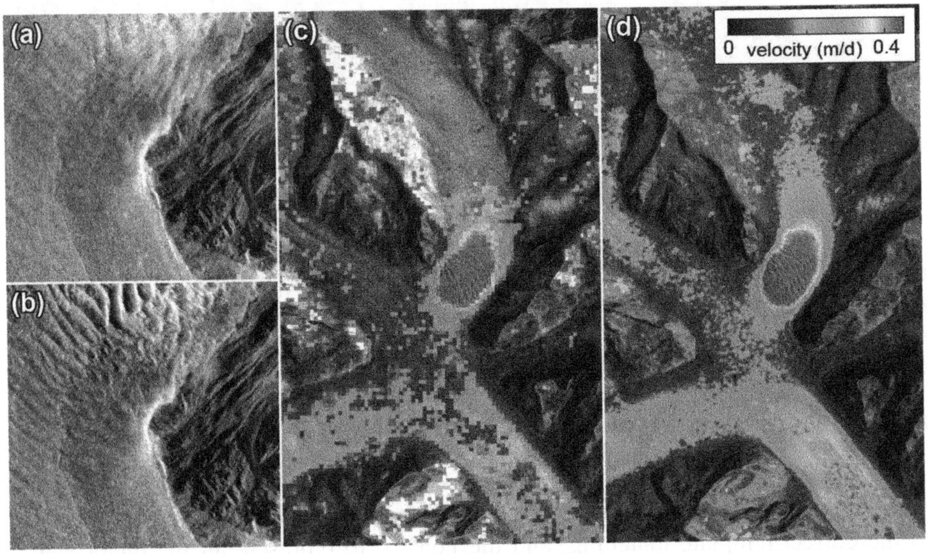

Figure 3.11. *(a) The temporal average of 10 SAR images (TerraSAR-X, ©DLR 2021) shows strong motion blur in the moving glacier. (b) After correcting for motion blur, the averaged images reveal many details of the glacier. (c) Velocity from intensity tracking of a single image pair (template size: 96 × 96 pixels, spacing: 32 pixels). (d) Velocity from autofocusing motion blur in a stack of 10 images (template size: 30 × 30 pixels). For a color version of this figure, see www.iste.co.uk/cavalie/images.zip*

3.2.3.2. *Autofocusing motion blur in image stacks*

Temporal multi-looking causes strong motion blur on moving surfaces [Figure 3.11a], which is why the averaging of image stacks before cross-correlation, studied by Meinhart *et al.* (2000) for PIV with optical imagery of low particle density, cannot be directly applied to SAR images. Instead, Leinss *et al.* (2021b) used an autofocus method to combine offset estimation with temporal multi-looking through stack averaging. With the aim of generating two locally coinciding image templates, they split a stack of N images into two sub-stacks each containing $N_1 \approx N_2 \approx N/2$

images. Then, they sampled different velocities by correcting for the time- and velocity-dependent displacement in every image. When the correct local velocity was sampled, temporal multi-looking of the sub-stacks created two speckle-reduced and motion-compensated SAR images revealing fine details that spatially coincided and generated a high cross-correlation score. For sampled velocities that did not match the true velocity, details in the averaged sub-stacks became blurred, resulting in a lower cross-correlation score. The method outperforms pairwise incoherent intensity tracking because the multi-looking process strongly reduces temporally uncorrelated speckle in SAR images.

The method requires a stack of N radar intensity images I_i, acquired at times $t_1, ..., t_N$, during which most surface features remain visible. The images must be coregistered with sub-pixel precision on stable terrain or the orbit accuracy must be precise enough to ensure sub-pixel registration. Varying time intervals Δt between acquisitions and minor gaps in the time series pose no problem.

To avoid motion blur in the time-averaged sub-stacks, the shift of surface features must be corrected with precisely known offset fields. In contrast, we can determine the offset field by trying to obtain a temporally multi-looked SAR image where motion blur is minimized. To achieve this, the (so far unknown) time-dependent offset field $(\delta x, \delta y)(x, y, t)$ can be approximated by the stationary velocity model

$$\left.\frac{\delta x}{\delta t}\right|_{xy,t^*} = u(x, y, t^*)$$
$$\left.\frac{\delta y}{\delta t}\right|_{xy,t^*} = v(x, y, t^*) \qquad [3.9]$$

where u, v describe the two components of the local velocity vector, $t^* \approx \frac{1}{N}\sum_i^N t_i$ is approximately the mean acquisition time of the stack and $\delta t = t - t^*$ is the time interval during which the offset $(\delta x, \delta y)$ occurs. Then, possible velocity values are sampled from a predefined distribution, for example a uniform distribution within a certain velocity range. The sampled velocity is used to estimate the time-dependent position $(x', y')(\delta t)$ of features located at (x, y) at time t^*. For the i-th image in the stack, the position at time $\delta t_i = t_i - t^*$ follows from the warping equations

$$x'_i = x + u(x, y)\,\delta t_i$$
$$y'_i = y + v(x, y)\,\delta t_i. \qquad [3.10]$$

For a spatially constant velocity field, the warping equations describe a simple shear of the two image stacks. For an arbitrary but stationary velocity field, we can track image features that deviate for large δt from a linear trajectory using finite-step time integration of equation [3.10]. Using the warping equations for time-dependent interpolation of the original stacks, the offset-corrected image stacks can be obtained. Averaging each corrected sub-stack results in two offset-corrected temporally multi-looked intensity images in which speckle is strongly reduced.

To estimate the surface velocity field (u, v), the motion blur of the temporally multi-looked sub-stacks is iteratively evaluated for each sampled velocity until the blur is minimized. The motion blur in each iteration is quantified using a method adapted from the "passive phase detection autofocus" method of photographic cameras: two bundles of light rays, each entering the camera at opposite sides of the lens, generate two sub-images, and the focal-length is determined from the offset between the sub-image pair via cross-correlation. Similarly, we calculate the correlation score between the images I'_k $(k = 1, 2)$ obtained by averaging the offset-corrected sub-stack:

$$I'_k(x, y, t^*) = \frac{1}{N_k} \sum_{i=k_0}^{k_N} I_i \left[x'(t_i), y'(t_i)\right], \qquad [3.11]$$

with $(k_0, k_N) = (1, N_1)$ for $k = 1$, $(k_0, k_N) = (N_1 + 1, N)$ for $k = 2$ and $N_k = k_N - k_0 + 1$. Already, with small N_k and a correctly chosen velocity, the images I'_k reveal surface structures that are hidden by speckle in single SAR images because the standard deviation of the intensity scales down with $1/\sqrt{N_k}$ (section 3.1.2). Due to reduced speckle, a relatively small template size can be used for the NCC. To compress the dynamic range and focus on small-scale features, I'_k should be converted to dB and a high-pass filter can be applied before calculating the NCC.

Each sampled velocity (u, v) shifts the images in the sub-stacks according to equation [3.10] and generates I'_1 and I'_2 to calculate an NCC value $\overline{\gamma}(u, v)$. The NCC score is maximized when the sampled velocity eliminates the motion blur of features in the stack, and more importantly, when the features revealed in I'_1 and I'_2 coincide. In practice, the sample velocity is applied to shift the entire image and the velocity that improves the local NCC score is stored. With sufficiently dense sampling, the optimal velocity for every image pixel can be found, which represents the estimated velocity field.

The velocity field can be iteratively refined, and thus the method is suitable to effectively track shear movements. The method requires that temporal variations in the velocity field be small, whereas there is no restriction on the spatial complexity of the observed flow patterns. In general, the uniform sampling of the velocity is computationally expensive as *each* iteration requires an interpolation of the entire image stack, the sum of the sub-stacks and the calculation of the NCC score. For effective velocity sampling, a pyramidal approach, as described by Vesecky *et al.* (1988), can be used.

As the method effectively correlates full-resolution speckle-reduced SAR imagery, the spatial resolution and also the coverage of the obtained velocity are significantly better compared to pairwise intensity cross-correlation, for which larger templates are required (compare Figure 3.11(c) vs. (d)). Compared to cross-correlation stacking, the method performs slightly better in robustness and accuracy. The reason is that in fact each image from one sub-stack is correlated with every image in the other sub-stack, resulting in $(N/2)^2$ cross-correlations. In contrast, for cross-correlation stacking, only a consecutive series of $N-1$ cross-correlations are evaluated. As a side product, the methods generate temporally multi-looked radar images that are free of motion blur (Figure 3.11(b)) and that can even be smoothly animated by an arbitrary choice of t^* and a running average over a large image series (Leinss *et al.* 2021a).

3.2.4. *Tracking from single and multiple orbits*

Most imaging methods project information from a 3D terrain into two image dimensions. SAR sensors project the scene content, and hence also the true 3D displacement, into the slant-range and azimuth geometry. Single-pass interferometric SAR systems are an example where displacements perpendicular to the slant range and azimuth can be observed, as in, for example, Leinss and Bernhard (2021). In this section, we discuss projections of displacements into the desired geometry and considerations relevant for the radar imaging geometry.

3.2.4.1. *Single-orbit tracking in slant-range coordinates*

The projection of the 3D terrain into the SAR imaging geometry depends on the terrain and on the incidence angle (see Figure 3.3). As long as the incidence angle is constant between two acquisitions, offset tracking reveals only slant-range and azimuth displacement. However, small deviations from the reference orbit cause terrain-dependent distortions in the image content, resulting in terrain-dependent offset fields. To avoid such terrain-dependent offsets, the satellite orbit must be exactly repeated for every acquisition. As this is not always feasible, such geometric distortion can be corrected by DEM-based coregistration (Sansosti *et al.* 2006; Nitti *et al.* 2010). However, for SAR offset tracking, we should at least use imagery from the same orbit, as image features show different backscatter characteristics at different incidence angles. Speckle and coherence tracking are most sensitive to small deviations, as speckle can already decorrelate strongly for deviations of a few hundreds of meters from the reference orbit (see also the critical baseline described in Chapter 4, section 4.2.2).

Offset tracking in the native slant-range/azimuth geometry has the advantage that no resampling for orthorectification is required, providing slightly better image quality. Another interpolation can be avoided when tracking is done without resampling for coregistration and instead the offset field is corrected for global and terrain-induced shifts through post-processing. In the latter case, however, cross-correlation stacking and autofocusing motion blur (see sections 3.2.3.1 and 3.2.3.2) require geometry-dependent corrections.

To obtain displacements in a map geometry, the 2D projection of the displacement can be orthorectified (see also Chapter 11, section 11.3.3.3). However, it is crucial to specify that slant-range (line-of-slight) and azimuth displacements are shown or that additional, possibly model-based, transformation into other displacement geometries is applied.

3.2.4.2. *Tracking in map coordinates*

Offset tracking can also be applied to orthorectified radar images to obtain results directly in map coordinates. The oblique imaging geometry makes a precise orthorectification crucial to avoid image artifacts. For precise orthorectification, a SAR image simulated from an external DEM (Small *et al.* 1998; Small 2011) can be used as a reference to align the reference SAR image with the DEM used for orthorectification (Wegmüller 1999). DEM-based coregistration of a SAR image stack to a common reference scene is an advantage, in addition to coregistration of the reference scene to the simulated reference.

Working in map coordinates can simplify offset tracking because of the horizontally constant pixel spacing. However, tracking in map coordinates implies the implicit assumption of displacements parallel to the DEM slope. First, this is because the 3D displacement is projected to the slant range and azimuth, and then the orthorectification transforms the apparent slant-range displacement through the DEM-based orthorectification lookup-table into a horizontal displacement. While for slope-parallel displacements only the horizontal component is measured, non-slope-parallel displacements can introduce large biases into the obtained velocity maps.

Despite a constant pixel spacing in the orthorectified imagery, the slope-dependent effective resolution can vary strongly over the scene, which makes the tracking precision dependent on the terrain slope (see Figure 3.3). Another problem relates to small-scale features in the orthorectification DEM, which can distort or wrinkle the orthorectified images when the features move in time (e.g. moving glacier crevasses, growing forest).

3.2.4.3. *Three-dimensional inversion*

Three-dimensional displacement estimation (including east, north and up components) is an important subject in SAR displacement measurement exploration in relation to both offset tracking and SAR interferometry (Wright *et al.* 2004; Pathier *et al.* 2006). Displacement maps showing only line-of-slight and possibly also only azimuth displacements can make the comparison with other sources of information difficult. Furthermore, interpretation can be difficult for those not familiar with SAR geometries. The common approach is thus to estimate the 3D displacement, referred to as the common terrestrial coordinate, from InSAR or offset-tracking displacement measurements obtained from different directions (i.e. range and azimuth directions, different incidence angles, different orbit directions).

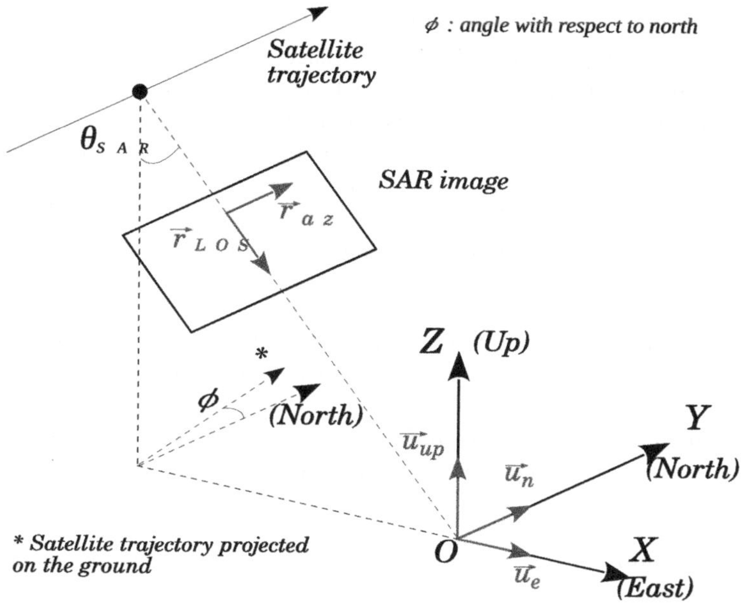

Figure 3.12. *Geometric model of 3D displacement and the displacement in the range and azimuth directions*

The problem can be resolved using a linear geometric model (Wright *et al.* 2004), as shown in Figure 3.12, and mathematically expressed as

$$\mathbf{r} = \mathbf{P}\mathbf{u} \tag{3.12}$$

where $\mathbf{r} = (r_{\text{LOS},1}, r_{\text{LOS},2}, ..., r_{\text{az},1}, r_{\text{az},2}, ...)^T$ denotes the vector of N displacement measurements in different directions (LOS, azimuth) obtained from offset tracking or InSAR, and $\mathbf{u}$ is the vector containing the true 3D displacement components to estimate. The matrix $\mathbf{P}$ is the $N \times 3$ projection matrix, where the rows define the projection either into the slant-range ($\mathbf{p}_{\text{LOS,i}}$) or azimuth direction ($\mathbf{p}_{\text{az,i}}$). The projections depend on the incidence angle θ_i and the azimuth heading angle ϕ_i of each acquisition geometry used. ϕ_i is defined as between the direction of the satellite trajectory and north. The projections are given by the unit row vectors

$$\begin{aligned} P_{\text{LOS,i}} &= \begin{pmatrix} -\cos\phi_i \sin\theta_i & \sin\phi_i \sin\theta_i & -\cos\theta_i \end{pmatrix} \\ P_{\text{az,i}} &= \begin{pmatrix} \sin\phi_i & \cos\phi_i & 0 \end{pmatrix} \end{aligned} \tag{3.13}$$

To solve equation [3.12], the least-squares approach is commonly used:

$$\mathbf{u} = (\mathbf{P}^T \mathbf{\Sigma_r}^{-1} \mathbf{P})^{-1} \mathbf{P}^T \mathbf{\Sigma_r}^{-1} \mathbf{r} \tag{3.14}$$

with $\mathbf{\Sigma_r}$ being the error covariance of displacement measurements in $\mathbf{r}$. Its main role consists of appropriately weighting displacement measurements of different uncertainties and propagating the uncertainty from $\mathbf{r}$ to $\mathbf{u}$. In many cases, a diagonal matrix is used for the sake of simplicity, assuming independence of each displacement measurement. In the case of measurements sharing the same reference/secondary images, the latter assumption may not hold and improved estimations of the covariance matrix $\mathbf{\Sigma_r}$ may be necessary (Teunissen and Amiri-Simkooei 2008). $\mathbf{\Sigma_r}$ can also be approximated by other quality indicator parameters, such as interferometric coherence or cross-correlation coefficients.

Since $\mathbf{u}$ is a vector with three components, there should be displacement measurements in at least three directions in r. Tracking results from descending and ascending passes are therefore often combined. However, outside of high latitudes, the azimuth directions of the ascending and descending passes are nearly collinear. In practice, they are often considered as the same contribution to the least-square solution.

Through equation [3.14], a large number of displacement measurements can be transformed into a single 3D displacement result. Besides the ease of interpretation and comparison, another main feature of interest lies in the reduction of the uncertainty. The uncertainty of the 3D displacement obtained is smaller than that of any displacement measurement in r. For a random uncertainty, the uncertainty decreases with the number of displacement measurements used. Therefore, the use of all available displacement measurements is preferred, knowing that the eventual interdependence between different measurements can result in an underestimation of the uncertainty of the 3D displacement. However, when systematic uncertainty (e.g. from orbit uncertainties) is present, fuzzy logic allows us to better describe the uncertainty due to data fusion and provides an upper bound for the uncertainty (Yan *et al.* 2012). The extraction of 3D displacement fields also simplifies the comparison of results between different sensors, models and field measurements (Yan *et al.* 2013).

3.2.5. *Summary*

This chapter provided a brief introduction to the SAR imaging method, the imaging geometry and the origin of radar speckle. The basic principles of offset tracking using the cross-correlation of pairs of image templates were introduced, followed by details of how to characterize the correlation score and the signal-to-noise ratio and how to efficiently calculate the normalized cross-correlation in the frequency domain. Depending on the temporal coherence of the speckle pattern in SAR imagery, different methods, such as coherence tracking, speckle tracking or intensity tracking, can be applied, with the latter requiring significantly larger image templates due to uncorrelated speckle noise. To make tracking more robust, stacks of cross-correlation functions or motion-compensated image stacks can be evaluated. We concluded this chapter by outlining the advantages and disadvantages of offset tracking using SAR imagery radar and map coordinates and describing a common method to combine offset estimates from multiple orbits and obtain 3D displacements.

3.3. Acknowledgments

The authors thank Anna Wendleder and Tazio Strozzi for their careful reviews and comments that helped to improve this chapter.

3.4. References

Altena, B. and Kääb, A. (2020). Ensemble matching of repeat satellite images applied to measure fast-changing ice flow, verified with mountain climber trajectories on Khumbu icefall, Mount Everest. *Journal of Glaciology*, 66(260), 905–915.

Balci, M. and Foroosh, H. (2006). Subpixel estimation of shifts directly in the Fourier domain. *IEEE Transactions on Image Processing*, 15(7), 1965–1972.

Bindschadler, R.A. and Scambos, T.A. (1991). Satellite-image-derived velocity field of an Antarctic ice stream. *Science*, 252(5003), 242–246.

Chen, J. and Katz, J. (2005). Elimination of peak-locking error in PIV analysis using the correlation mapping method. *Measurement Science and Technology*, 16(8), 1605–1618.

De Zan, F. (2014). Accuracy of incoherent speckle tracking for circular Gaussian signals. *IEEE Geoscience and Remote Sensing Letters*, 11(1), 264–267.

Debella-Gilo, M. and Kääb, A. (2011). Sub-pixel precision image matching for measuring surface displacements on mass movements using normalized cross-correlation. *Remote Sensing of Environment*, 115(1), 130–142.

Dehecq, A., Gourmelen, N., Trouvé, E. (2015). Deriving large-scale glacier velocities from a complete satellite archive: Application to the Pamir–Karakoram–Himalaya. *Remote Sensing of Environment*, 162, 55–66.

Derauw, D. (1999). DInSAR and coherence tracking applied to glaciology: The example of Shirase Glacier. *Proceedings of the FRINGE*, ESA.

Fitch, A.J., Kadyrov, A., Christmas, W.J., Kittler, J. (2002). Orientation correlation. *Proceedings of 13th British Machine Vision Conference*, 1–10.

Foroosh, H., Zerubia, J.B., Berthod, M. (2002). Extension of phase correlation to subpixel registration. *IEEE Transactions on Image Processing*, 11(3), 188–200.

Funning, G.J., Parsons, B., Wright, T.J., Jackson, J.A., Fielding, E.J. (2005). Surface displacements and source parameters of the 2003 Bam (Iran) earthquake from Envisat advanced synthetic aperture radar imagery. *Journal of Geophysical Research: Solid Earth*, 110(B9).

Gray, A.L., Mattar, K.E., Vachon, P.W., Bindschadler, R., Jezek, K.C., Forster, R., Crawford, J.P. (1998). InSAR results from the RADARSAT Antarctic mapping mission data: Estimation of glacier motion using a simple registration procedure. *IEEE International Geoscience and Remote Sensing Symposium*, 3, 1638–1640.

Heid, T. and Kääb, A. (2012). Evaluation of existing image matching methods for deriving glacier surface displacements globally from optical satellite imagery. *Remote Sensing of Environment*, 118, 339–355.

Hoge, W. (2003). A subspace identification extension to the phase correlation method. *IEEE Transactions on Medical Imaging*, 22(2), 277–280.

Hu, X., Wang, T., Liao, M. (2013). Measuring coseismic displacements with point-like targets offset tracking. *IEEE Geoscience and Remote Sensing Letters*, 11(1), 283–287.

Joughin, I. (2002). Ice-sheet velocity mapping: A combined interferometric and speckle-tracking approach. *Annals of Glaciology*, 34, 195–201.

Karvonen, J., Cheng, B., Vihma, T., Arkett, M., Carrieres, T. (2012). A method for sea ice thickness and concentration analysis based on SAR data and a thermodynamic model. *The Cryosphere*, 6(6), 1507–1526.

de Lange, R., Luckman, A., Murray, T. (2007). Improvement of satellite radar feature tracking for ice velocity derivation by spatial frequency filtering. *IEEE Transactions on Geoscience and Remote Sensing*, 45(7), 2309–2318.

Leinss, S. and Bernhard, P. (2021). TanDEM-X: Deriving InSAR height changes and velocity dynamics of Great Aletsch Glacier. *IEEE Journal of Selected Topics in Applied Earth Observervation and Remote Sensing*, 14, 1–18.

Leinss, S., Li, S., Frey, O. (2021a). Flow animation of Great Aletsch Glacier generated by autofocusing temporally multilooked SAR time series. Video, ETH Research Collection.

Leinss, S., Li, S., Frey, O. (2021b). Measuring glacier velocity by autofocusing temporally multilooked SAR time series. *IEEE International Geoscience and Remote Sensing Symposium*, 5493–5496.

Leprince, S., Barbot, S., Ayoub, F., Avouac, J.-P. (2007). Automatic and precise orthorectification, coregistration, and subpixel correlation of satellite images, application to ground deformation measurements. *IEEE Transactions on Geoscience and Remote Sensing*, 45(6), 1529–1558.

Li, M., Zhang, L., Shi, X., Liao, M., Yang, M. (2019). Monitoring active motion of the Guobu landslide near the Laxiwa hydropower station in China by time-series point-like targets offset tracking. *Remote Sensing of Environment*, 221, 80–93.

Li, S., Leinss, S., Hajnsek, I. (2021). Cross-correlation stacking for robust offset tracking using SAR image time-series. *IEEE Journal of Selected Topics in Applied Earth Observations and Remote Sensing*, 14, 4765–4778.

Lucchitta, B.K. and Ferguson, H.M. (1986). Antarctica: Measuring glacier velocity from satellite images. *Science*, 234(4780), 1105–1108.

Meinhart, C.D., Wereley, S.T., Santiago, J.G. (2000). A PIV algorithm for estimating time-averaged velocity fields. *Journal of Fluids Engineering*, 122(2), 285–289.

Michel, R., Avouac, J.-P., Taboury, J. (1999). Measuring ground displacements from SAR amplitude images: Application to the Landers earthquake. *Geophysical Research Letters*, 26(7), 875–878.

Nitti, D.O., Hanssen, R.F., Refice, A., Bovenga, F., Nutricato, R. (2010). Impact of DEM-assisted coregistration on high-resolution SAR interferometry. *IEEE Transactions on Geoscience and Remote Sensing*, 49(3), 1127–1143.

Oliver, C. and Quegan, S. (2004). *Understanding Synthetic Aperture Radar Images*. SciTech Publishing, Raleigh.

Olmsted, C. (1993). Alaska SAR Facility – Scientific SAR User's Guide. Document.

Pathier, E., Fielding, E.J., Wright, T.J., Walker, R., Parsons, B.E., Hensley, S. (2006). Displacement field and slip distribution of the 2005 Kashmir earthquake from SAR imagery. *Geophysical Research Letters*, 33(20).

Paul, F., Bolch, T., Kääb, A., Nagler, T., Nuth, C., Scharrer, K., Shepherd, A., Strozzi, T., Ticconi, F., Bhambri, R., Berthier, E., Bevan, S., Gourmelen, N., Heid, T., Jeong, S., Kunz, M., Lauknes, T.R., Luckman, A., Merryman Boncori, J.P., Moholdt, G., Muir, A., Neelmeijer, J., Rankl, M., Van Looy, J., Van Niel, T. (2015). The glaciers climate change initiative: Methods for creating glacier area, elevation change and velocity products. *Remote Sensing of Environment*, 162, 408–426.

Rosen, P.A., Hensley, S., Joughin, I.R., Fuk, K.L., Madsen, S.N., Rodriguez, E., Goldstein, R.M. (2000). Synthetic aperture radar interferometry. *Proceedings of the IEEE*, 88(3), 333–382.

Sansosti, E., Berardino, P., Manunta, M., Serafino, F., Fornaro, G. (2006). Geometrical SAR image registration. *IEEE Transactions on Geoscience and Remote Sensing*, 44(10), 2861–2870.

Scambos, T.A., Dutkiewicz, M.J., Wilson, J.C., Bindschadler, R.A. (1992). Application of image cross-correlation to the measurement of glacier velocity using satellite image data. *Remote Sensing of Environment*, 42(3), 177–186.

Schaum, A. and McHugh, M. (1991). Analytic methods of image registration: Displacement estimation and resampling. Technical report, Naval Research Lab, Washington DC.

Small, D. (2011). Flattening gamma: Radiometric terrain correction for SAR imagery. *IEEE Transactions on Geoscience and Remote Sensing*, 49(8), 3081–3093.

Small, D., Holecz, F., Meier, E., Nuesch, D. (1998). Absolute radiometric correction in rugged terrain: A plea for integrated radar brightness. *IEEE International Geoscience and Remote Sensing Symposium*, 330–332.

Stone, H., Orchard, M., Chang, E.-C., Martucci, S. (2001). A fast direct Fourier-based algorithm for subpixel registration of images. *IEEE Transactions on Geoscience and Remote Sensing*, 39(10), 2235–2243.

Strozzi, T., Luckman, A., Murray, T., Wegmüller, U., Werner, C.L. (2002). Glacier motion estimation using SAR offset-tracking procedures. *IEEE Transactions on Geoscience and Remote Sensing*, 40(11), 2384–2391.

Teunissen, P.J. and Amiri-Simkooei, A. (2008). Least-squares variance component estimation. *Journal of Geodesy*, 82(2), 65–82.

Tong, X., Ye, Z., Xu, Y., Gao, S., Xie, H., Du, Q., Liu, S., Xu, X., Liu, S., Luan, K., Still, U. (2019). Image registration with Fourier-based image correlation: A comprehensive review of developments and applications. *IEEE Journal of Selected Topics in Applied Earth Observations and Remote Sensing*, 12(10), 4062–4081.

Trouvé, E., Nicolas, J.-M., Maitre, H. (1998). Improving phase unwrapping techniques by the use of local frequency estimates. *IEEE Transactions on Geoscience and Remote Sensing*, 36(6), 1963–1972.

Vesecky, J.F., Samadani, R., Smith, M.P., Daida, J.M., Bracewell, R.N. (1988). Observation of sea-ice dynamics using synthetic aperture radar images: Automated analysis. *IEEE Transactions on Geoscience and Remote Sensing*, 26(1), 38–48.

Wegmüller, U. (1999). Automated terrain corrected SAR geocoding. *IEEE International Geoscience and Remote Sensing Symposium*, 3, 1712–1714.

Werner, C., Wegmüller, U., Strozzi, T., Wiesmann, A. (2005). Precision estimation of local offsets between pairs of SAR SLCs and detected SAR images. *IEEE International Geoscience and Remote Sensing Symposium*, 7, 4803.

Wright, T.J., Parsons, B.E., Lu, Z. (2004). Toward mapping surface deformation in three dimensions using InSAR. *Geophysical Research Letters*, 31(1).

Yan, Y., Mauris, G., Trouvé, E., Pinel, V. (2012). Fuzzy uncertainty representations of co-seismic displacement measurements issued from SAR imagery. *IEEE Transactions on Instrumentation and Measurement*, 61(5), 1278–1286.

Yan, Y., Pinel, V., Trouvé, E., Pathier, E., Perrain, J., Bascou, P., Jouanne, F. (2013). Coseismic slip distribution of the 2005 Kashmir earthquake from SAR amplitude image correlation and differential interferometry. *Geophysical Journal International*, 193(1), 29–46.

Zitová, B. and Flusser, J. (2003). Image registration methods: A survey. *Image and Vision Computing*, 21, 977–1000.

4

SAR Interferometry: Principles and Processing

Philippe DURAND[1], Nadine POURTHIÉ[1], Céline TISON[1] and Giorgio GOMBA[2]

[1] *CNES, Toulouse, France*

[2] *German Aerospace Center (DLR), Oberpfaffenhofen, Germany*

4.1. Introduction

Synthetic aperture radar interferometry (InSAR) combines two SAR single-look-complex (SLC) images to exploit the geometrical information contained in the phase difference. The phase of the electromagnetic wave describes, among other contributions, the wave propagation path. This access to the geometry is very valuable for describing the topography of the illuminated terrain. By combining more than two phases, InSAR allows the measurement of the topography changes by minimizing the phase noise.

To obtain accurate measurements of the wave path difference, the conventional InSAR technique computes the normalized complex correlation coefficient called the coherence. The coherence argument provides the wrapped phase image (value ranging from 0 to 2π) called the interferogram. It is related to the acquisition geometry and topographic/displacement information. The coherence modulus (value ranging from 0 to 1) provides for each pixel an indicator of the similarity of the two SAR complex images. It is related to the phase noise and characterizes the geometric phase reliability, as explained later in this chapter.

Surface Displacement Measurement from Remote Sensing Images,
coordinated by Olivier CAVALIÉ and Emmanuel TROUVÉ.
© ISTE Ltd 2022.

The inversion of the interferometric phase delivers direct information on the surface topography, provided that the satellite orbit and a reference geoid are accurately defined. Furthermore, the measurement of the variations of topography gives a surface displacement map, leading to the concept of differential interferometry. Differential interferometry (DInSAR) is the combination of two interferograms or one interferogram and a reference digital elevation model (DEM)[1]. The DInSAR method enables movement detection at a fraction of the radar wavelengths (i.e. a few millimeters).

In this chapter, interferometry principles are explained. The limits of the method are also discussed (coherence loss, relief impact in shadows and layovers, atmospheric artifacts, etc.). A particular emphasis is put on the atmospheric impacts, which are a major issue for displacement observations. In fact, displacement detection requires the combination of interferograms acquired at different times, sometimes with quite a large gap. The meteorological conditions are thus often different. The water content of the atmosphere disturbs the radar phase, which can mislead the DInSAR interpretation. The interferometric processing steps are illustrated on the basis of Differential Interferometric Automated Process in the ORFEO Tool Box (DiapOTB) processors. This open source interferometric toolbox, made available by CNES in the framework of the Orfeo ToolBox (OTB), is one of the available interferometric software packages. The main processing steps are common to all the approaches, with a key point being the registration. Intrinsic decorrelation due to the native geometry and surface changes can affect the quality of the interferogram, and the remaining quality improvement comes from the registration accuracy of both images.

4.2. Principle and limits of SAR interferometry

The phase of SLC images depends on the geometric path of the electromagnetic waves. Interactions with the backscattering surface or volume (forest, snow/ice, etc.) can also modify it so that the phase of each pixel is tainted by speckle noise due to the physical properties and spatial distribution of the elementary targets within the corresponding resolution cell.

The main idea of interferometry is to combine two SAR images acquired with very similar geometries. To measure the topography, a baseline (i.e. a geometrical

1. A DEM is a model providing knowledge of the ground terrain, a "bare Earth" elevation model, on a geolocalized regular grid, unmodified from its original data source, which is in theory free of vegetation, buildings and other "non-ground" objects (generic term for a superset including both a digital terrain model (DTM) and digital surface model (DSM) with objects), but which in practice has some artifacts. A DEM is a digital image in which each pixel has a value corresponding to its altitude above sea level. It can be derived by digitizing elevation data from topographic maps or, directly, from stereo imagery, interferometric synthetic aperture radar or light detection and ranging (LiDAR).

difference between the two sensor positions) is necessary to create slightly different viewing angles in the acquisitions, as in an optical stereo configuration. Note that the magnitude of the B/H ratio is much lower in SAR imagery than in optical imagery. The accuracy of the estimated topography benefits from a large baseline. The limits for the baseline/height (B/H) ratio for InSAR were introduced in Chapter 1, section 1.1.2. However, this geometrical difference has to be small enough to limit the speckle noise decorrelation, leading to an important trade-off between noise level and geometric accuracy. Indeed, speckle is a deterministic phenomenon: its realization will be similar in both images if the surface has not changed and the geometries of the images are close enough. For the measurement of displacement by differential interferometry using an available DEM, this geometric difference should be as small as possible: the differential interferogram will be less affected by DEM errors and the phase noise from different speckle realizations.

4.2.1. *Geometrical information*

Most interferograms are computed from two SAR images acquired by the same sensor with two different positions in space and time. In some cases, interferometry is performed with two antennas, with one emitting and both receiving simultaneously, either on the same platform, as in the Shuttle Radar Topographic Mission (SRTM), or on two platforms flying on simultaneous parallel orbits, such as the TerraSAR-X/TanDEM-X satellites. This configuration is referred to as the bistatic case.

In the following, we only consider the monostatic case: acquisition of two separate images by the same sensor at two different dates (also called repeat-pass interferometry). The bistatic case involves the same geometric behavior as the monostatic one, except that the travel path difference is divided by two since only the return path is different. The bistatic case has the advantage of not being affected by atmospheric disturbances and ground surface changes in the quasi-simultaneous bistatic acquisitions. All the other processing issues are the same.

Figure 4.1 provides the InSAR geometry and the notations used in the following text. The geometric distance between the two positions of the sensor, called the "baseline", B, is a fundamental parameter. Its importance for the signal quality is discussed in section 4.2.2. In this section, we first provide the major equations for the understanding of interferometry. $B_\perp$ is the orthogonal baseline after projection onto the direction perpendicular to the range vector. $B_\perp$ and the incidence angle θ are the main factors in the InSAR geometric equations.

The interferometric phase is linked to the differential geometric path of the electromagnetic wave:

$$\phi = \frac{4\pi}{\lambda}(R_2 - R_1) \qquad [4.1]$$

where λ is the wavelength and R_2 and R_1 the range distance between one point and the two satellite positions. The differential geometric path ϕ can be decomposed into several contributions:

$$\phi = \phi_{orb} + \phi_{topo} + \phi_{dis} + \phi_{aps} + \phi_{noise} \quad [4.2]$$

with ϕ_{orb} being the orbital phase (flat terrain phase referenced on a geoid), ϕ_{topo} the topographic phase, ϕ_{dis} the phase due to surface displacement, ϕ_{aps} the differential atmospheric phase screen (APS) and ϕ_{noise} the differential phase noise.

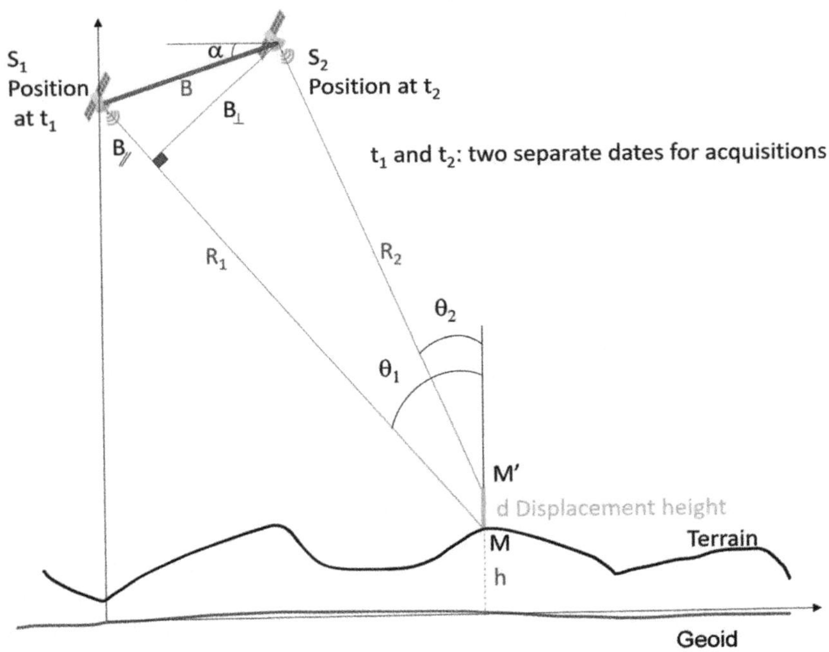

Figure 4.1. *Interferometry geometry and the notation used in this chapter. For a color version of this figure, see www.iste.co.uk/cavalie/images.zip*

The expressions of ϕ_{orb}, ϕ_{topo} and ϕ_{dis} are as follows (Rosen *et al.* 2000), with $B_\perp$ being the orthogonal baseline and θ the incidence angle:

$$\phi_{orb} = \frac{4\pi}{\lambda}\frac{B_\perp \Delta R}{R\tan\theta} \quad [4.3]$$

$$\phi_{topo} = \frac{4\pi}{\lambda}\frac{B_\perp h}{R\sin\theta} \quad [4.4]$$

$$\phi_{dis} = \frac{4\pi d_{LOS}}{\lambda} \quad [4.5]$$

where d_{LOS} is the projection of the displacement on the radar line-of-sight (LOS); for a vertical displacement d illustrated in Figure 4.1, $d_{LOS} = d\cos\theta$.

ϕ_{orb} is estimated and removed by considering the terrestrial geoid surface and the orbit directly in the registration process. For surface displacement applications, ϕ_{topo} is removed by considering a reference DEM or by using a second interferogram. The quality of the orbit, the geoid and the DEM are essential for the final accuracy of the estimation. For displacement analysis, the focus is placed on ϕ_{dis}. For TOPSAR and spotlight modes, a specific phase correction is needed to avoid the misalignment variations (the squint angle varies during the acquisition). This correction, commonly called deramping or reramping, should be performed before the coregistration of the two images.

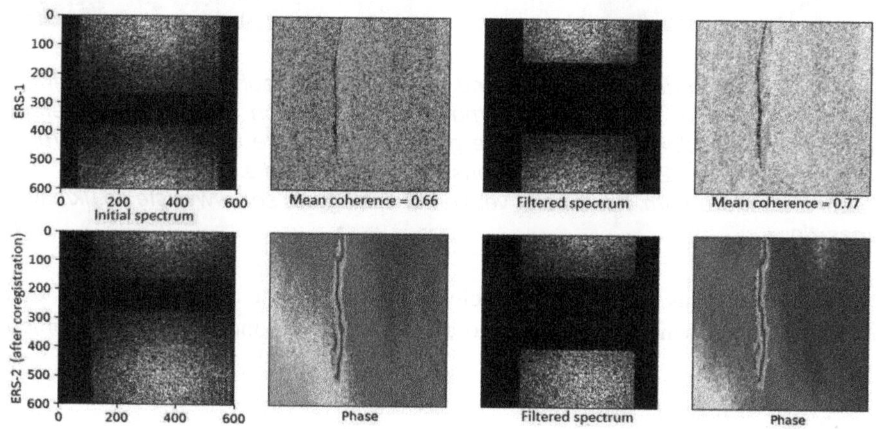

Figure 4.2. *Illustration of frequency shift in the interferometric pair. The spectra are filtered to retain the common part and enhance the coherence at the price of a loss in resolution. ERS-1/ERS-2 interferogram (May 2 and 3, 1995, Isla Incahuasi in the center of the Salar de Uyuni, Bolivia). For a color version of this figure, see www.iste.co.uk/cavalie/images.zip*

The registration also induces a slight frequency shift between the two images, leading to a decorrelation term ρ_{geo}. This is due to the ground projection of the frequency (Fornaro and Guarnieri 2002). Figure 4.2 shows that retaining only the common part of the two image spectra after the coregistration (common band filtering) can increase the InSAR coherence. For ERS satellites, the frequency shift was very important and could not be ignored. With the new SAR system, this shift is minimized. The InSAR processor may even skip this correction to preserve the spatial resolution, which is the case for DiapOTB. The impact of the decorrelation is acceptable in these cases.

Figure 4.3 illustrates the wrapped phase in an interferogram computed from a Sentinel-1 Stripmap pair, first when no corrections are applied, and second when topographic and orbital fringes are removed, leaving only displacements and residual tropospheric delays.

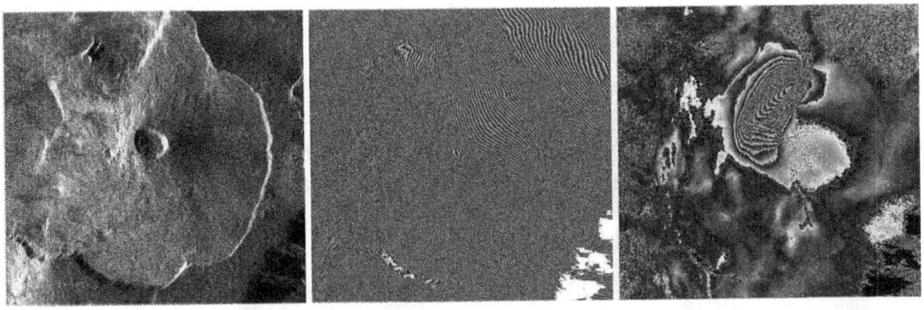

Figure 4.3. *An interferogram with and without topographic and orbital fringes (Sentinel-1 Stripmap, Piton de la Fournaise area, Reunion Island). The differential interferogram very clearly highlights the ground motion due to the volcanic eruption on September 11, 2016. Left: intensity image, middle: interferogram, right: differential interferogram. For a color version of this figure, see www.iste.co.uk/cavalie/images.zip*

From the previous definitions, it is clear that the altitude of ambiguity $h_{2\pi}$ is the altitude difference which corresponds to a complete 2π round of the interferometric phase :

$$h_{2\pi} = \frac{\lambda R \sin\theta}{2B_{\perp}} \quad [4.6]$$

Then, knowing the phase difference $\delta\phi$, the height difference δh can be deduced as:

$$\delta h = \frac{\delta\phi}{2\pi} h_{2\pi} \quad [4.7]$$

As an example, if a 100 m perpendicular baseline is used with $\theta = 35°$, a 2π change of the interferometric phase corresponds to an altitude difference ($h_{2\pi}$) of about 131 m for Sentinel-1 (C-band, R = 826 km), 66 m for CSK (X-band, R = 741 km) and 54 m for TSX (X-band, R = 614 km).

Since the phase of the interferogram is measured modulo 2π, the elevation is the observed modulo $h_{2\pi}$. Here, $h_{2\pi}$ gives the sensitivity of the DEM used. The smaller the $h_{2\pi}$ is, the better the DEM. If the DEM is old (or not accurate) and topographic changes occur before image acquisition, then fringes proportional to the changes will appear. Each fringe will represent an error of $h_{2\pi}$ in the interferogram. This will cause errors in the phase unwrapping and may be misinterpreted as a displacement. In this case, a small $h_{2\pi}$ (large baseline) is a disadvantage. $h_{2\pi}$ should be chosen as not too small with respect to the DEM variations. Phase unwrapping methods are discussed in Chapter 6.

As a consequence, the uncertainty of h with respect to ϕ and B is equal to:

$$\frac{\partial h}{\partial \phi} = -\frac{h_{2\pi}}{2\pi} \qquad [4.8]$$

$$\frac{\partial h}{\partial B} = -\frac{R \sin\theta \tan(\theta - \alpha)}{2B} \qquad [4.9]$$

In addition, the digitization of the signal implies a quantification noise in the interferometric phase. The impact of this noise on the height accuracy depends directly on $h_{2\pi}$.

In order to measure surface displacement, differential interferograms are computed from at least three SAR images or two SAR images and a DEM. Differential interferometry aims to remove the contribution of the DEM and obtain only the topographic variations.

4.2.2. *How to choose an interferometric pair?*

The baseline B between the two acquisitions should be carefully chosen. A first criterion is respect of the *critical baseline*. Inside the pixel resolution cell, there is a geometrical difference between the closest part of the pixel and the farthest part of the pixel. This creates a phase difference inside the pixel (Figure 4.4). If this difference is higher than $\lambda/2$ in two acquisitions, the speckle due to distributed scatterers within the resolution cell will not be correlated for the two acquisitions. In contrast to noise reduction in intensity images where the decorrelation of speckle is the key, the conventional InSAR approach seeks speckle correlation to erase the speckle phase. Note that this constraint disappears when a single dominant scatterer is present in the resolution cell, as in the permanent scatterer (PS) approach (see Chapter 5). The critical baseline is thus the maximum baseline, which guarantees a correlation of the speckle within the pixel:

$$B_{\perp_c} = \frac{\lambda}{2} \frac{R \tan\theta}{\Delta r} \qquad [4.10]$$

At the pixel level, the interferometric condition is:

$$L \sin\theta_1 - L \sin\theta_2 < \frac{\lambda}{2} \qquad [4.11]$$

The limit of this critical baseline for InSAR processing is when the common part of the two SLC spectra becomes null.

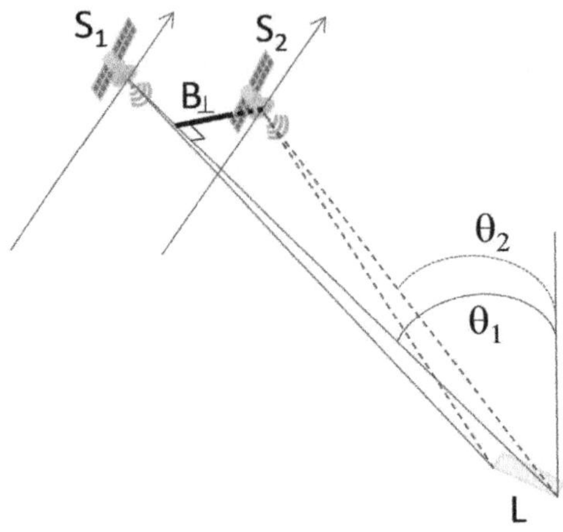

Figure 4.4. *Condition between view angles to guarantee internal correlation inside the pixel. $B_\perp$ should not be too large in order to guarantee that the internal path delay is below $\frac{\lambda}{2}$. For a color version of this figure, see www.iste.co.uk/cavalie/images.zip*

Therefore, the baseline should not be too large in order to meet the requirement with respect to $B_{\perp_c}$.

For Sentinel-1, $\Delta r = 2.33$ m, $R = 826$ km, $\theta = 35°$ and $B_{\perp_c} = 6884$ m, which largely exceeds a fixed orbital tube of 100 m. The orbital tube ensures ground-track coverage repetitiveness and consequently ensures repeat-pass interferometric compatibility between SAR images acquired on the same orbit. The size of the orbital tube may also have an impact on the interferometric performance in terms of azimuth spectral decorrelation and azimuth coregistration accuracy under the presence of squint. These effects require special consideration for ScanSAR, spotlight and TOPSAR modes (Prats-Iraola *et al.* 2015).

When using an interferogram dedicated to DEM generation, the second constraint on the baseline is the ambiguity altitude. From equation [4.8], the smaller the $h_{2\pi}$ is, the more precise the estimation of h is. However, $h_{2\pi}$ should not be overly small compared to the local topography in order to avoid too many phase wrappings, which can result in narrow fringes or even signal aliasing when the phase rotation between adjacent pixels is higher than π. The eligible SLC pairs are thus the ones checking the two conditions on $B_\perp$: $B_\perp$ below the critical baseline and $B_\perp$ leading to a suitable $h_{2\pi}$. In other words, the common part of the spatial spectrum must be large enough.

With C-band wavelengths, baselines lengths between 30 m and 70 m are useful for surface change detection and baselines smaller than 30 m are needed to study the movements of the Earth's surface. For the generation of an InSAR DEM, the optimal perpendicular baseline is in the range of 150–300 m. However, the best result is achieved by using a combination of interferograms with different $h_{2\pi}$. Indeed, interferograms with small baselines can be used to help unwrap interferograms with high baselines.

4.2.3. *Phase and coherence estimation*

The phase difference between two coregistered SLC images s_1 and s_2 can be obtained at each pixel (k, l) by the Hermitian product $s_1(k,l)s_2^*(k,l)$. However, to reduce the phase noise and to estimate the coherence, spatial averaging is necessary. If Ω denotes the averaging window (boxcar, adaptive, etc.), the interferogram ϕ and its coherence ρ are estimated as (Rosen *et al.* 2000):

$$\rho e^{i\phi} = \frac{\sum_{(k,l)\in\Omega} s_1(k,l)s_2^*(k,l)}{\sqrt{\sum_{(k,l)\in\Omega} |s_1(k,l)|^2}\sqrt{\sum_{(k,l)\in\Omega} |s_2(k,l)|^2}} \quad [4.12]$$

By analogy with the processing of intensity images, this initial spatial averaging is (abusively) called complex multi-looking and the number of SLC pixels averaged is called the number of looks L. This formula implies that the L looks are fully decorrelated. This strong hypothesis is not always true, as the impulse response of the instrument and the nature of the ground may induce some spatial correlations between pixels. Nonetheless, modeling the inter-pixel correlation is not straightforward and the number of looks is often assimilated as the number of averaged pixels at native SAR resolution.

Despite this initial complex multi-looking, the phase and the coherence values estimated in equation [4.12] are still noisy. The noise level depends on the true coherence level D and the number of independent looks L. The phase statistics are ideally described by a probability distribution function (PDF) of the interferometric phases given by equation [4.13] (Tupin *et al.* 2014) and illustrated in Figure 4.5.

$$p(\phi|D,L) = \frac{(1-D^2)^L}{2\pi}\frac{1}{2L+1}\,{}_2F_1\left(2, 2L; L+\frac{3}{2}; \frac{1+D\cos(\phi)}{2}\right) \quad [4.13]$$

where ${}_2F_1$ represents a hypergeometric function.

The noise characteristics of interferometric phases can play an important role in InSAR performance. For example, these statistics can be used for maximum likelihood estimation of different parameters of interest, such as deformation, deformation rate and topography, and can also be used for uncertainty (precision and reliability) descriptions of these products.

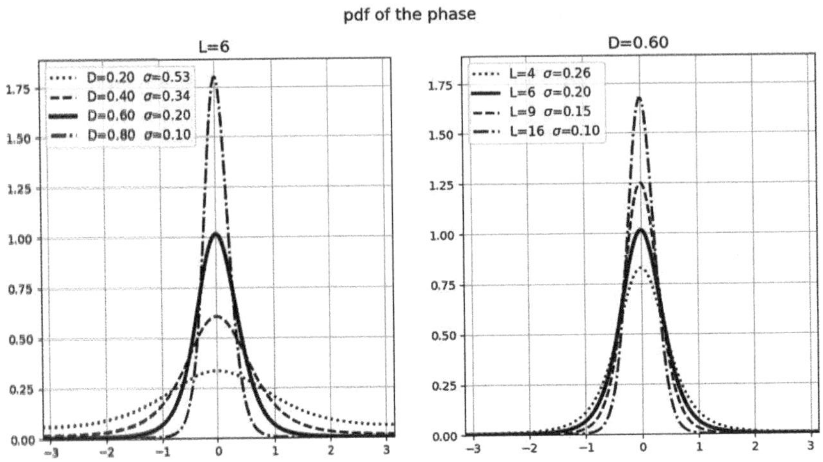

Figure 4.5. *Phase probability density functions (PDFs) according to the level of coherence D and the number of independent looks L*

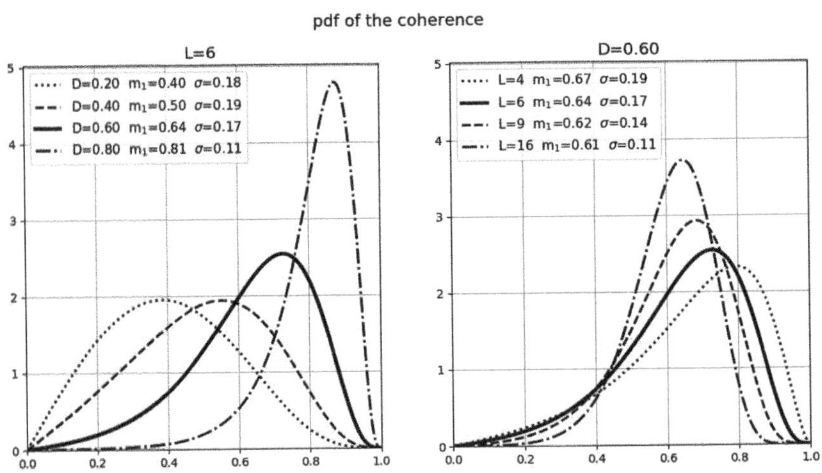

Figure 4.6. *Left: coherence probability density functions ($L = 6$); right: the estimated coherence for different numbers of looks*

The coherence value is between 0 and 1; the closer to 1, the more reliable the ϕ is. Equation [4.12] provides an estimate of the coherence value ρ, but this estimate is biased and its variance depends on the true coherence D and the number of looks L, as illustrated in Figure 4.6. The coherence PDF is given by equation [4.14] and the first- and second-order moments by equations [4.15] and [4.16], respectively (Tupin *et al.* 2014):

$$p(\rho|D,L) = 2(L-1)(1-D^2)^L \rho(1-\rho^2)^{L-2}\, {}_2F_1(L,L;1;\rho^2 D^2) \quad [4.14]$$

$$m_1 = (1-D^2)^L \frac{\sqrt{\pi}\,\Gamma(L)}{2\,\Gamma\left(\frac{1}{2}+L\right)}\, {}_3F_2(L,L,\frac{3}{2};1,L+\frac{1}{2};D^2) \quad [4.15]$$

$$m_2 = (1-D^2)^L \frac{1}{L}\, {}_3F_2(L,L,2;1,L+1;D^2) \quad [4.16]$$

Moreover, coherence estimation implies spatial averaging of the data inside a given estimation window (equation [4.12]). This estimation technique relies on the hypothesis that all the signals involved in the estimation process are locally stationary. When this is not the case, biased coherence values result. For example, if phase values due to the topography are not properly compensated for, the estimated coherence value is underestimated.

4.2.4. *Loss of coherency: several reasons*

Low-coherence areas in InSAR results show low degrees of similarity between echoes into these areas, and consequently phase difference with low coherence does not completely represent the distance difference between echoes. In addition, coherence can reflect whether the terrain changes during the interval between the two illuminations, and it can also be used as an index of SAR image classification. The quality of an interferogram can be measured by estimating the coherency of the data, but this coherency is affected by many factors, such as Doppler centroid difference, the baseline and atmospheric effects. Accordingly, the coherence can be written as (Rosen *et al.* 2000):

$$\rho = \rho_{temp} \times \rho_{geo} \times \rho_{vol} \times \rho_{Doppler} \times \rho_{cor} \times \rho_{2\pi} \times \rho_{quant} \times \rho_{SNR} \quad [4.17]$$

with ρ_{temp} being the temporal decorrelation due to surface changes, ρ_{geo} the geometrical decorrelation due to the perpendicular baseline $B_\perp$, ρ_{vol} the volumic decorrelation, $\rho_{Doppler}$ the Doppler decorrelation due to different squints in the acquisitions, ρ_{cor} the decorrelation due to coregistration errors, $\rho_{2\pi}$ the decorrelation due to the $h_{2\pi}$ value, ρ_{quant} the decorrelation due to signal quantization and ρ_{SNR} the decorrelation due to the image SNR.

Coherency drops may be due to various phenomena, mainly atmospheric content or drastic changes in the soil (vegetation growing, a change in moisture conditions, surface modification, impact of wind on trees or bare soils), as these change the scattering element geometry or dielectric properties, reducing the interferometric correlation. Low correlation values (< 0.3) significantly reduce interferogram quality, and the measurement of topography or motion is then difficult.

The longer the time intervals are between image acquisitions, the more likely it is that the coherence will drop. For differential interferometry, the reference image has to be carefully chosen to assure the best coherence between dates. In Figure 4.7, we can clearly see a loss of coherence in a pair separated by 12 days compared with a four-month-interval case.

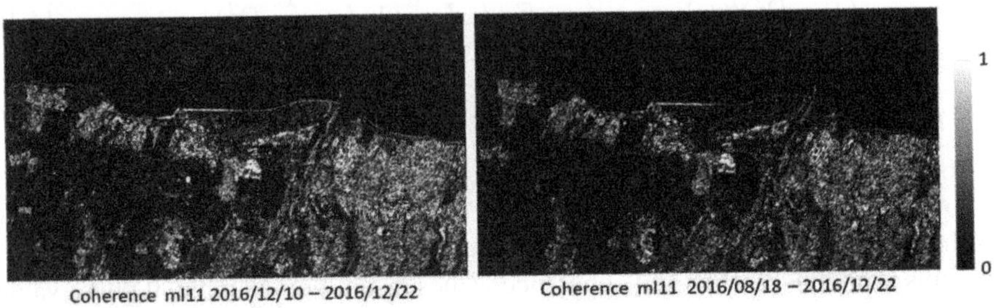

Figure 4.7. *Coherence images computed with a 12 day interval (left) and a 4 month interval (right) from Sentinel-1 images of the Saint Denis area (Réunion Island). The coherence is higher with the 12 day pair, especially for natural surfaces for which it is otherwise quite low. For human-made structures, the coherence can stay high if there is no modification*

In practice, low coherence values are often due to a change in the observed surface (e.g. vegetation growing, impact of wind on the surface, movements), atmospheric conditions or a very large difference in the geometry (baseline too large; see section 4.2.2). One way to optimize the coherence is to privilege a small baseline and small time spacing between the images.

4.2.5. *Other InSAR limitations*

One limitation of using the conventional InSAR technique is temporal and/or geometrical decorrelations in areas with large deformation rates and/or with dense vegetation.

For DEMs with steep slopes, layovers and shadows may occur. Layovers are areas where the radar backscattered signal from the ground and that from the relief are mixed (see Chapter 3, section 3.1.1). This occurs when ground and relief are at the same range distance from the sensor. For instance, in mountainous or urban areas, there are many layovers because of the high steepness of the terrain. Unmixing is a very tricky operation, and layovers are mostly considered as lost information. Shadows are due to areas inaccessible to the radar signal, mainly in mountainous regions where mountains hide the ground behind them. In such conditions, interferograms cannot provide information. As a result, ground movement monitoring in mountainous regions is always challenging. However, these are regions highly susceptible to earthquakes and thus ground motions.

Conventional SAR interferometry does not work with an InSAR monostatic configuration over water (especially smooth open water) because the surface returns little backscattered energy to the SAR antenna (specular reflection) and it is constantly changing (low coherence). There are some methods that can combine multiple interferograms and potentially increase the signal strength of river areas, so that the absolute phase information of the water level is obtained to provide spatio-temporal dynamics of the river levels. Special modes and wavelengths, such as near-nadir Ka-band InSAR, allow strong water/land radiometric contrast (typically of the order of 10 dB) and very high interferometric coherence on water. The Ka-band radar interferometer (KaRIn) of Surface Water and Ocean Topography (SWOT), the NASA/CNES wideswath altimetry mission scheduled to launch at the end of 2022 (Fjortoft *et al.* 2014), is a bistatic system especially designed for interferometry over oceans and in-land waters.

DEM accuracy influences the final precision of deformation monitoring, especially when processing DInSAR with only one image pair. The DEM grid resolution, the DEM height precision and the DEM's being up-to-date with regard to the InSAR images processed should be considered. The theoretical limit is $\lambda/4$ for proper deformation monitoring. When the dataset includes a lot of images (typically more than 20 with more widely spread perpendicular baselines), the InSAR results are less impacted by DEM errors (Lazecky *et al.* 2015).

4.3. Atmospheric corrections

4.3.1. *Compensation for tropospheric delays*

Due to the presence of air in the atmosphere, electromagnetic waves travel at lower speeds than in the near-vacuum that prevails at the satellite altitude. As a result, the back-and-forth travel time of a radar pulse emitted by a SAR satellite experiences a delay, so that the total travel time cannot simply be converted into a geometric distance between the satellite and the ground target. The total excess *atmospheric delay* experienced by the radar wave typically amounts to an apparent one-way zenithal excess length of approximately 3 meters at sea level altitude and temperate latitudes. Unfortunately, due to the dynamic nature of the Earth's atmosphere, this atmospheric delay varies temporally, laterally and vertically. As a result, SAR interferograms are commonly affected by atmospheric artifacts, manifesting as distortions in the fringe patterns, eventually affecting the quality of the resulting geodetic measurements (Massonnet and Feigl 1995; Zebker *et al.* 1997).

In the microwave range, the speed of light is mainly affected by air pressure (P), air temperature (T) and water vapor pressure (e). The total one-way tropospheric delay (in meters) from the satellite to a ground pixel at elevation z can be expressed as (Baby *et al.* 1988; Doin *et al.* 2009):

$$\delta L^{s}_{LOS}(z) = \frac{10^{-6}}{cos(\theta)} \left(\underbrace{\frac{k_1 R_d}{g_m}(P(z) - P(z_{ref}))}_{\text{Hydrostatic delay}} + \underbrace{\int_{z}^{z_{ref}} \left(\left(k_2 - \frac{R_d}{R_v} k_1\right) \frac{e(z)}{T(z)} + k_3 \frac{e(z)}{T(z)^2} \right) dz}_{\text{Wet delay}} + \underbrace{\int_{z}^{z_{ref}} \left(k_4 W_{cl} \right) dz}_{\text{Liquid water delay}} \right) \quad [4.18]$$

where θ is the local incidence angle, $R_d = 287.05$ J.kg^{-1}.K^{-1} and $R_v = 461.495$ J.kg^{-1}.K^{-1} are, respectively, the dry air and water vapor specific gas constants, g_m is a weighted average of the gravity acceleration between z and z_{ref}, P is the dry air partial pressure in Pa, e is the water vapor partial pressure in Pa, T is the temperature in K and W_{cl} is the cloud (liquid) water content. The constants are $k_1 = 0.776$ K.Pa^{-1}, $k_2 = 0.716$ K.Pa^{-1}, $k_3 = 3.75 \cdot 10^3$ K^2.Pa^{-1}v and $k_4 = 1.45 \times 10^3$ m^3.kg^{-1}. In the above formula, the additional frequency-dependent contribution of the ionosphere (see section 4.3.2) is ignored.

As shown in Figure 4.8, the hydrostatic delay (HD) depends on the air pressure profile for the altitude, which follows an exponential law, whereas the wet delay (WD) relates to the contribution of air moisture integrated vertically through the atmosphere, which is mainly influenced by the lower troposphere ($z < 2$–5 km). While the former term dominates in the absolute value against the latter (several meters for HD vs. several tens of centimeters for WD), their variations in time are comparable (a few tens of centimeters). As interferometry is sensitive to temporal variations of the absolute delay, both the hydrostatic and wet delays substantially contribute to atmospheric artifacts.

However, these contributions to the total delay differ significantly in terms of spatial distribution. Spatial variations of atmospheric pressure are generally of longer wavelengths (hundreds of kilometers) than variations in temperature and air moisture. Air moisture varies over characteristic distances ranging from hundreds of meters to tens of kilometers, with very large local variations. The resulting interferometric (i.e. differential) delay is then a combination of very long wavelength variations of hydrostatic delay and local variations of wet delay. Considering that an interferogram is a measurement of the relative apparent travel time, signals of characteristic wavelengths longer than the footprint of an image are actually not detectable, hence the prevalence of wet delay in the interferometric phase. Local variations in delay can be approximated by second-order stationary Gaussian noise in each interferogram individually and are random in time.

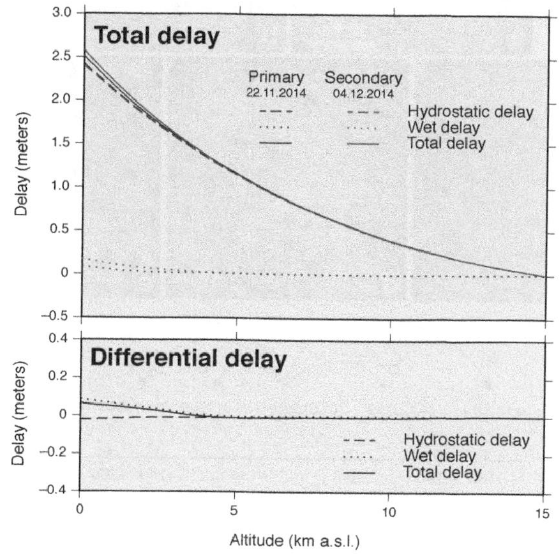

Figure 4.8. *Integrated atmospheric delay (equation [4.18]) predicted by the European Centre for Medium-Range Weather Forecasts (ECMWF) ERA-40 reanalysis for two different dates in 2014 (blue: November 11, 2014; red: December 4, 2014) for a location in Albania (41.75° N; 21.75° E). For a color version of this figure, see www.iste.co.uk/cavalie/images.zip*

Indeed, the typical duration of a SAR acquisition is usually of the order of approximately 1 minute (a typical integration time for a given target is of the order of approximately 1 second), whereas the revisit interval between two successive acquisitions is generally several days (six days for the Sentinel-1 system). A typical decorrelation time for zenith tropospheric delays measured by GPS is of the order of six hours, reflecting the dynamic nature of the atmospheric circulation locally (Emardson *et al.* 2003). Therefore, a single acquisition only captures a "snapshot" of the distribution of atmospheric heterogeneities. These features are subsequently advected and dispersed well before the next acquisition. The atmospheric artifacts resulting from this process are referred to as *turbulent* atmospheric artifacts. As shown in Figure 4.9, turbulent atmospheric artifacts, generally related to air moisture, are characterized by a broad spectrum of spatial correlation lengths (originating from the spatial scale of heterogeneities at any given acquisition time) but appear random in time (Hannsen 2001).

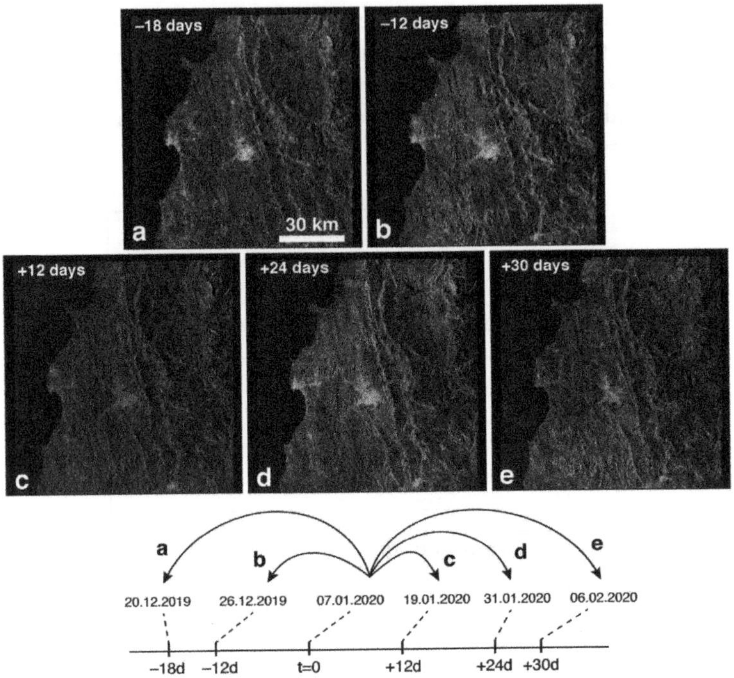

Figure 4.9. *Sentinel-1 interferograms sharing the same primary acquisition over the central part of Albania. The short time span of each interferogram ensures that the fringes visible in the interferograms are primarily due to atmospheric effects, mainly caused by atmospheric turbulence. The primary acquisition was chosen for its stable atmospheric state, so that all interferograms would reflect the state of the atmosphere of the secondary acquisition. The spatially correlated nature of water vapor heterogeneity is particularly evident in (a) and (e). For a color version of this figure, see www.iste.co.uk/cavalie/images.zip*

Various strategies for correcting these artifacts have been implemented. First, independent remote sensing observations of the water vapor distribution in the observed scene can be acquired by spectrometers operating in one or several absorption windows of water vapor. For instance, the MERIS instrument onboard the ENVISAT satellite, operating in visible and infrared wavelengths, allowed for mapping of the water vapor distribution at approximately 1 km spatial resolution, synchronously with the SAR acquisitions of the ASAR instrument. Using assumptions on the vertical distribution of water vapor in the column, and under cloud-free conditions, these maps could then be used to predict the wet delay in the radio domain and correct interferograms (Puysségur *et al.* 2007; Li *et al.* 2012). Unfortunately, recent SAR satellites are usually not equipped with such ancillary instruments. Alternatively, the integration of total zenithal delays routinely estimated from the analysis of GNSS data also allows for prediction of the artifacts that affect InSAR data acquired over a GNSS network (Onn and Zebker 2006). However, this approach is limited by the existence and accessibility of still scarce GNSS data and does not provide regular lateral and vertical sampling of the atmospheric column due to the constraints on the installation of GNSS stations. All these methods rely on external datasets that are not always available and for which spatial resolution generally does not compare well with that of SAR acquisitions.

Other strategies do not rely on external data. First, we can consider that interferometric delay at a given pixel depends on the pixel elevation, since the atmosphere is, to the first order, a stratified medium, as shown in Figure 4.10(a). This effect has long been recognized as a major source of uncertainty when attempting to interpret interferometric fringes in terms of ground deformation in regions characterized by large topographic features, such as volcanoes (e.g. Delacourt *et al.* 1998) or mountain ranges (e.g Elliott *et al.* 2008). The covariation between the interferometric phase and elevation is the result of the temporal variation of the altitude dependency of the atmospheric delays in equation 4.18, and, to a lesser degree, to its lateral variations. Hence, evaluating a correlation between the interferometric phase and elevation is an easy first-order correction (e.g. Cavalié *et al.* 2007; Jolivet *et al.* 2012). However, such an approach neglects local turbulent delays and can bias estimates of ground deformation if the latter also correlates with topography. Averaging of a large number of independent acquisitions, in order to cancel out these artifacts, has also been proposed. This can be achieved by stacking (thereby reinforcing any underlying steady signal, such as interseismic deformation) (Peltzer *et al.* 2001), performing a least-squares multi-temporal inversion (Berardino *et al.* 2002) or applying a low-pass temporal filter (Hooper *et al.* 2007). It has been shown, however, that the uneven temporal sampling of turbulent behavior and its subsequent low-pass filtering (or averaging) can lead to aliasing, which has significantly biased the estimates of ground deformation (Doin *et al.* 2009).

Another radically different approach consists of attempting to predict atmospheric delays using numerical weather forecast models. The rationale behind this approach relies once again on the observation that atmospheric artifacts often follow topographic features. The relation between the altitude and the delay often follows a nonlinear function, which can be empirically estimated or directly predicted from weather forecasts (Doin *et al.* 2009; Jolivet *et al.* 2011). As shown in the inset of Figure 4.10(b), for two sites located about 200 km apart, the differential delay is predicted to approximately vary linearly with elevation, but with a proportionality coefficient differing by about 30%. Mapping the predicted delays on the topography (Figure 4.10(d)) makes it possible to construct a synthetic atmospheric phase screen that can correct the observed interferogram (Figure 4.10(c)).

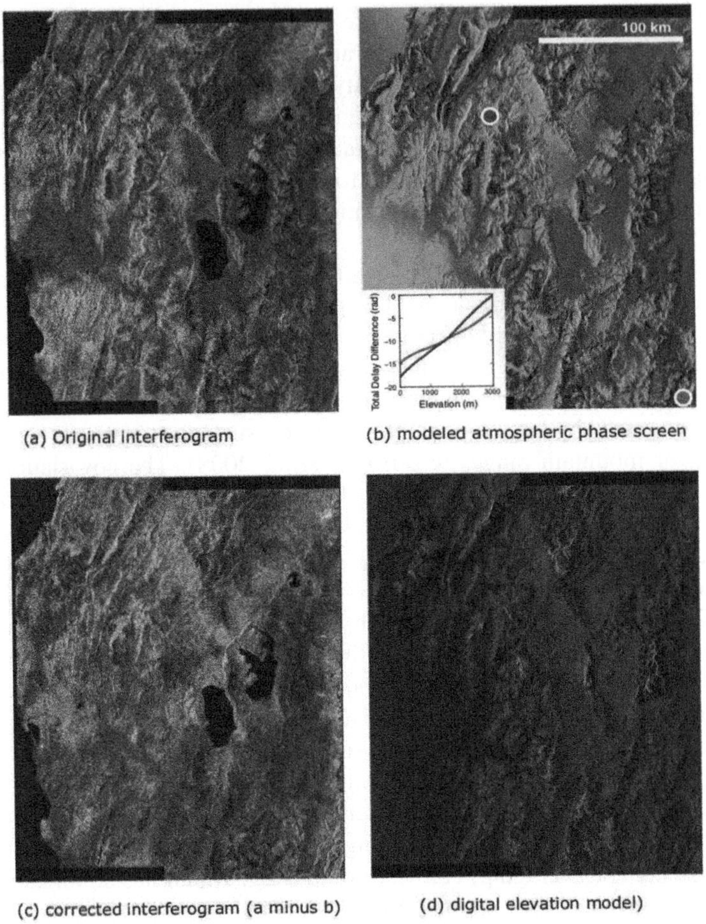

Figure 4.10. *Correction of stratified tropospheric delays using the ERA-5 meteorological analysis. This Sentinel-1 interferogram covers Albania (ascending track 175; image 1: November 22, 2014; image 2: December 4, 2014). Two sub-swaths have been assembled (iw1 and iw2). The covered area is approximately 150 km wide (across-track) and 220 km long (along-track). The elevation in the scene ranges from sea level to more than 2000 m above sea level. The blue and red dots in (b) show the locations of the ERA-5 delays shown in the inset. North is at the bottom. For a color version of this figure, see www.iste.co.uk/cavalie/images.zip*

The most accurate models available to describe the state of the troposphere at the relevant resolution consist of weather forecast reanalyses, which rely on the assimilation of a wealth of remote sensing and *in situ* meteorological observations, such as the ERA-40 and ERA-5 reanalyses of European Centre for Medium-Range Weather Forecasts (ECMWF) data (Dee *et al.* 2011) or the National Centers for

Atmospheric Prediction (NCEP) and the National Center for Atmospheric Research (NCAR) reanalyses of National Oceanic and Atmospheric Administration (NOAA) data (Kalnay *et al.* 1996). In terms of temporal resolution, the ERA-5 model of the ECMWF is currently available at a one-hour resolution, which is a significant improvement compared to the six-hour resolution of the previous-generation ERA-40 reanalysis. However, these reanalyses are only available after a latency of several weeks to months. Near-real-time solutions, such as ERA-5T, or the HRES forecasts from ECMWF data, are currently considered as surrogates for correcting InSAR data for applications requiring rapid processing (Yu *et al.* 2018).

4.3.2. *Estimation of and compensation for ionospheric propagation delays*

4.3.2.1. *Introduction*

The ionosphere is the portion of the Earth's upper atmosphere where ions and electrons are present with sufficient density to significantly affect the propagation of radio waves (Davies 1990). The slant total electron content (TEC) is the integrated electron density between the satellite and the target, along a tube of one square meter cross-section following the radar line-of-sight (LOS). TEC levels are higher at the equator than at higher latitudes because the ionization of the atmosphere gases depends on the solar radiation, which is stronger at lower latitudes. The activity of the sun, as well as the ionosphere, follows an 11 year cycle, with the addition of seasonal and daily variations. The most frequent perturbations of the smooth background ionosphere are equatorial and high-latitude scintillation, auroral arcs and traveling ionospheric disturbances.

The refractive index for a wave in the ionosphere is inversely proportional to its frequency, since the wave propagation speed depends on the frequency and the ionosphere is dispersive. The wave speed then needs to be distinguished into phase and group velocity, which can be shown to be equal but opposite. The main effects of the ionosphere on a traversing microwave are delay of the wave packet, advance of the phase and rotation of the polarization angle. The wave propagation in the ionosphere takes longer than in vacuum; the one-way delay is:

$$\Delta t_{iono} = \frac{K}{cf_0^2} TEC, \qquad [4.19]$$

where f_0 is the carrier frequency, c is the speed of light in vacuum and $K = 40.28\,\text{m}^3/\text{s}^2$. While the pulse envelope is retarded, the pulse phase is advanced by the same amount:

$$\Delta\phi_{iono} = -\frac{4\pi K}{cf_0} TEC = -4\pi f_0 \Delta t_{iono}. \qquad [4.20]$$

The rotation of the polarization angle, i.e. the Faraday rotation, depends on the strength of the magnetic field $\boldsymbol{B}$ and the angle formed with the radar wave propagation vector $\hat{\boldsymbol{\kappa}}$:

$$\Omega = \frac{Ke\,\boldsymbol{B}\cdot\hat{\boldsymbol{\kappa}}}{cmf^2}TEC, \qquad [4.21]$$

where e and m are the electron charge and mass, respectively. Additional ionospheric effects are caused by the small-scale perturbations of the ionosphere along the satellite orbit. In the azimuth direction, the synthetic aperture acts as a moving window filter. The average TEC within the aperture determines the range phase advance and group delay. The linear TEC variation shifts the focused data with respect to their correct positions by changing their zero-Doppler point. Residual nonlinear variations blur the azimuth impulse response. The nonlinear phase advance within the signal bandwidth is approximately quadratic and blurs the range impulse response. However, for typical ionospheric states and carrier frequencies higher than P-band, the blurring is minimal and can be ignored.

Ionospheric errors of many fringes and shifts of many meters can be quite common in C- and L-band interferometric and motion tracking results. It is therefore necessary to estimate and remove the ionospheric effects.

4.3.2.2. *Ionosphere estimation and compensation*

Absolute ionospheric levels can only be estimated with low precision from SAR data. Range blurring caused by the dispersive propagation in the ionosphere can be used to estimate the absolute TEC with very low precision (Meyer 2010). If quad-pol acquisitions are available, the Faraday rotation angle can be measured from the images and converted to a TEC value, even if the accuracy of this method is limited by the inaccurate knowledge of the magnetic field (Meyer and Nicoll 2008; Kim and Papathanassiou 2010). Alternatively, GNSS-based TEC maps can give the absolute TEC level for a SAR observation, but these values are typically biased by the very low resolution of the maps. In conclusion, a precise compensation of the errors caused by the absolute ionosphere levels, for example geolocalization or absolute motion, may not be possible.

In contrast, the compensation of differential and relative errors between SAR images is possible. The mutual ionospheric azimuth shifts can be corrected by the coregistration and resampling steps, which recover the coherence but do not separate the ground motion from the ionospheric shifts. In order to then correct the interferogram, it is necessary to estimate the ionospheric phase variation in the scene. A method that has been demonstrated to be successful is the split-spectrum method (Gomba *et al.* 2016, 2017). It is based on the dispersive property of the ionosphere: it can separate the ionospheric phase variation from other contributions and the estimated ionosphere can then be removed from the interferogram. Bandpass filtering of the range spectrum is used to obtain interferograms at different carrier frequencies, and then the ionospheric contribution is estimated. An example of the results of the method is shown in Figure 4.11.

Gomba and De Zan (2017) combine the split-spectrum method with the azimuth shifts estimated during the coregistration to improve the estimation results. The split-spectrum method can also be applied to a stack of images to correct the ionospheric errors in the measurement of ground motion, as demonstrated by Fattahi *et al.* (2017).

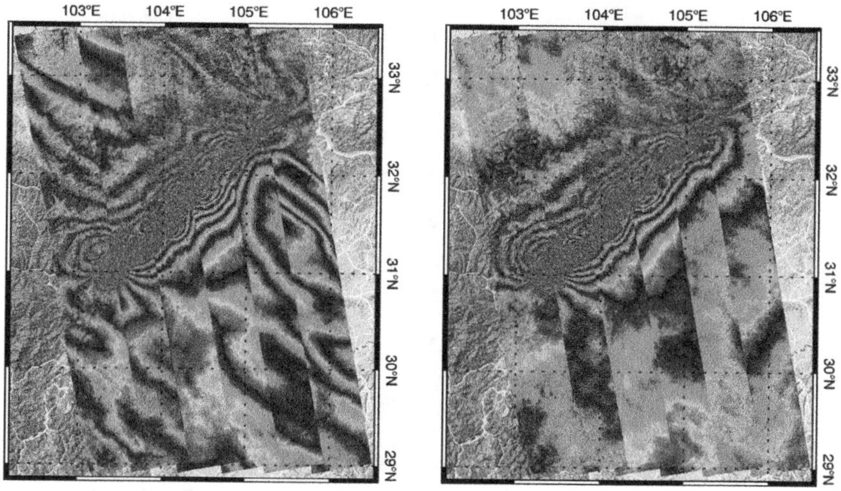

Figure 4.11. *The 2008 Wenchuan Earthquake measured with ALOS interferograms, before and after applying the split-spectrum method (Gomba* et al. *2016). For a color version of this figure, see www.iste.co.uk/cavalie/images.zip*

4.4. InSAR processing chains

This section details the main processing steps in DInSAR processing, which are common to almost all InSAR chains (e.g. image focusing, resampling with the same acquisition geometry, common-band filtering, interferogram and coherence map generation, filtering of fringes, etc.). After discussing these key steps, we illustrate them with an earthquake case using the DiapOTB open-source software (Durand *et al.* 2021). A quick overview of the existing InSAR software solutions is provided at the end of this section.

4.4.1. *Main steps in InSAR processing*

Figure 4.12 presents the main steps in DInSAR computation.

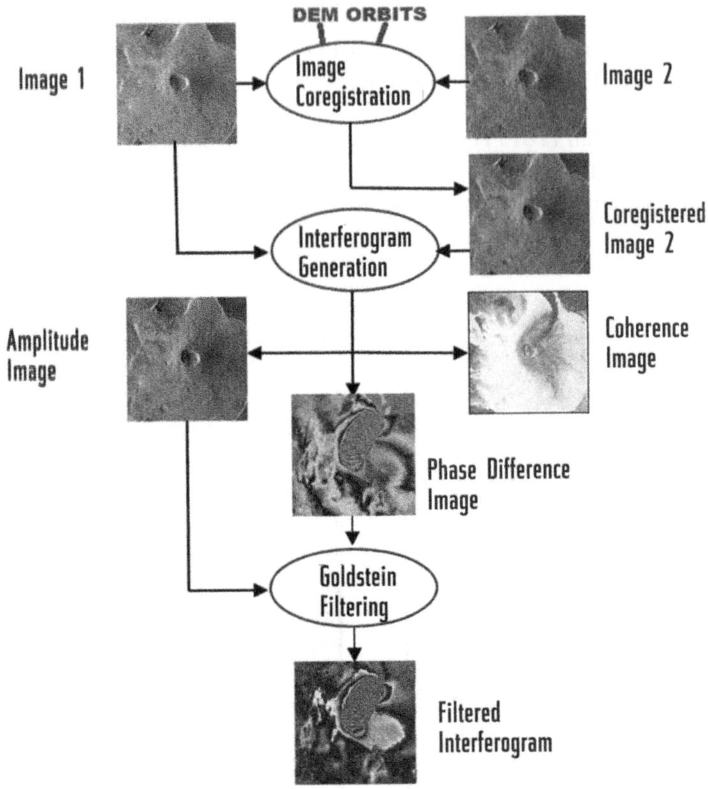

Figure 4.12. *DInSAR data processing diagram with Sentinel-1 images of the Piton de la Fournaise (La Réunion, France)*

4.4.1.1. *Coregistration*

The key step is the coregistration of the secondary image onto the reference image. The precision is a strong driver of the final coherence. For an appropriate interferogram, the coregistration should be better than a tenth of pixel.

The DEM and the orbits predict a deformation grid, which can be compared to a sparse grid obtained from local correlation with actual images: it may fail at places, revealing actual changes between images. The comparison of the two grids allows us to transfer the exact datum to the predicted distortion grid, which becomes accurate down to a few hundredths of a pixel, both locally and globally.

The primary radar image used as a reference image must be registered in absolute geographic coordinates. It is better for both images to be radar-compressed with the same SAR processor. They must share the same Doppler centroid, which should be set to the average of the optimal value found for each image.

In an ideal case of perfectly parallel orbits and aligned acquisitions, the coregistration would only need to compensate for the differing geometry due to the different viewing angle, by means of a proper cross-track stretching of one image. In practice, the coregistration should also account for orbit crossings/skewing, different sensor attitudes, different sampling rates (perhaps due to different pulse repetition frequencies (PRFs), sensor velocities, etc.) and along- and across-track shifts. The secondary image is resampled so that the final interferogram is in the same (slant-range, azimuth) reference geometry as the reference image. Range oversampling can sometimes be used in high-quality interferogram generation.

4.4.1.2. *Coherence and interferogram computation*

Once the two images are coregistered with the accuracy of a fraction of a pixel, the interferogram is generated by cross-multiplying pixel by pixel using the reference image with the complex conjugate of the secondary image, as previously explained in Figure 4.12.

The interferometric phase is the argument of this product or the phase difference between the images. The phase noise can be estimated from the interferometric SAR pair by means of the local coherence, which provides a useful measure of the interferogram quality (SNR). The local coherence is the cross-correlation coefficient of the SAR image pair estimated over a small window of a few pixels in range and azimuth, after the compensation for all the deterministic phase components (mainly due to the terrain elevation). The coherence map of the scene is then formed by computing the magnitude of the local coherence on a moving window that covers the whole SAR image.

4.4.1.3. *Topographic fringe removal*

For differential interferometry, the topographic contribution is eliminated by subtracting the fringe pattern calculated from the elevation model, removing any "predictable" fringes and leaving only those related to displacement and/or DEM improvement.

4.4.1.4. *Geocoding and mosaicking*

Geocoding and mosaicking are the very last steps in the interferogram generation chain, used to obtain a product in a "standard format" and to compare the results with maps and other results in the area of interest.

4.4.1.5. *Filtering and unwrapping*

As a final post-processing step, a Goldstein filter (Goldstein and Werner 1998) can be applied to the local interferogram power spectrum in order to increase the signal-to-noise ratio by using a fast Fourier transform on the interferogram patches. The patches are overlapped and a triangular weighted sum of the overlapping patches is used.

Most InSAR applications (e.g. topographic mapping and deformation monitoring) use the phase unwrapping approach in order to solve the 2π ambiguity in the wrapped phase (see Chapter 6).

4.4.2. Illustration of the DInSAR processing chain: a case study with DiapOTB

In this section, the earthquake that occurred on November 11, 2019, near the village of Le Teil in the Rhône River valley (Ardèche, France) is considered to illustrate the processing chain. The magnitude of the earthquake was 4.9, producing an unexpected surface rupture with ground displacement, caused by the reactivation of an ancient fault (La Rouvière) formed during an extensional tectonic period about 20–30 million years ago. Figure 4.13 show the location of Le Teil village in an IW ascending scene.

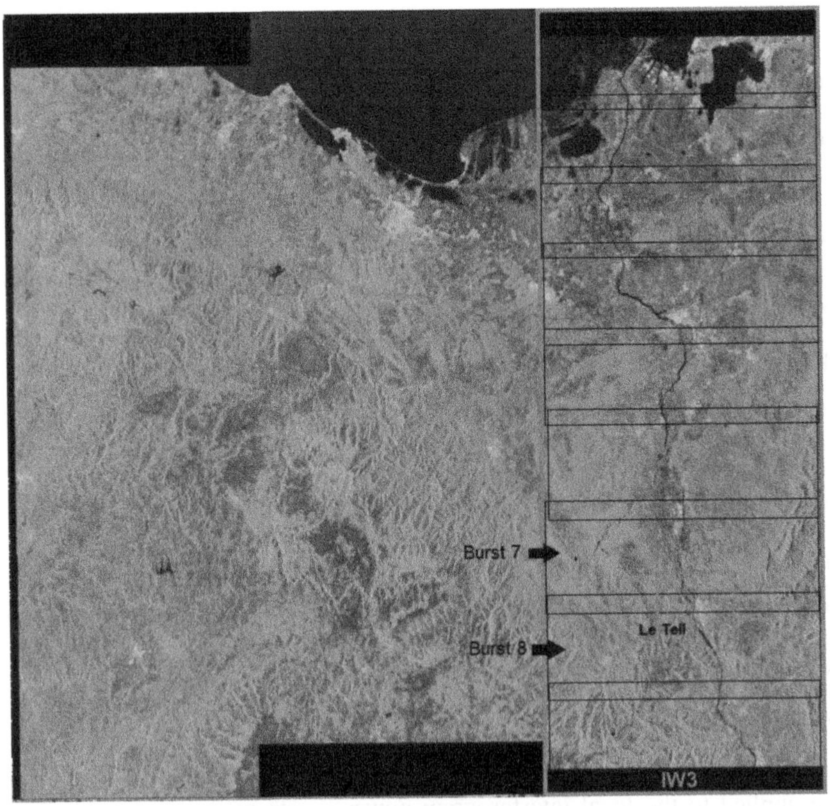

Figure 4.13. *Sentinel-1 acquisition (ascending mode geometry, north is at the bottom): the Le Teil area is covered in the IW3 swath with two bursts. For a color version of this figure, see www.iste.co.uk/cavalie/images.zip*

Since 2018, the new DiapOTB chain developed and provided by the CNES has offered a set of open-source filters and applications as part of the OTB. This section explains the different steps in this software, with illustrations from the Le Teil earthquake. Based on the heritage of DIAPASON, the legacy CNES interferometric processing package, DiapOTB is a complete DInSAR chain designed to work as an OTB remote module or as a stand-alone module.

DiapOTB is able to process Sentinel-1 images, both in stripmap and TOPSAR modes, and COSMO SkyMed images, both in stripmap and spotlight modes; by the end of 2021, DiapOTB will also handle TerraSAR-X and PAZ images.

4.4.2.1. *Detailed description of steps in the DiapOTB chain*

For the TOPSAR mode, interferometric processing is undertaken at the burst level in each DiapOTB step, as shown in the flowchart in Figure 4.14.

1) *Pre-processing steps*: multi-look images are produced by time-domain averaging of the SLC image samples. An independent pre-processing application is available in DiapOTB to estimate the altitude of ambiguity between a couple of images (in order to choose the best interferometric pairs). This step is not needed with Sentinel-1 images because interferometry is the main native acquisition mode (narrow orbital tube). For the TOPSAR mode, the bursts of the SLC images are extracted and a deramping function is applied (a reramping function will be needed after the final coregistration of the images before the computation of the interferogram).

2) *Deformation grid for reference and other images*: an accuracy better than 1/100 pixels is required for TOPSAR InSAR generation to ensure a continuous phase in the interferometric product. The slope of azimuth phase ramps induced by imperfect synchronization of reference and secondary images is proportional to the temporal lag between burst center times in the two acquisitions.

3) *Coregistration of the two SLCs* (secondary image onto reference geometry): for Sentinel-1 IW TOPS mode, data coregistration and DInSAR are supported.

4) *Interferogram creation* (a raw one or an interferogram with topographic phase suppression): for the TOPSAR mode, each interferogram is computed at the burst level and at the end recomposed at the swath level with a reramp function (an option of the deramp step). A spectral diversity method (enhanced spectral diversity (ESD)) can exploit double-difference interferograms in the burst overlap areas in order to suppress the phase discontinuities at burst edges.

5) *Post-processing steps*: DiapOTB includes an interferogram filtering step that uses the Goldstein method (Goldstein and Werner 1998). Terrain and ellipsoid-corrected geocoding from range-Doppler to map coordinates are included. Well-suited interpolation algorithms are applied for the resampling step. The geocoding of images in ground-range geometry is also supported.

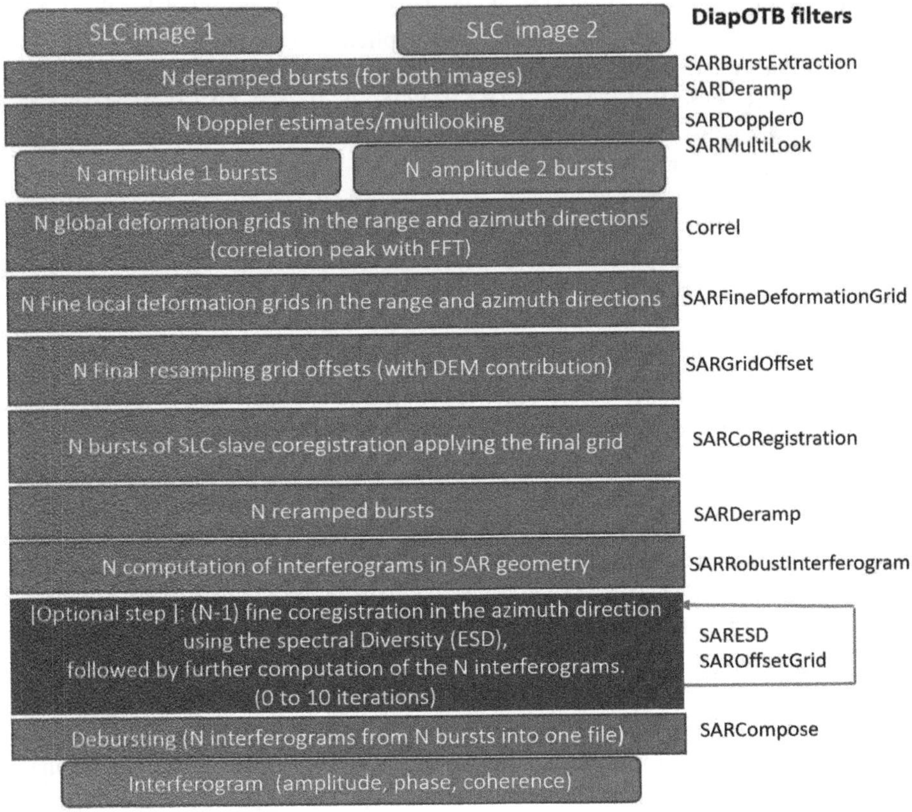

Figure 4.14. *Main DiapOTB steps for the TOPSAR processing mode. For a color version of this figure, see www.iste.co.uk/cavalie/images.zip*

A python script allows the launch of the computation of interferometric time series. It includes a registration module for all the images with respect to a reference image. Then, the mean amplitude image, the interferogram and the coherence images are computed for each image pair. A DEM, from SRTM tiles, for example, overlapping the image is automatically generated and used as input. The interferograms are thus differential interferograms with respect to the DEM.

4.4.2.2. *Example of processing for the Le Teil earthquake*

A couple of Sentinel-1 wide interferometric images from just before and after the event, both in ascending and descending orbital directions in case of foreshortening in the area of interest, were selected and downloaded from the PEPS French portal (Figure 4.15):

S1A-IW-SLC-1SDV-20191031T173917-20191031T173944-029706-03627B-52F8.SAFE
S1A-IW-SLC-1SDV-20191112T173917-20191112T173944-029881-03689D-753A.SAFE
S1B-IW-SLC-1SDV-20191031T054322-20191031T054349-018715-023463-F95B.SAFE
S1B-IW-SLC-1SDV-20191112T054322-20191112T054349-018890-023A0A-BF8C.SAFE

The pre-earthquake image was from October 31, 2019, and the post-earthquake image was from November 12, 2019 (12 days baseline).

Figure 4.15. *Data selection in the PEPS portal for Sentinel-1 data*

DiapOTB needs metadata to process images; data must be stored in its native repositories (.SAFE for Sentinel-1 data) and formats (a tif file for Sentinel-1 data). Access to elevation data (a DEM) for the target area is also required for DInSAR processing. The DEM accuracy for the back-geocoding coregistration and terrain correction processing must be carefully chosen because phase is highly correlated with terrain topography. DiapOTB can automatically extract the required SRTM hgt tiles.

The orbit state vectors provided in the metadata of a SAR product are generally not accurate and can be refined with the precise orbit files, which are available hours to weeks after the generation of the product.

The altitude of ambiguity was estimated with the DiapOTB SARAltambig tool for the target area (latitude: 44.518, longitude: 4.671, height: 136 m), and the results led as expected to a small perpendicular baseline:

– ascending couple: incidence angle = 43.72 deg, $H_{2\pi}$ = -1375.61 m, $B_\perp$ = 11.52 m;

– descending couple: incidence angle = 44.65 deg, $H_{2\pi}$ = -256.91 m, $B_\perp$ = 62.72 m.

Thus, the selection is suitable for DInSAR, the topography contribution will be minimized and the coherence will be convenient for InSAR results.

This example uses VV polarization images; the processing would have been the same with VH ones. The region of interest is situated in the first sub-swath, bursts 6–8 of 9. The practical run can be restricted to these bursts.

Running the DiapOTB chain involves checking (and eventually changing) the default configuration parameters for each of the steps in a .json file for the input of the DiapOTB IW python script (see *https://gitlab.orfeo-toolbox.org/remote_modules/diapotb/-/wikis/ProcessingChains*).

Sometimes, input processing parameters must also be defined and optimized for the InSAR image pair, depending on the sensor and the final measurement accuracy: for example, the azimuth and range multi-look factor, windows and grid sizes, etc.

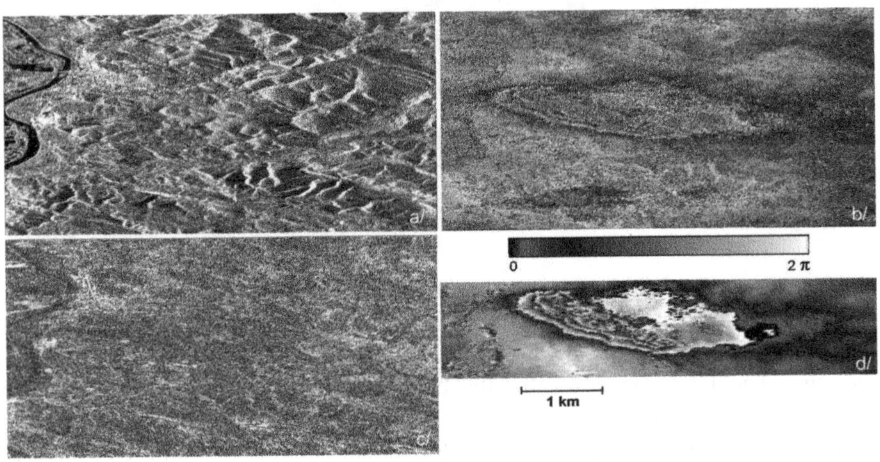

Figure 4.16. *Le Teil earthquake (France, November 11, 2019): DiapOTB DInSAR raw results with SAR geometry (extracts for (a) amplitude, (b) phase before filtering, (c) coherence and (d) filtered phase) from descending Sentinel-1 acquisitions. For a color version of this figure, see www.iste.co.uk/cavalie/images.zip*

The outputs of the processing are the coregistered secondary image with the same geometry as the reference image, then the raw interferogram with the three bands with SAR geometry: amplitude, phase and coherence images and the optional filtered interferogram (three bands). Figure 4.16 shows the resulting raw interferogram for the descending path with SAR geometry near Le Teil town. In Figure 4.16(c), the raw phase is still noisy. In Figure 4.16(d), the raw phase is filtered and displacement fringes appear clearly.

Figure 4.17 shows the DiapOTB results of the DInSAR processing (after Goldstein filtering and projection over the SRTM DEM tile) for a couple of Sentinel-1 images

(October 31, 2019, and November 12, 2019) acquired in the two directions, from ascending and descending orbits.

We can clearly see the low coherence in the ascending path couple and thus the bad quality of the DInSAR phase image. In the area of the earthquake, the mean coherence is 0.68 (standard deviation is 0.23). The loss of coherence can be explained mainly by tropospheric artifacts (the clouds between the two acquisitions were not the same). The coherence image is much better for the descending path couple: the mean coherence is 0.89 and the standard deviation is 0.17. The fringes of the interferogram can then be analyzed by scientists.

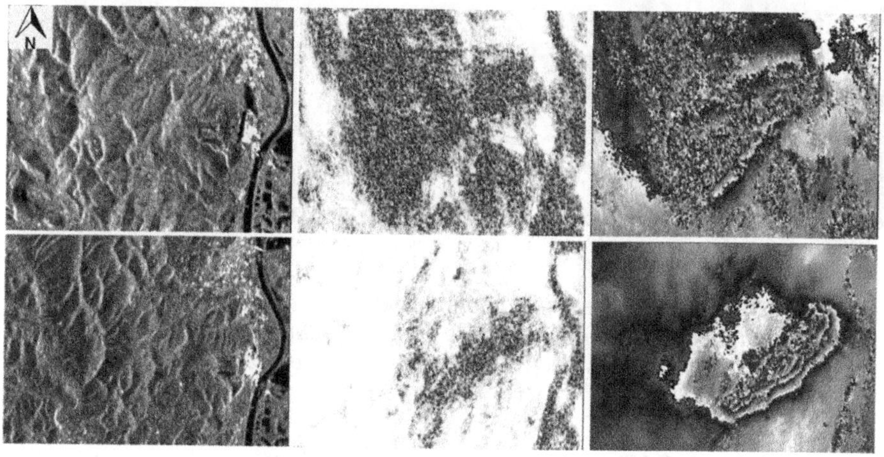

Figure 4.17. *Le Teil earthquake DiapOTB DInSAR results with ground geometry (extracts for amplitude, coherence, phase from left to right) from ascending (top) and descending (bottom) Sentinel-1 acquisitions. For a color version of this figure, see www.iste.co.uk/cavalie/images.zip*

Ritz *et al.* (2020) have reported all the analyses conducted on this earthquake. The seismological and interferometric Sentinel-1 observations indicate that the earthquake occurred at a very shallow focal depth on a southeast-dipping reverse fault. An interferogram with a good coherence (descending orbits) reveals a surface rupture and up to 15 cm uplift of the hanging wall along a northeast–southwest trending discontinuity with a length of about 5 km. Together, these lines of evidence suggest that the Oligocene La Rouvière fault was reactivated.

Figure 4.18 shows a final interferometric chart (wrapped phase) obtained using Sentinel-1 synthetic-aperture radar data pre- and post-event. This type of chart helps scientists a great deal in their geological deductions. The black lines correspond to faults after the geological map. The white line defines the about 5 km long northern section of the La Rouvière Fault (LRF) matching with the InSAR discontinuity.

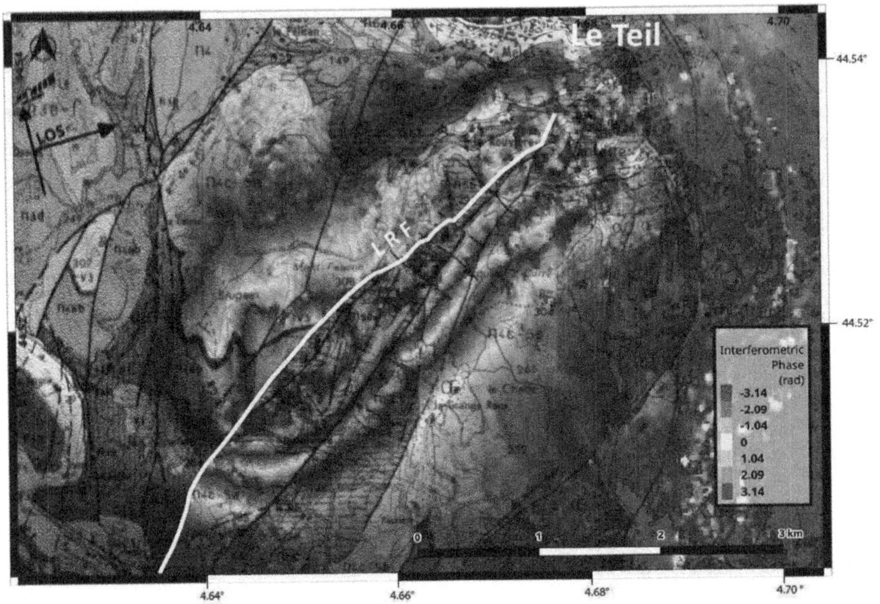

Figure 4.18. *Le Teil earthquake (figure from Ritz* et al. *(2020)): the Sentinel-1 ascending interferogram reveals the surface rupture. For a color version of this figure, see www.iste.co.uk/cavalie/images.zip*

4.4.3. *Available InSAR software: a quick overview*

With the large amount of freely available SLC SAR data, such as Sentinel-1 products, InSAR processing has become accessible even to non-radar specialists through the use of scientific or commercial software running automatic steps controlled by simple parameter initialization. In this section, some of the main available software packages for interferometry are listed. Some are under commercial licenses, such as GAMMA or ENVI/Sarscape; others have open licenses. Most of the free software packages for SAR interferometric processing have been developed by research institutes that work in the field of SAR interferometry. They are copyrighted and mainly for internal use, but they are also available to and shared with external SAR communities, supporting the research carried out at the respective institutes. Most of them are regularly maintained for different platforms and support new SAR mission data. The authors do not intend to advertize, nor to produce an exhaustive listing of the current InSAR software packages.

In Table 4.1, we only provide a short overview of the most current InSAR toolboxes available for the users. For more details on the software, refer to each referenced URL.

Name	Developed by	License	Capabilities	Sensors
DIAPASON Massonnet 1997	CNES	Not distributed anymore	Standard DInSAR	ERS, ENVISAT, JERS, ALOS, ALOS-2, RSAT, TSX, CSK, S1
DiapOTB https://gitlab.orfeo–toolbox.org/remote modules/diapotb.git	CNES	OpenSource OTB	Standard DInSAR	S1, CSK, TSX
ENVI (Sarscape) harrisgeospatial.com/sarscape	RSI	Commercial	Advanced DInSAR	ERS, JERS, RSAT, ENVISAT, S1, CSK, TSX
GAMMA http://www.gamma-rs.ch/	GAMMA	Commercial	Advanced DInSAR	All (including new constellation missions)
NSBAS https://github.com/IPGP/NSBAS-API	IPGP, ISTER, LGL	Free license	Standard DInSAR	S1, ALOS, ALOS-2
ROIPAC geohazards-tep/InSAR-ROIPAC	Berkeley University	Free license	Advanced DInSAR	ERS, JERS, RSAT, ENVISAT, CSK, TSX, S1
SNAP S1TBX https://sentinel.esa.int/web/sentinel/toolboxes/sentinel-1	ESA	Free license	Advanced DInSAR	ERS, ENVISAT, S1 Basic tools for: ALOS, ALOS-2, CSK, JERS, KOMPSAT-5, TSX, PAZ, RSAT, RCM, Seasar, ICEYE

Table 4.1. *Main DInSAR software packages*

All the current software packages offer the entire InSAR processing chain, from SLC data to InSAR and DInSAR basic products. Some of them can undertake further steps in order to produce digital elevation models, displacement maps and land-use maps. They can include a wide range of functions – from image visualization, to DEM import and interpolation, to cartographic and geodetic transforms such as multi-looking, coregistration, despeckling, geocoding and radiometric calibration, mosaicking, feature extraction, segmentation and classification.

These chains are suited for geophysical monitoring of natural hazards such as earthquakes, volcanoes and landslides. They are also useful in structural engineering, particularly for the monitoring of subsidence and structural stability.

4.5. Conclusion

The main scientific objective of DInSAR is the characterization and quantification of underlying processes to enable the creation of predictive models that can precisely map the deformation of the Earth's surface due to tectonic, volcanic and glacial processes. The main applications are as follows:

– understanding strain changes leading to and following major earthquakes;

– characterizing magma movements to predict volcanic eruptions;

– assessing the impact of ice-sheet and glacier-system dynamics on sea level rise and characterizing temporal variability;

– characterizing land surface motion related to subsidence, hydrology and coastal processes.

The second part of this book explains some of the InSAR applications in detail.

For surface topography mapping, a pre-existing digital elevation model (DEM) can be used. With a slight difference in radar position, changes in topography cause a distortion in the phase, much like a stereographic image. Using the known baseline distance, the height of the surface can be derived from multiple observations and a DEM created. The topography can then be corrected, and the phase and the image pixels rearranged to correspond to latitude and longitude in order to produce a correct "geocoded" projection.

When the contributions to the signal's phase of the ground surface, orbital effects and topography are removed, what remains in the image is the change in surface movement or deformation. Differential InSAR is not a tool for accurate pixelic local displacement measurements, but it is useful in monitoring progressing movements over large areas. With regard to the identification of linear and nonlinear deformation patterns and the abillity to follow time histories of movement for every radar target of a large dataset, a significant development of conventional InSAR is the use of permanent scatterer interferometry (PSI), as explained in the next chapter dedicated to multi-temporal InSAR approaches.

4.6. Acknowledgments

The authors thank Raphaël Grandin and Romain Jolivet for writing section 4.3.1 and Jean-Marie Nicolas for the careful review and comments that helped to improve this chapter.

4.7. References

Baby, H.B., Gole, P., Lavergnat, J. (1988). A model for the tropospheric excess path length of radio waves from surface meteorological measurements. *Radio Science*, 23, 1023–1038.

Berardino, P., Fornaro, G., Lanari, R., Sansosti, E. (2002). A new algorithm for surface deformation monitoring based on small baseline differential SAR interferograms. *IEEE Transactions on Geoscience and Remote Sensing*, 40, 2375–2383.

Cavalié, O., Doin, M.-P., Lasserre, C., Briole, P. (2007). Ground motion measurement in the Lake Mead Area, Nevada, by differential synthetic aperture radar interferometry time series analysis: Probing the lithosphere rheological structure. *Journal of Geophysical Research: Solid Earth*, 112(B3), 1–18.

Davies, K. (1990). *Ionospheric Radio*. IET, London.

Dee, D.P., Uppala, S.M., Simmons, A., Berrisford, P., Poli, P., Kobayashi, S., Andrae, U., Balmaseda, M., Balsamo, G., Bauer, D.P. *et al.* (2011). The ERA-Interim reanalysis: Configuration and performance of the data assimilation system. *Quarterly Journal of the Royal Meteorological Society*, 137(656), 553–597.

Delacourt, C., Briole, P., Achache, J. (1998). Tropospheric corrections of SAR interferograms with strong topography. *Application to Etna*, 25, 2849–2852.

Doin, M.-P., Lasserre, C., Peltzer, G., Cavalié, O., Doubre, C. (2009). Corrections of stratified tropospheric delays in SAR interferometry: Validation with global atmospheric models. *Journal of Applied Geophysics*, 69(1), 35–50.

Durand, P., Pourthié, N., Tison, C., Usseglio, G. (2021). DiapOTB: A new open source tool for differential SAR interferometry. *EUSAR On Line Conference*, 29 March–1 April.

Elliott, J.R., Biggs, J., Parsons, B., Wright, T.J. (2008). InSAR slip rate determination on the Altyn Tagh Fault, northern Tibet, in the presence of topographically correlated atmospheric delays. *Geophysical Research Letters*, 35.

Emardson, T.R., Simons, M., Webb, F.H. (2003). Neutral atmospheric delay in interferometric synthetic aperture radar applications: Statistical description and mitigation. *Journal of Geophysical Research*, 108(B5), 1–8.

Fattahi, H., Simons, M., Agram, P. (2017). InSAR time-series estimation of the ionospheric phase delay: An extension of the split range-spectrum technique. *IEEE Transactions on Geoscience and Remote Sensing*, 55(10), 5984–5996.

Fjortoft, R., Gaudin, J., Pourthié, N., Lalaurie, J., Mallet, A., Nouvel, J., Martinot-Lagarde, J., Oriot, H., Borderies, P., Ruiz, C., Daniel, S. (2014). KaRIn on SWOT: Characteristics of near-nadir Ka-band interferometric SAR imagery. *IEEE Transactions on Geoscience and Remote Sensing*, 52(4), 2172–2185.

Fornaro, G. and Guarnieri, A.M. (2002). Minimum mean square error space-varying filtering of interferometric SAR data. *IEEE Transactions on Geoscience and Remote Sensing*, 40(1), 11–21.

Goldstein, R. and Werner, C. (1998). Radar interferogram filtering for geophysical applications. *Geophysical Research Letters*, 25(21), 4035–4038.

Gomba, G. and De Zan, F. (2017). Bayesian data combination for the estimation of ionospheric effects in SAR interferograms. *IEEE Transactions on Geoscience and Remote Sensing*, 55(11), 6582–6593.

Gomba, G., Parizzi, A., De Zan, F., Eineder, M., Bamler, R. (2016). Toward operational compensation of ionospheric effects in SAR interferograms: The split-spectrum method. *IEEE Transactions on Geoscience and Remote Sensing*, 54(3), 1446–1461.

Gomba, G., Gonzalez, F.R., De Zan, F. (2017). Ionospheric phase screen compensation for the Sentinel-1 TOPS and ALOS-2 ScanSAR modes. *IEEE Transactions on Geoscience and Remote Sensing*, 55(1), 223–235.

Hannsen, R.F. (2001). *Radar Interferometry: Data Interpretation and Error Analysis*. Springer-Verlag, Heidelberg.

Hooper, A., Segall, P., Zebker, H.A. (2007). Persistent scatterer interferometric synthetic aperture radar for crustal deformation analysis, with application to Volcán Alcedo, Galápagos. *Journal of Geophysical Research: Solid Earth*, 112(B7), 1–21.

Jolivet, R., Grandin, R., Lasserre, C., Doin, M.P., Peltzer, G. (2011). Systematic InSAR tropospheric phase delay corrections from global meteorological reanalysis data. *Geophysical Research Letters*, 38(17), 1–6.

Jolivet, R., Lasserre, C., Doin, M.-P., Guillaso, S., Peltzer, G., Dailu, R., Sun, J., Shen, Z.-K., Xu, X. (2012). Shallow creep on the Haiyuan Fault (Gansu, China) revealed by SAR interferometry. *Journal of Geophysical Research: Solid Earth*, 117(B6), 1–8.

Kalnay, E., Kanamitsu, M., Kistler, R., Collins, W., Deaven, D., Gandin, L., Iredell, M., Saha, S., White, G., Woollen, J. *et al.* (1996). The NCEP/NCAR 40-year reanalysis project. *Bulletin of the American Meteorological Society*, 77(3), 437–472.

Kim, J.S. and Papathanassiou, K.P. (2010). Faraday rotation estimation performance analysis. *8th European Conference on Synthetic Aperture Radar (EUSAR)*, Aachen.

Lazecky, M., Bakon, M., Hlavacova, I. (2015). Effect of DEM inaccuracy on precision of satellite InSAR results. *GIS Ostrava Conference*, Ostrava.

Li, Z., Xu, W., Feng, G., Hu, J., Wang, C., Ding, X., Zhu, J. (2012). Correcting atmospheric effects on InSAR with MERIS water vapour data and elevation-dependent interpolation model. *Geophysical Journal International*, 189(2), 898–910.

Massonnet, D. (1997). Producing ground deformation maps automatically: The diapason concept. *IEEE International Geoscience and Remote Sensing Symposium*, 3–8 August, Singapore.

Massonnet, D. and Feigl, K.L. (1995). Discrimination of geophysical phenomena in satellite radar interferograms. *Geophysical Research Letters*, 22(12), 1537–1540.

Meyer, F. (2010). A review of ionospheric effects in low-frequency SAR – Signals, correction methods, and performance requirements. *Geoscience and Remote Sensing Symposium*, 25–30 July.

Meyer, F. and Nicoll, J. (2008). Prediction, detection, and correction of Faraday rotation in full-polarimetric L-band SAR data. *IEEE Transactions on Geoscience and Remote Sensing*, 46(10), 3076–3086.

Onn, F. and Zebker, H.A. (2006). Correction for interferometric synthetic aperture radar atmospheric phase artifacts using time series of zenith wet delay observations from a GPS network. *Journal of Geophysical Research: Solid Earth*, 111(B9), 1–16.

Peltzer, G., Crampé, F., Hensley, S., Rosen, P. (2001). Transient strain accumulation and fault interaction in the Eastern California shear zone. *Geology*, 29, 975–978.

Prats-Iraola, P., Rodriguez-Cassola, M., Zan, F.D., Scheiber, R., Lopez-Dekker, P., Barat, I., Geudtner, D. (2015). Role of the orbital tube in interferometric spaceborne SAR missions. *IEEE Geoscience and Remote Sensing Letters*, 12(7), 1486–1490.

Puysségur, B., Michel, R., Avouac, J.-P. (2007). Tropospheric phase delay in interferometric synthetic aperture radar estimated from meteorological model and multispectral imagery. *Journal of Geophysical Research Solid Earth*, 112(B11), B05419.

Ritz, J., Baize, S., Ferry, M., Larroque, C., Audin, L., Delouis, B., Mathot, E. (2020). Surface rupture and shallow fault reactivation during the 2019 Mw 4.9 Le Teil earthquake, France. *Communications Earth and Environment*, August, 1(10).

Rosen, P., Hensley, S., Joughin, I., Li, F., Madsen, S., Rodriguez, E., Goldstein, R. (2000). Synthetic aperture radar interferometry. *Proceedings of the IEEE*, 88, 333–382.

Tupin, F., Inglada, J., Nicolas, J.-M. (2014). *Remote Sensing Imagery*. ISTE Ltd, London, and Wiley, New York.

Yu, C., Li, Z., Penna, N.T., Crippa, P. (2018). Generic atmospheric correction model for interferometric synthetic aperture radar observations. *Journal of Geophysical Research: Solid Earth*, 123(10), 9202–9222.

Zebker, H.A., Rosen, P.A., Hensley, S. (1997). Atmospheric effects in interferometric synthetic aperture radar surface deformation and topographic maps. *Journal of Geophysical Research: Solid Earth*, 102(B4), 7547–7563.

5

Advanced Methods for Time-series InSAR

Dinh Ho Tong Minh[1], Ramon Hanssen[2], Marie-Pierre Doin[3] and Erwan Pathier[3]

[1] *UMR-TETIS, INRAE, France*
[2] *Delft University of Technology, The Netherlands*
[3] *ISTerre, Grenoble, France*

5.1. Introduction

Time-series InSAR analysis is a well-known technique that is used to map large-scale surface subsidence with high accuracy from space. There are two main reasons to move from traditional InSAR (i.e. based on a single image pair) to time-series InSAR: (1) it would enable the study of temporarily extended and varying phenomena and (2) it would improve the precision of surface displacement measurements. Indeed, for deformation processes that develop slowly (e.g. land compaction, interseismic deformation, reservoir pressurization, slow-moving landslides or ice flows), this technique allows us to exploit the redundancy in InSAR time-series stacks to mitigate noise and nuisance signals such as atmospheric effects, residual topography and decorrelations.

While many time-series InSAR methods have been developed in the last 20 years, most of them share similar characteristics so that they can be categorized into two types of techniques based on how they account for signal decorrelations (Ho Tong Minh *et al.* 2020). The first category of techniques is based on distributed

Surface Displacement Measurement from Remote Sensing Images,
coordinated by Olivier Cavalié and Emmanuel Trouvé.
© ISTE Ltd 2022.

scatterers (DSs) for deformation monitoring. Distributed targets occur in natural environments (meadows, fields, bare soil) where many similarly bright scatterers contribute to the information existing in a resolution cell. A common way of reducing signal decorrelations is to select interferograms with short spatial and temporal baselines (small baseline (SB) techniques). Distributed targets are plentiful, facilitating densely sampled deformation maps. However, deformation measurements of distributed targets are often lower quality and therefore require spatial filtering. The second approach is permanent/persistent scatterer (PS) InSAR techniques, which use individual scatterers that dominate the signal from within a resolution cell to track deformation through time. The well-known scattering stability is due to the fact that point-like targets are often associated with stable human-made structures such as buildings, poles and gratings, resulting in very low decorrelations. Consequently, PS techniques provide high-quality deformation information at point target locations. However, the density of PS targets is often low in non-urban areas, resulting in a sparse distribution of usable points. To overcome this, a recent advanced technique allows us to combine both PSs and DSs to overcome the sparsity of identified points for estimation (the persistent scatterer–distributed scatterer (PSDS) technique).

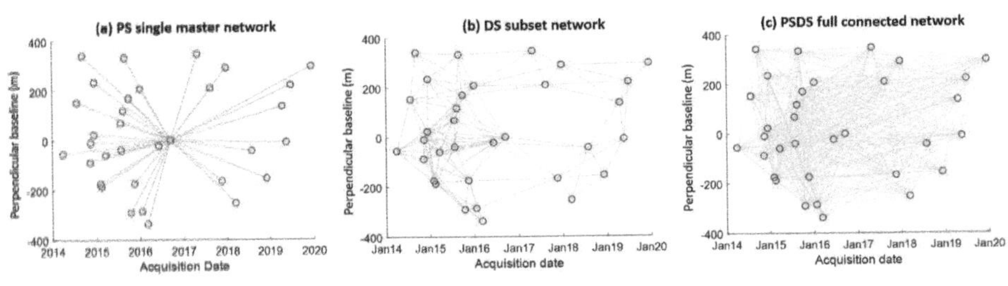

Figure 5.1. *Examples of interferogram networks generated by using COSMO SkyMed data (provided by the Italian Space Agency under research project ID226) from over the Ho Chi Minh City area. (a) The single master network with permanent/persistent scatterer (PS) techniques. (b) The subset network consisting of interferograms with small temporal and spatial baselines with the small baseline subset (SBAS) technique. (c) The full network consisting of all possible interferograms with the persistent scatterer–distributed scatterer (PSDS) technique*

To perform an InSAR time-series analysis, it is necessary to reconstruct the evolution of the unwrapped phase between each acquisition date. If we rely on interferograms, it is therefore necessary for all dates to be connected by interferograms. A common way to represent these connections is a graph in which the interferograms are represented by edges that link the nodes of the network, which are the SAR images plotted according to their acquisition time and perpendicular baseline. Figure 5.1 shows three examples of interferogram networks related to the different time-series techniques mentioned above. If some acquisition dates are not connected by the interferogram network, it is necessary to make more or less strong assumptions to reconstruct the phase evolution. An example is the so-called "stacking" method, which makes the strong assumption that, in each pixel, the phase evolution is linear with time. This relies on the sum of independent (i.e. not connected) interferograms to estimate a velocity map (e.g. Wright *et al.* 2001). However, this assumption is not acceptable in most cases, and other approaches that require access to the unwrapped phase of each interferogram are necessary.

This chapter describes the two main families of time-series InSAR techniques (SB and PSDS). It is organized as follows: the mathematical background is provided in section 5.2; methodology advancements are summarized in section 5.3; the SB technique is presented in section 5.4; and finally the PSDS technique is presented in section 5.5.

5.2. Background of time-series InSAR analysis

Let us assume that N SAR images are available. The images are co-registered on a reference/common grid. Let us assume that phase terms due to orbit and terrain topography have been estimated and compensated for. The residual phase of the unwrapped differential interferogram φ^n, for each acquisition n, is composed of the phase components contributed by residual topography, deformation, atmosphere and noise (Hanssen 2001):

$$\varphi^n = \varphi^n_{topo} + \varphi^n_{dis} + \varphi^n_{aps} + \varphi^n_{noise} + 2k\pi \qquad [5.1]$$

where φ^n_{topo} is the residual topographic phase; φ^n_{dis} is the phase due to the surface displacement; φ^n_{aps} is the atmospheric phase screen (APS), representing the contribution of weather conditions; φ^n_{noise} is the phase noise due to all other possible contributions, such as temporal decorrelation, mis-coregistration, uncompensated spectral shift decorrelation, orbital errors, soil moisture and thermal noise; and k is an integer ambiguity number.

For a point p, the residual topographic phase can be written as (Hanssen 2001):

$$\varphi^n_{p,topo} = k^z_{p,n}\varepsilon^z_p \qquad [5.2]$$

where $k^z_{p,n} = \frac{4\pi B^{\perp}_n}{\lambda sin\theta_p R_{n,p}}$ is the height-to-phase factor; ε^z_p is the residual topography or the elevation error; $B^{\perp}_n$ is the perpendicular baseline of the n-th image with respect to the reference image; θ_p is the local incidence angle; λ is the carrier radar wavelength; and $R_{n,p}$ is the target–satellite distance.

Without loss of generality, the phase component due to the displacement at a point p can be usefully split into two components:

$$\varphi^n_{p,dis} = k^t_n v_p + \mu_{NL} \qquad [5.3]$$

where v_p is the mean light-of-sight (LOS) velocity of the target; $k_n^t = \frac{4\pi t_n}{\lambda}$ is the time-to-phase factor; t_n is the temporal baseline; and μ_{NL} is the phase term due to a possible nonlinear motion.

The APS φ_{aps}^n is usually not modeled because it depends on weather conditions, but it can be reduced significantly by considering phase differences between the nearby points thanks to its spatial covariance properties (Emardson *et al.* 2003). The noise term φ_{noise}^n contains all the other phase contributions.

Most often, a linear model for the residual topography and a constant velocity are assumed. More specifically, given $N-1$ unwrapped differential interferograms, for each pixel, using equation [5.1], we can write a system of $N-1$ equations with two unknowns:

$$\varphi^n = k_n^z \varepsilon^z + k_n^t v + w^n + 2k\pi \quad [5.4]$$

where $w^n = \mu_{NL} + \varphi_{aps}^n + \varphi_{noise}^n$ is the residual phase. The problem in equation 5.4 will be linear if the integer ambiguity number k is known. Unfortunately, φ^n can be only observed as a wrapped phase (i.e. k is also an unknown) (Ghiglia and Pritt 1998; Hanssen 2001). PS and PSDS approaches can be carried out directly from the wrapped phase, whereas the phase input in SB analyses should be unwrapped.

5.3. A review

5.3.1. *Small baseline methods for time-series analysis*

The objective of this processing strategy is to establish a time series for ground displacement with the largest possible spatial coverage and density for different types of land cover. Achieving this objective depends essentially on the ability to obtain the unwrapped phase over the maximum number of pixels at the scale of a network of interferograms connecting the different dates of acquisition of the time series. This capacity quickly becomes limited in the presence of strong phase gradients and decorrelation noise (see Chapter 6). In natural environments, dominated by DSs, geometric and temporal decorrelation are two main limiting factors (see Chapter 4). The idea of the SB approach is therefore to establish a network of interferograms for which these decorrelation effects are minimized.

In early radar missions such as Radarsat-1 and JERS-1, the large variability of the perpendicular baseline and the viewing squint angle, due to poor orbital control and yaw steering capabilities (see Chapter 1), made geometrical decorrelation the main limitation. For C-band ERS and Envisat satellites that were better yaw steered but remained in an orbital tube with a diameter larger than 1 km, the perpendicular baseline was still a critical criterion to be minimized when the first InSAR time-series analysis started by the end of the 1990s. This explains the term "small baseline"; however, it is important to keep in mind that other criteria are applied to select interferograms, especially those that aim to minimize the temporal decorrelation. Today, with recent missions such as Sentinel-1, which have a narrower orbital tube, the perpendicular baseline tends to be a less restrictive criterion in the network design. In practice, computation time and data volume must sometimes also be taken into account in the case of long time series in order to limit the number of interferograms in the network while keeping good redundancy.

Early attempts to reconstruct ground displacement time series from a small baseline network of unwrapped interferograms used a pixel-by-pixel least-squares adjustment (e.g. Usai *et al.* 1999; Beauducel *et al.* 2000). This numerical method is limited when the system to solve is under-determined (rank deficiency), which can occur when the unwrapped phase is not available for all the pixels of all the interferograms of the network.

It should be recalled that even if the interferogram network connects all the dates, at the individual pixel level, the situation may be different. For example, in the case of a pixel located in a noisy environment, which cannot be reliably unwrapped over several interferograms, there may be several missing links in the network at the pixel level (Figure 5.2), possibly resulting in disconnected sub-networks called subsets.

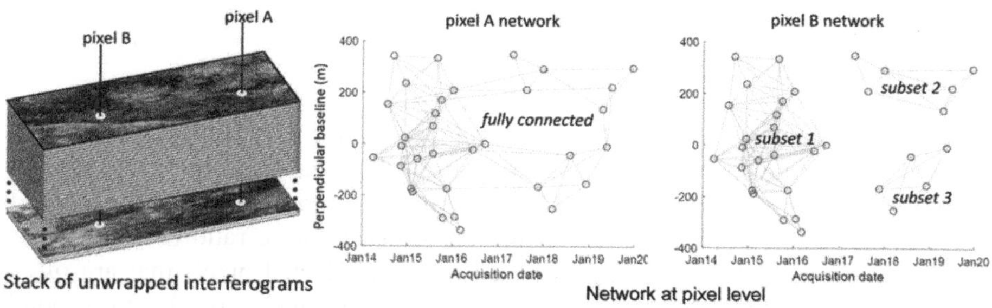

Figure 5.2. *Two examples of networks at the pixel level starting from the network at the image level shown in Figure 5.1(b). The network of pixel A is fully connected despite some missing links with respect to the interferogram network at the image level (i.e. there are some interferograms in which pixel A is not unwrapped, but all dates are connected through a single network). The network of pixel B shows disconnected subsets due to missing links. In this case, recovering the full time series requires some assumptions and suitable inversion methods to reconnect the subsets. For a color version of this figure, see www.iste.co.uk/cavalie/images.zip*

To tackle this problem, the small baseline subset (SBAS) technique uses a singular value decomposition (SVD) to solve the under-determined system (Berardino *et al.* 2002). More recently, several other implementations have been proposed to reconstruct the time series with different degrees of sophistication, including regularization constraints, the weighting of interferograms, solving for phase evolution and topographical residual error at the same time, the combination of information from low- and high-resolution interferograms through an iterative approach and correcting for unwrapping errors (Schmidt and Bürgmann 2003; Mora *et al.* 2003; Lanari *et al.* 2004; Fornaro *et al.* 2009; López-Quiroz *et al.* 2009; Doin *et al.* 2011; Jolivet *et al.* 2013; Yunjun *et al.* 2019). One of these implementations is discussed in more detail in section 5.4.

Compared to other methods such as PS or those that exploit all possible interferograms (see section 5.5), SBAS methods are confronted with the question of how to choose the optimal interferogram network. This choice is often based on empirical considerations about the quality of interferograms (with the difficulty that this may vary from region to region) and practical considerations about computational performance. This possible compromise makes SBAS methods competitive with respect to other methods, especially when very large areas (> millions of km^2) need to be covered. To reach such coverage objectives, the problems of massive processing should also be taken into account (e.g. Casu *et al.* 2014; Manunta *et al.* 2019). Multi-looking is then commonly used to improve performance (e.g. Morishita *et al.* 2020). However, studies on phase closure, detailed in Chapter 6, have shown that this can lead to biases in the measurement of surface displacements, which must be taken into account.

5.3.2. *From PS to PSDS*

Permanent scatterer interferometry, developed in the late 1990s, is the first InSAR technique involving analysis directly from the measured wrapped phase in equation [5.4] (Ferretti *et al.* 2000, 2001). Parameters can be estimated through a search of the solution space, provided that the signal-to-noise ratio (SNR) is high enough. Consequently, the analysis is based only on selected pixels that are often characterized by highly coherent targets (i.e. low σ^2_{noise}), which typically correspond to human-made objects present over an urban city but which are less available in non-urban areas. Such stable targets, called permanent scatterers (Ferretti *et al.* 2000), have a dominant scatterer within the resolution cell and can be identified by a criterion relating to amplitude stability through time. Several approaches have been presented in previous studies for the performance of InSAR analysis over non-urban scenes (e.g. non-cultivated land, desert and debris areas) where the PS assumption cannot be retained by considering DS (Berardino *et al.* 2002; Lanari *et al.* 2004; Fialko 2006). In Figure 5.3, examples of PS and DS targets are shown.

The main aim is to achieve better performance by increasing the density of identified points by combining PS and DS targets. This combination has been proposed and studied by many authors (Fornaro *et al.* 2006; Rocca 2007; Ferretti *et al.*

2008, 2011; Hooper 2008). In Hooper (2008), a hybrid multi-temporal processing scheme was proposed, where two approaches (PS and SBAS) were applied to the same data input and then SBAS results were inverted by an SVD to be combined with PS. However, in the work of Fornaro *et al.* (2006), Ferretti *et al.* (2008), Ferretti *et al.* (2011) and Rocca (2007), the combination of PS and DS is carried out using a maximum likelihood (ML) estimation framework. In this approach, an optimized process for the parameters of interest can be implemented by exploiting target statistics, which can be represented by a coherence matrix. The advantage of this technique is that the criteria, which determine the weight of each interferogram in the estimation process, are directly derived from coherences through a rigorous mathematical approach. With the ML optimization, the parameters of interest can be asymptotically unbiased and of minimum variance. However, ML techniques require reliable information on target statistics (i.e. a coherence matrix) to determine the weight in the estimation algorithm.

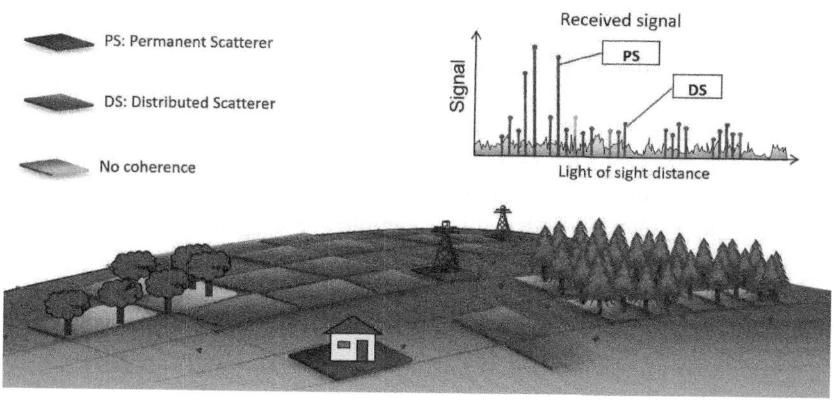

Figure 5.3. *Examples of permanent and distributed scatterers. For a color version of this figure, see www.iste.co.uk/cavalie/images.zip*

Rocca (2007) and Ferretti *et al.* (2011) have proposed a two-step ML estimation process. In the first step, the ML estimation is applied to all the $N(N-1)/2$ interferograms available from N images, in order to squeeze the best estimates of the $N-1$ interferometric phases, which is called SqueeSAR. This ML step was first known as phase linking (Guarnieri and Tebaldini 2008) or as the phase triangulation algorithm (PTA; Ferretti *et al.* 2011). This step is efficient and has a low computational burden. In the second step, the separation of the contributions of decorrelation noises from the parameters of interest can be processed as in the PS algorithm. Nowadays, the term "PSDS" refers to group techniques that exploit both PS and DS targets for time-series InSAR analysis.

5.4. The SBAS technique

5.4.1. *Principle and definition of a "small baseline" network of interferograms*

As shown in section 5.3, there is a continuum of methods from PS to small baseline subset techniques. Here, we focus on an end-member small baseline method that is determined by:

– the definition of a "small baseline" network of interferograms;

– multi-looking and spatial filtering of differential interferograms;

– spatial unwrapping of spatially continuous interferograms after atmospheric and/or DEM error corrections;

– time-series reconstruction of phase delay maps with time from a redundant network of interferograms;

– source separation between atmospheric and deformation contributions.

While the exact order, combination and iteration of these steps depend on specific implementations, all of them are present in most SBAS algorithms. We here focus on the NSBAS chain organization of these steps (Doin *et al.* 2011), but we will mention some of the other SBAS chains. Note that the main difference with the PSDS technique, described above as an extension of the PS algorithms, is the assumption of a mostly continuous interferometric image sampled on a regular grid, as opposed to a discrete punctual sampling of space.

The basic rationale behind the SB technique is to recognize the non-closure behavior of the filtered unwrapped interferometric phase:

$$\varphi_{AC} \neq \varphi_{AB} + \varphi_{BC} \quad [5.5]$$

where A, B and C are three acquisitions and φ_{AC}, φ_{AB}, φ_{BC} are the unwrapped interferometric phases between A and C, A and B, and B and C, respectively (see Chapter 6). Non-closures are derived from three aspects: (a) range spectral filtering of the non-overlapping part of the spectrum between two single look complexes (SLCs) involved in the interferogram formation, (b) spatial averaging and filtering of complex values $A_1 A_2 \exp(j(\phi_2 - \phi_1))$ in the presence of decorrelation noise that reaches π, and (c) path-dependent unwrapping ambiguities.

Due to these three sources of non-closure, φ_{AC} brings complementary information to φ_{AB} and φ_{BC}. A key for the success of a particular SBAS implementation relies on how the information retrieval from each individual interferogram is maximized while minimizing the noise propagation, when the time series is reconstructed from a network of interferograms. For example, if A and C are taken during the same season, say autumn, while B is taken during spring, φ_{AC} might be more coherent than $\varphi_{AB} + \varphi_{BC}$, despite AB and BC being separated by a shorter time span than AC. In general, however, the longer the temporal or perpendicular baseline, the larger the phase noise in the full-resolution wrapped interferograms.

SB techniques thus try to optimize the choice of interferometric pairs by minimizing their perpendicular baseline, temporal baseline, seasonal difference, etc.

One source of decorrelation arises from the perpendicular baseline, which is mitigated by range spectral filtering. Range spectral filtering consists of removing the non-overlapping part of the range spectrum before interferogram computation. The relative apparent shift in the radar wavelength, as sensed by DSs on ground, between two acquisitions, depends on the terrain slope and the perpendicular baseline between two acquisitions (Gatelli *et al.* 1994). The overlapping fraction of the spectrum depends on the frequency shift and the range bandwidth. For ERS and Envisat data, for example, a perpendicular baseline of about 1,100 m leads to a complete disappearance of the spectrum overlap and thus a complete decorrelation of DSs. With adequate range spectrum filtering adapted to a local terrain slope (Davidson and Bamler 1999), interferograms with baselines up to 500 m can be obtained with a good preservation of the coherence over DSs. For ALOS interferograms, interferograms with very large perpendicular baselines can remain coherent due to a much larger range bandwidth than for ERS or Envisat. In the case of a restricted orbital tube and large bandwidth, such as Sentinel-1, the benefit of range spectral filtering is limited, and setting a small perpendicular baseline for interferograms is no longer the most critical criterion when defining the small baseline interferometric network.

A second source of decorrelation includes changes in land cover, soil moisture, etc. As time separation increases, the probability of the phase noise for each pixel going beyond π increases, either because of the complete change of a dominant scatterer or because of progressive transformation of multiple backscattering interactions within a resolution cell. As the phase is only known modulo 2π, a large phase noise produces a complete loss of information. Multi-looking and spatial filtering of small baseline interferograms are thus used to reduce phase noise before they are combined to reconstruct interferograms with any arbitrary temporal baseline. Therefore, SBAS methods can be considered as one way of solving for the ambiguity of phase noise in interferograms with large temporal baselines.

After multi-looking, filtering and unwrapping, the network of interferometric phases is inverted into time series. The temporal reconstruction consists of solving the overdetermined, but also possibly under-determined, set of equations:

$$\sum_{l} G_{kl}\delta\varphi_l = \varphi_{ij}; \text{ with } G_{kl} = 1 \text{ for } i \leq l < j; \text{otherwise } G_{kl} = 0, \qquad [5.6]$$

where $\delta\varphi_l$ are the unknown phase increments from acquisitions l to $l+1$ and φ_{ij} is the kth unwrapped interferometric phase between date i and date j. If the above system is under-determined, or $\text{rank}(G^TG) < N - 1$, with N being the number of images, it means that the interferometric network is made up of several ($= N - \text{rank}(G^TG)$) independent subsets of images (Berardino *et al.* 2002; López-Quiroz *et al.* 2009). The relative information between independent subsets is missing. In order to reconstruct the time series, we must "invent" the missing links between subsets based on a predefined criterion. This additional criterion is the origin of the "subset" addition to the SBAS acronym. Although one may make an educated guess about the missing information (see, for example, López-Quiroz *et al.* (2009)), the disruption of the interferometric network into subsets will increase the uncertainty in the inverted time series, especially if subsets present no overlap in time.

To conclude, SBAS methods make it possible, using multi-looking and filtering of the interferometric phase in small baseline interferograms, to monitor the phase delay of a possibly evolving gathering of objects on the ground, with changing combinations of weights between backscatterers in the chosen neighborhood. To optimize the choice of the interferometric network, we must consider (1) a certain level of redundancy that will ensure a varied sampling of the above-mentioned combinations, (2) a perpendicular baseline that is small enough to preserve a spectrum overlap of at least half the range bandwidth, (3) a temporal baseline that enhances coherence and (4) network continuity from the first acquisition to the last one.

5.4.2. *Multi-looking and filtering, atmospheric or DEM error corrections and unwrapping*

As mentioned above, multi-looking and filtering are essential steps in SBAS techniques. Multi-looking is often performed on the complex interferogram, including the backscattering amplitude of the reference and secondary SLCs. As a result, high-intensity point scatterers, when present in the neighborhood, act as a huge weight on the multi-looked interferogram. Alternatively, in natural environments, the amplitude of the interferogram can be replaced by a measure of phase local stability, such as the collinearity defined by Pinel-Puyssegur *et al.* (2012), before multi-looking to increase the SNR of the multi-looked phase (Daout *et al.* 2018). There are many filtering methods (e.g. Goldstein and Werner 1998). Both multi-looking and filtering must preserve steep phase gradients, if there are any. When the latter occur along slopes in foreshortening areas, for example, or in incoherent areas, they produce inconsistencies in interferometric fringe patterns, which lead to unwrapping errors. This is why it is judicious to flatten the wrapped interferometric phase by, for example, correcting atmospheric delays and DEM errors before filtering and unwrapping (Doin *et al.* 2015; Daout *et al.* 2017; Maubant *et al.* 2020).

The component of the APS that can easily be corrected before filtering and unwrapping is the stratified atmospheric delay, i.e. the atmospheric contribution that is first-order dependent on elevation (see Chapter 4 and also Hanssen (2001) and

Doin *et al.* (2009)). The stratified atmospheric correction can be evaluated empirically using the wrapped phase versus the elevation relationship (Grandin *et al.* 2012; Doin *et al.* 2015) or it can be predicted using global atmospheric models (Doin *et al.* 2009; Jolivet *et al.* 2011) together with GNSS zenith delays if present (Yu *et al.* 2018). In any case, it helps to reduce phase gradients in mountainous areas (Grandin *et al.* 2012) and thus facilitates filtering and unwrapping. The local DEM error correction can also be evaluated empirically using the relationship between the wrapped phase and the perpendicular baseline. This estimation can be performed for PSs or DSs (Ferretti *et al.* 2001; Hooper 2008) or for the whole multi-looked data stack (Ducret *et al.* 2014). It reduces the local phase scatter and again improves the multi-looking, filtering and unwrapping steps. Figure 5.4 shows the impact of corrections on wrapped interferograms, which can improve subsequent unwrapping.

When deformation gradients are very large and the deformation pattern is mostly constant through time, as in the case of subsidence in Mexico City (López-Quiroz *et al.* 2009) or permafrost dynamics Daout *et al.* (2017), it can be useful to obtain a preliminary deformation pattern map that can be scaled and removed before unwrapping and added back after unwrapping. Similarly, some authors prefer to remove a somewhat complex, empirically adjusted, stratified APS before unwrapping to efficiently remove steep phase gradients in mountainous areas, adding it back after unwrapping in order to make sure that there is no deformation leakage (Maubant *et al.* 2020). All this helps in solving unwrapping ambiguities that arise from abrupt phase gradients and/or from the presence of phase noise. The SBAS algorithm of Lanari *et al.* (2004) also presents an iterative unwrapping procedure where the first low-pass deformation time series obtained and the DEM correction map are used to correct wrapped interferograms and refine the unwrapping.

Finally, most of the SBAS methods rely on the spatial unwrapping of interferograms using different classes of methods (see Chapter 6): branch cut (Zebker and Lu 1998), global optimization (Chen and Zebker 2002) or spatial phase integration based on coherence (Grandin *et al.* 2012). The interferometric phase must be spatially continuous to connect all the parts of an interferogram without remaining 2π ambiguities. Spatial unwrapping is a critical step in InSAR processing, even for small baseline interferograms, especially in vegetated, mountainous or snowy regions. Note that in SBAS techniques this step replaces the spatial integration of velocity differences Δv_{pq} and residual wrapped phase differences Δw_{pq}^n measured along the arcs pq of a network of PSDS points. The retrieved integrated residual phase, w^n, of PSDSs for each time step n can also be affected by unwrapping errors due to possible large phase gradients related to the atmosphere (across atmospheric fronts or steep terrains) or nonlinear deformation between PSDS locations.

Once the interferograms have been unwrapped, they must be referenced by setting the average phase to zero in a given area. Unwrapped interferograms can also be flattened along the range, x, and azimuth, y, directions by adjusting a phase ramp in the form: $\varphi_{ij}^{est} = a_{ij}^{est}x + b_{ij}^{est}y + c_{ij}^{est}$. In this case, a_{ij}^{est} and b_{ij}^{est} must be used to invert the time series a_l^{inv} and b_l^{inv} for each time step l and then to reconstruct $a_{ij}^{inv} = a_j^{inv} - a_i^{inv}$ and $b_{ij}^{inv} = b_j^{inv} - b_i^{inv}$. This ensures that the ramp correction is consistent within the interferogram network rule of thumb (Biggs *et al.* 2007). Alternatively, GPS data, within the InSAR footprint and converted to the LOS, can be used to fix the phase ramp.

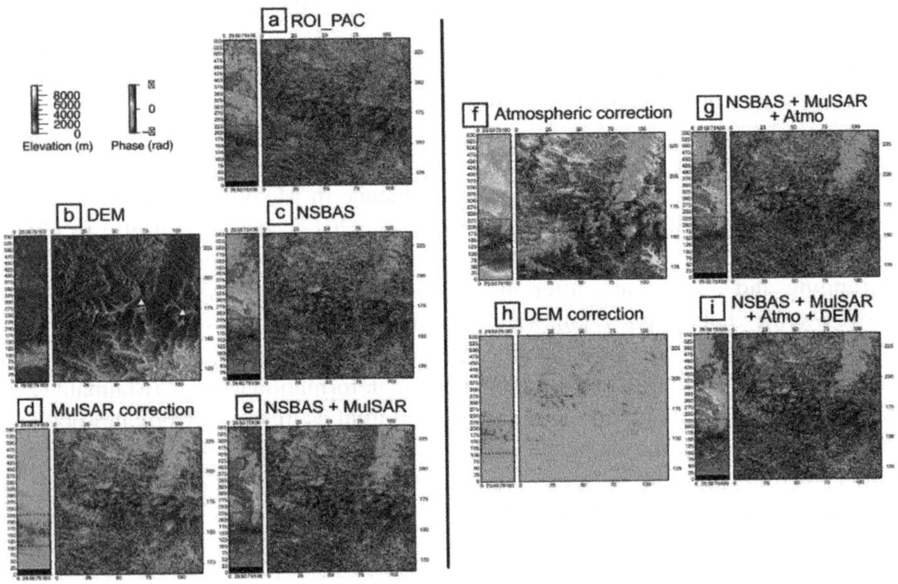

Figure 5.4. *Examples showing how successive corrections on the wrapped phase help later unwrapping for a long temporal baseline Envisat interferogram (3.1 years) with a large perpendicular baseline (236 m) across the Himalayan range, i.e. in an area with strong relief and vegetation. The complete interferogram together with a zoom is shown at each correction step. The figure was reproduced from Grandin* et al. *(2012). (a) Multi-looked interferogram processed with ROI-PAC. (b) Shuttle Radar Topography Mission (SRTM) DEM in radar geometry. (c) Interferogram processed with NSBAS (Doin* et al. *2011). (d) MulSAR correction (Pinel-Puyssegur* et al. *2012). (e) Interferogram after the MulSAR correction. (f) Tropospheric delay correction computed from the ERA-Interim atmospheric model (Jolivet* et al. *2011). (g) Interferogram after atmospheric correction. (h) DEM correction calculated from the combination of (wrapped) coherent interferograms with large perpendicular baselines and small temporal baselines (Ducret* et al. *2014). (i) Interferogram after DEM correction. For a color version of this figure, see www.iste.co.uk/cavalie/images.zip*

5.4.3. *Reconstruction of the time series*

We first focus here on the resolution of the set of equations [5.6], the only aim of which is to reconstruct the complete phase delay time series, including the residual atmospheric contribution. The linear system is written here for pedagogical purposes:

$$\begin{pmatrix} \varphi_{1,3} \\ \varphi_{2,5} \\ \vdots \\ \varphi_{i,j} \\ \vdots \\ \varphi_{N-1,N} \end{pmatrix} = \begin{pmatrix} 1 \ 1 \ 0 0 \ 0 \ 0 0 \cdots 0 \\ 0 \ 1 \ 1 1 \ 0 \ 0 0 \cdots 0 \\ \vdots \ \ddots 0 \, 1 \cdots 1 \, 0 \ \ddots \ \vdots \\ 0 \cdots 0 \, 1 \cdots 1 \, 0 \cdots 0 \\ \vdots \ \ddots 0 \, 1 \cdots 1 \, 0 \ \ddots \ \vdots \\ 0 \ 0 \ 0 0 \cdots 0 0 \ 1 \ 1 \end{pmatrix} \begin{pmatrix} \delta\varphi_1 \\ \vdots \\ \delta\varphi_l \\ \vdots \\ \delta\varphi_{N-1} \end{pmatrix}$$

where φ_{ij} are the M measured interferometric phases, N is the number of acquisitions and the $M \times (N-1)$ matrix is the design matric G. It can be solved pixel by pixel by least squares or by singular value decomposition if the rank of $G^T G$ is lower than $N-1$ (Berardino *et al.* 2002):

$$\delta\varphi^{inv} = (G^T G)^{-1} G^T \varphi \qquad [5.7]$$

In order to determine how the data (φ) or model phase increments ($\delta\varphi_{inv}$) can be weighted by covariance matrices on the data C_d or on the model C_m in the estimation process, we must recall the reasons why the set of equations [5.6] will not be perfectly respected. The residuals $\varphi - G\delta\varphi^{inv}$, in principle, contain only non-closure associated with multi-looking and filtering of phase decorrelation noise and unwrapping errors. Therefore, we can safely assume that there is no built-in covariance between the estimations of φ_{AB}, φ_{BC} and φ_{AC}, although they share common dates. The covariance, C_d, is thus to the first-order diagonal. Similarly, as the model phase increments contain two physical signals, deformation and atmospheric delay, there is no reason to include a temporal correlation length in the non-diagonal elements of C_m. Finally, we recommend a simple weighting of each equation in [5.6] by the amplitude of the estimated phase decorrelation noise, σ_φ^{ij}, of each interferogram.

In the case of good data redundancy (evaluated by the data resolution matrix), and if unwrapping errors affect less than about one third of interferograms, the residuals between original and reconstructed interferograms, $r_{ij} = \varphi_{ij} - \sum_l G_{kl}\delta\varphi_l^{inv}$ (with the interferogram ij corresponding to the kth interferogram), can be used to detect and correct unwrapping errors (Doin *et al.* 2011). We must first obtain r_{ij} by solving equation [5.7] and then by rule of thumb use these residuals to down-weight the set of equations by $1/(\text{abs}(r_{ij}) + r_0)$, where r_0 corresponds to a typical phase noise standard deviation, about 0.4 rad. Then, the residuals obtained with the new solution $\delta\varphi^{inv2}$ will point towards interferograms containing unwrapping errors. As a result, $+/-2\pi$ can be added to the "wrong" φ_{ij} values and a new solution can be computed.

A few iterations allow us to converge to a consistent network of corrected unwrapped phases, provided that the number of initial errors affect only less than one fourth (by rule of thumb) the number of interferometric links.

Finally, even if the built interferometric network does not isolate any image subsets, unwrapping may not have reached all the parts of interferograms, such that, for a given pixel, $G^T G$ might not be invertible. This is particularly common if there is snow during a part of the year, for example. In this case, singular value decomposition can be used, which will set the mean phase increment between the reconstructed phase through time of the different subsets to zero. However, this is not always the optimal way to reconstruct the missing information between subsets. Alternatively, we can implement constraints to provide the necessary interpolation across subsets using two equivalent approaches:

– in the first, we compute the average velocity v, then subtract its contribution from each interferogram, $\varphi_{ij} - v \times (t_j - t_i)$, and iterate on the time-series reconstruction. Afterwards, the offset between subsets obtained by SVD will correspond to the average velocity times the temporal separation of subsets. It is possible to write somewhat more complex formulations based on this principle;

– in the second, we add constraints to the reconstructed phase with supplementary equations that model the phase behavior with time and/or the perpendicular baseline. The system then becomes invertible without having to resort to SVD.

In the following, we focus on the second case, i.e. the NSBAS inversion technique (Doin *et al.* 2011). The new constraints are applied in parallel to the first set of equations (equation [5.6]), meaning that we must solve additional variables, such as v, the acceleration a or a coefficient proportional to the perpendicular baseline $B^{\perp}$, while still solving for the phase increments $\delta\varphi^{inv}$ (López-Quiroz *et al.* 2009):

$$\begin{cases} \omega_{ij}\left(\sum_{n=i}^{n=j-1} \delta\varphi_n^{inv}\right) = \omega_{ij}\varphi_{ij} \\ W\Omega_1 c = 0 \\ W\Omega_l\left(\sum_{n=1}^{n=l-1} \delta\varphi_n^{inv} - v(t_l - t_1) - a(t_l - t_1)^2 - \alpha(B_l^{\perp} - B_1^{\perp}) - c\right) = 0 \end{cases} \quad [5.8]$$

where t_l and $B_l^{\perp}$ are the time and perpendicular baseline of acquisition l, respectively; l is greater than 2 in the last set of equations; and c is a constant. The weight W of the new constraint lines must be low enough for them not to affect the phase time-series reconstruction φ_l^{inv} in the case of $G^T G$ being inversible. The weight ω_{ij} should depend on interferogram coherence: we can, for example, set it to be inversely proportional to the temporal baseline or, more empirically, inversely dependent on a measure of the local interferogram coherence. However, the weight Ω_l should be set to be inversely correlated with the amplitude of the APS of each acquisition l. Therefore, the coefficients v, a, α and c are constrained mostly from acquisition dates with limited residual atmospheric contributions.

The deformation term may not be well represented by a linear term plus an acceleration. We can then decide to choose another parsimonious dictionary of functions, with a step (for earthquake), an exponential (for relaxation), trigonometric functions (for seasonal fluctuations), etc. Alternatively, for a very general inversion procedure, we can also decide to simultaneously invert for a smooth deformation contribution, with N additional variables, φ_l^{def}, with l being the image acquisition (Doin *et al.* 2011):

$$\begin{cases} \omega_{ij}\left(\sum_{n=i}^{n=j-1}\delta\varphi_n^{inv}\right) = \omega_{ij}\varphi_{ij} \\ W\Omega_1\varphi_1^{def} = 0 \\ W\Omega_l\left(\sum_{n=1}^{n=l-1}\delta\varphi_n^{inv} - \varphi_l^{def} - \alpha(B_l^{\perp} - B_1^{\perp})\right) = 0 \\ W\frac{\partial^2\varphi^{def}}{\partial t^2} = 0 \end{cases} \qquad [5.9]$$

The constraint $W\frac{\partial^2\varphi^{def}}{\partial t^2} = 0$ can be replaced by introducing a covariance between modeled variables φ_l^{def} through non-zero diagonal terms in C_m (Jolivet *et al.* 2013).

Finally, for present-day SAR missions with systematic and massive acquisition strategies (such as Sentinel-1, every 6 or 12 days), resolving the above system of equations by progressive assimilation, as done by Dalaison and Jolivet (2020), strongly reduces the computational burden, while also making it possible to evaluate uncertainties.

Note that, in the NSBAS approach, in contrast to the traditional SBAS method (Lanari *et al.* 2004), the priors on φ_l^{def} are only used for regularization to fill the missing links, but the inverted phase time series φ_l^{inv} includes both the residual APS (after the correction of the stratified tropospheric effect) and the deformation, and it is not filtered in time. It can therefore be considered as a measurement with a quantified measurement error and it is obtained without an *a priori* constraint on deformation. This then makes it possible to proceed to geophysically driven source separation.

5.4.4. *Source separation, atmosphere versus deformation*

Source separation on the total phase delay time series (hereafter referred to as the "data cube") should in principle allow us to (1) extract the deformation associated with different sources and discard the atmospheric phase screens and (2) quantify errors associated with this extraction. The source separation can be blind or aided by supplementary geophysical information. Blind methods allow us to explore datasets and prove the existence of a given geophysical signal, while geophysically aided extraction is more efficient for quantifying a giving process and giving an uncertainty in the result. In any case, adding a term proportional to the perpendicular baseline separates the residual contribution from the DEM error and helps in the retrieval of other geophysical signals.

The source separation problem can be solved with different approaches depending on the scientific goal. The "classical" blind method is to isolate APS by applying successive combinations of a high-pass filter to time followed by a low-pass filter to space for individual acquisitions (Berardino *et al.* 2002). The smoothed time series after the elimination of APS is the standard product given by many software packages, for example the P-SBAS approach (Lanari *et al.* 2020). However, possible leakages between APS and deformation, especially at large scales, cannot be easily evaluated using this approach. Furthermore, low-pass filtering a time series with uneven sampling, or sampled at a frequency lower than the oscillations of the signal, can produce strong aliasing (Doin *et al.* 2009). The phase delay data cube can also be explored using principal component analysis (Chaussard *et al.* 2014) or independent component analysis (ICA) (Ebmeier 2016; Maubant *et al.* 2020). The main contributions to the phase delay are then identified without introducing any external knowledge. For example, Maubant *et al.* (2020) could, using ICA, identify both the stratified atmospheric delay and the tenuous transient deformation signal, of a duration of a few months, associated with a slow slip event along the Pacific subduction zone in Mexico. In this case, the stratified APS showed a high-frequency component in time superimposed on a seasonal fluctuation, and, as a result, broke the basic assumption classically made for isolating APS. The validity of ICA source separation could then be checked using independent constraints from GNSS tropospheric and deformation measurements, as well as from the ERA-Interim atmospheric model from the European Centre for Median-Range Weather Forecasting (ECMWF).

Another type of approach involves seperating the sources using a regression analysis, assuming that deformation can be expressed by temporal functions (linear, seasonal, step, etc.). The advantages of this simple method are its robustness in the case of low SNR, the quantification of associated uncertainties and the possibility of weighting each acquisition based on the amplitude of its APS content. Note that retrieving only the LOS velocity field at the mm/yr accuracy is already a challenge at large scales. This is especially true when the expected deformation partly correlates with elevation, as is the case for some major faults (e.g. the Altyn Tagh fault (Elliott *et al.* 2008; Daout *et al.* 2018)) or volcanoes (Beauducel *et al.* 2000). Therefore, the uncertainty in the retrieved velocity field in mountainous areas cannot be discussed without a thorough error analysis of the stratified atmospheric correction. An in-depth presentation of how to derive the uncertainties in stratified atmospheric correction was proposed in Daout *et al.* (2018). Most regression analyses are performed on a pixel-by-pixel basis. However, Jolivet and Simons (2018) introduced a method that includes both spatial and temporal functions as well as a spatially correlated covariance matrix on the data.

Finally, source decomposition can be aided by modeling or a prior knowledge of the shape of the deformation in time, $T(t)$, or in space, $S(x, y)$. For example, assuming that the deformation associated with the load variation of the Siling Co lake is in the form $T(t) * S(x, y)$, a shape $S(x, y)$ associated with known water level fluctuations can be retrieved or, inversely, the temporal behavior $T(t)$ associated with the model shape of load-induced deflection can be estimated (Doin *et al.* 2015). The advantages again are the possibility of taking into account the noise properties of each acquisition and giving an error bar on $T(t)$ or $S(x, y)$. In addition, including modeling predictions for the east, north and up components, $M_i(x, y)$, for different sources, i, also makes it possible to merge the time series obtained with different viewing angles using the unit LOS projection vector $u_{LOS}(x, y)$. This can be done by solving:

$$\phi(x, y, t) = \sum_i M_i(x, y) u_{LOS}(x, y) * T_i(t), \qquad [5.10]$$

with possible regularization applied on $T_i(t)$ (Grandin *et al.* 2010).

To conclude, the strength of the SBAS method is its ability to both adapt to the natural environment and provide a "data cube", with the unwrapping error automatically being detected and corrected, including both deformation and residual atmospheric contributions and thus free from priors for the temporal or spatial behavior of deformation. Source separation adapted to the geophysical signal of interest can then be very efficient and accompanied by uncertainty estimates. In the following, we describe the PSDS method, which maximizes the retrieval of ground deformation from selected pixels or groups of neighboring pixels.

5.5. The PSDS technique

5.5.1. *The PS algorithm*

Let us now consider the phase difference between two neighboring PSs (e.g. targets p and q), which can be written as (Ferretti *et al.* 2000; Ho Tong Minh *et al.* 2020):

$$\Delta\varphi_{pq}^n = W\left\{k_n^z \Delta\varepsilon_{pq}^z + k_n^t \Delta v_{pq} + \Delta w_{pq}^n\right\} \qquad [5.11]$$

where W is the wrapping operator; $\Delta\varepsilon_{pq}^z$ is the relative elevation error between the targets p and q; Δv_{pq} is the relative constant velocity; and $\Delta w_{pq}^n = \mu_{pq,NL}^n + \varphi_{pq,aps}^n + \varphi_{pq,noise}^n$ is the difference between the observed and modeled phase for the points p and q in the n-th interferogram.

The residue phase variance $\sigma_{\Delta w}^2$ is the sum of three independent contributions, σ_μ^2, σ_{aps}^2 and σ_{noise}^2. Since a pair of neighboring PSs is not too far apart, low σ_{aps}^2 (i.e. less than $0.1\ rad^2$) can be expected for short distance (1 km) (Ferretti *et al.* 2000). In addition, the nonlinear motion of neighboring targets is highly correlated, resulting in a low σ_μ^2. Finally, by definition, a low noise level is expected at the PS pixels. Thus, $\sigma_{\Delta w}^2$ remains low. In this case, it is possible to estimate $\Delta\varepsilon^z$ and Δv by optimizing a figure of merit with a high degree of accuracy. In fact, $\Delta\varepsilon^z$ and Δv can be jointly determined as the values that maximize the complex ensemble coherence in time $\hat{\gamma}_{pq}$, written as (Ferretti *et al.* 2000):

$$\hat{\gamma}_{pq} = \frac{1}{N-1} \sum_{n=1}^{N-1} exp\left(i\Delta w_{pq}^{n}\right) \quad [5.12]$$

under the condition:

$$\Delta w_{pq}^{n} \leq \pi. \quad [5.13]$$

The absolute value of the coherence is in the range of 0 to 1. A value of 1 means a perfect match between the modeled phase and the observed phase. The ensemble coherence of PS pairs lower than a given threshold (e.g. 0.75) is usually discarded to maintain the high degree of accuracy.

In this way, Δw_{pq}^{n}, $\Delta \varepsilon_{pq}^{z}$ and Δv_{pq} can be calculated for every PS pair. Note that large distances (i.e. greater than 1 km) for a PS pair should be discarded to guarantee the conditions of equation [5.13]. Then, the variables ε^{z} and v can be calculated for PS targets by integrating their relative values, $\Delta \varepsilon_{pq}^{z}$ and Δv_{pq}. Most often, the computation is performed with respect to a reference pixel that is assumed to have known elevation and motion. By integrating the relative residual unwrapped phase difference Δw_{pq}^{n} from equation [5.11] over the whole network of PS pixels, we can obtain the unwrapped residual phase delays, w_{p}^{n}, for each acquisition n and each PS p. Once the unwrapped phase of w_{p}^{n} is available, the atmospheric component can be separated by applying spatial–temporal filtering. After the atmospheric delays are estimated at the PS positions, a spatial interpolation can be applied to obtain the APS at the full resolution of differential interferograms.

5.5.2. *PS selection*

The main limitation of the PS approach outlined in section 5.5.1 lies in the sparsity of the grid for the selected PS. To overcome this limitation, the interpolated APSs are subtracted from differential interferograms, and additional PS candidates can be identified by applying a phase stability criterion. The computations described in section 5.5.1 are repeated for the new and denser PS network.

However, this PS technique has low performance in areas where the initial PS identification (i.e. based on the amplitude stability) does not suffice to cover the whole area of interest. This can occur for non-urban areas, especially for datasets suffering from amplitude calibration problems. The first solution to this problem was proposed by Hooper *et al.* (2004), who identified persistent scatterers based directly on the phase stability criterion rather than the amplitude stability. When compared to amplitude-based algorithms, this method has been shown to yield a denser point grid on rock areas. Nowadays, group techniques analyzing the time-series phase change of stable targets are referred to as persistent scatterer interferometry (PSI).

The gain performance of PSI is based on the selection of stable targets and their ability to remove the APS artifacts. The PSI technique offers a highly precise phase estimate despite temporal and geometrical decorrelations. Consequently, many related techniques have been investigated in the literature, for example interferometric point target analysis (Werner *et al.* 2003), coherent target monitoring (der Kooij 2003), the Stanford method for persistent scatterers (StaMPS) (Hooper *et al.* 2004), the spatial temporal unwrapping network (STUN) (Kampes 2006), persistent scatterer pairs (Costantini *et al.* 2008) and quasi-persistent scatterers (Perissin and Wang 2012). These techniques share similar properties in seeking modified approaches in order to improve the PS technique. The StaMPS algorithm makes no assumption of a prior deformation model and includes a three-dimensional phase unwrapping approach (Hooper and Zebker 2007; Hooper 2008). The STUN algorithm includes an integer least-squares estimator to solve equation [5.11] that allows us to drop the assumption made in equation [5.13]. Many methods have been proposed to add more reliable PS candidates (e.g. spectral phase diversity (Werner *et al.* 2003), a phase stability criterion (Hooper *et al.* 2004), maximum likelihood (Shanker and Zebker 2007)), resulting in a denser network and better estimation of phase measurements. The linear deformation model can be extended to consider thermal expansion (Monserrat *et al.* 2011) or to take into account seasonal motion phenomena (Colesanti *et al.* 2003). The signal model given in equation [5.1] makes an assumption of a dominant scatterer within the resolution cell. In practice, there could be more than one dominant point, as often is the case for layover due to buildings and towers. SAR tomography, which enables us to resolve layovers (Gini *et al.* 2002), can be exploited to relax this assumption. Siddique *et al.* (2016) demonstrated that the SAR tomography approach can improve deformation areas affected by layover.

The PSI method requires stable targets over a long time series. This results in a loss of information in some cases where stable points can be decorrelated in the majority of the low-coherence interferograms (Hooper *et al.* 2011). In such cases, the PS assumption can be relaxed by considering targets that are stable only in a subset of the dataset, for example temporary PSs (Ferrero *et al.* 2004; Hooper *et al.* 2011) and quasi-PSs (Perissin and Wang 2012). Finally, Ferretti *et al.* (2011) proposed an ML framework to consider both stable PS and slow decorrelation DS targets within the same PSI process. Hence, the PSDS algorithm is an extension of the PSI technique.

5.5.3. *DS selection*

DS targets are difficult to handle due to their low temporal coherence (Zebker and Villasenor 1992), which results in a low SNR. To enhance their SNRs, a number of pixels sharing the same statistical behavior can be exploited to make them emulate quasi-PSs. Groups of spatially close pixels that behave in a similar way (i.e. the same statistical behavior in the amplitude time series) must be identified. In practice, they can be selected by applying a two-sample Kolmogorov–Smirnov test (assuming a level of significance) to the amplitude time series of the considered pixel and all its neighbors. Pixels with a similar cumulative probability distribution are considered as "brothers". Consequently, a brotherhood can be created, which results in a statistically homogeneous pixel (SHP) family. We can then compute the sample coherence matrix ($\hat{\Gamma}$) by taking advantage of the SHP family.

Decorrelating targets poses a challenge to time-series InSAR techniques. To overcome this challenge, ML estimation has been proposed to retrieve the linked phase time series (Rocca 2007; Guarnieri and Tebaldini 2008). As a result, the now commonly used PSDS technique outperforms PSI, particularly in non-urban areas (Ferretti *et al.* 2011). Consequently, this approach has become the main technique for surface deformation applications (Goel and Adam 2014; Ho Tong Minh *et al.* 2015, 2019; Cao *et al.* 2016; Cohen-Waeber *et al.* 2018).

Many approaches have been developed to improve the selection of the SHP family: the Anderson–Darling test (Goel and Adam 2014), fast statistically homogeneous pixel selection (Jiang *et al.* 2015), mean amplitude difference (Spaans and Hooper 2016), t-tests (Shamshiri *et al.* 2018) and similar time-series interferometric phase (Narayan *et al.* 2018) have been proposed to increase the number of DS points. These techniques aim to mitigate the bias in the sample coherence. A DS involves an assumption that there are independent small scatterers with a uniform scattering mechanism. This assumption can be relaxed by accounting for two or more scattering mechanisms (Cao *et al.* 2016; Engelbrecht and Inggs 2016). Engelbrecht and Inggs (2016) showed the enhancement of deformation measurements in agricultural areas when exploiting the multiple scattering mechanism approach with L-band ALOS PalSAR.

5.5.4. *Phase linking*

In SB approaches, signal decorrelations lead to the definition of interferograms with short spatial and temporal baselines. With the ML estimation, all the $N(N-1)/2$ interferograms available from N SLC images are jointly exploited in a complex coherence matrix. In fact, thanks to the redundancy of the phase differences between all given pairs, the phase can be improved by exploiting the consistency of phase closure of triplet information (i.e. the phases of a loop of three time instants I_1-I_2, I_2-I_3, I_3-I_1, which would be zero) (De Zan *et al.* 2015). The coherence matrix ($\hat{\Gamma}$) can fully characterize the target statistics and is used to invert for linked phases ($\hat{\boldsymbol{\lambda}}$) as follows (Ferretti *et al.* 2011; Ho Tong Minh *et al.* 2020):

$$\hat{\boldsymbol{\lambda}} = argmax_{\lambda} \left\{ \Lambda^H \left(|\hat{\Gamma}|^{-1} \circ \hat{\Gamma} \right) \Lambda \right\} \qquad [5.14]$$

where $\Lambda = exp(j\boldsymbol{\lambda})$; ^{H} is the Hermitian conjugation; $\circ$ is the Hadamard entry-wise product; $\hat{\boldsymbol{\lambda}} = [\vartheta^1, \vartheta^2, ..., \vartheta^N]$; and ϑ^n is the estimated phase. It is noteworthy that the coherence matrix $\hat{\Gamma}$ has an Hermitian symmetry that can be calculated using equation [4.12] for the SHP families. The estimated phase $\hat{\boldsymbol{\lambda}}$ quality can be assessed using PTA coherence, which can be defined as (Ferretti *et al.* 2011):

$$\gamma_{PTA} = \frac{2}{N(N-1)} Re \sum_{n=1}^{N} \sum_{k=n+1}^{N} exp\left(i(\phi^{nk} - \vartheta^n + \vartheta^k)\right) \qquad [5.15]$$

where ϕ^{nk} is the phase term corresponding to the off-diagonal elements of the coherence matrix $\hat{\Gamma}$. If the γ_{PTA} coherence is above a certain threshold (e.g. 0.5), a DS point with the linked phase values $\hat{\boldsymbol{\lambda}}$ will be jointly processed with the PS using the PSI technique, as described in section 5.5.1.

The linked phase is the key step in the PSDS technique, resulting in many studies focused on improving its performance. Cao *et al.* (2015) proposed equal-weighted and coherence-weighted approaches for phase estimation. Samiei-Esfahany *et al.* (2016) extended the PSI STUN algorithm (Kampes 2006; based on the least-squares principle for integer numbers) for slowly decorrelating targets. The component extraction and selection SAR (CAESAR) technique (Fornaro *et al.* 2015) can be exploited to decompose different scattering components from the coherence matrix (Cao *et al.* 2016). Ansari *et al.* (2017, 2018) proposed a sequential estimator for DS interferometry. Finally, Ho Tong Minh and Ngo (2021) introduced a compressed SAR (ComSAR) interferometric algorithm in the PSDS framework, particularly useful for matching in the analysis of big InSAR data.

5.5.5. *Tools supporting PS and PSDS processing*

Today, a variety of software packages support PS processing, but there are only two PSDS tools available. The first package was developed by TRE-Altamira, who named it SqueeSAR (Ferretti *et al.* 2011). The TomoSAR platform created by D. Ho Tong Minh was originally developed at Politecnico di Milano, but subsequent development has taken place at the CESBIO and INRAE (Ho Tong Minh and Ngo 2017). Table 5.1 reports some well-known tools and their capabilities.

Although a Matlab license is required, the StaMPS code is the only open source tool available (Hooper *et al.* 2004). StaMPS can be used for PS or small baseline analysis. It allows the combination of PSs and DSs with a hybrid processing chain where two techniques (PS and SBAS) are applied, and then SBAS results are inverted by an SVD to be combined with PSs.

Name	Year	Main code	Main creator	Note	Support
Gamma	1995	C	C. Werner	Commercial	PS, SBAS
TRE-Altamira	1999	n/a	A. Ferretti	Service demand	SBAS, PSDS
StaMPS	2004	Matlab	A. Hooper	Open source	PS, SBAS
SARSCAPE	n/a	C	n/a	Commercial	PS, SBAS
SARPROZ	2009	Matlab	D. Perissin	Collaboration	PS
TomoSAR	2014	C	D. Ho Tong Minh	Service collaboration	SBAS, PSDS
DePSI	2014	Matlab	F. Van Leijen	Collaboration	PS

Table 5.1. *Overview of PS and PSDS time-series software packages*

Author contributions

Dinh Ho Tong Minh and Ramon Hanssen contributed to the sections on permanent/persistent scatterer and distributed scatterer InSAR techniques. Marie-Pierre Doin and Erwan Pathier contributed to the sections on small baseline methods. All authors read and approved the final manuscript.

5.6. Acknowledgments

We thank Fabio Rocca, Ekbal Hussain, Romain Jolivet and Raphaël Grandin for their constructive comments that improved the quality of this chapter.

5.7. References

Ansari, H., De Zan, F., Bamler, R. (2017). Sequential estimator: Toward efficient InSAR time series analysis. *IEEE Transactions on Geoscience and Remote Sensing*, 55(10), 5637–5652.

Ansari, H., De Zan, F., Bamler, R. (2018). Efficient phase estimation for interferogram stacks. *IEEE Transactions on Geoscience and Remote Sensing*, 56(7), 4109–4125.

Beauducel, F., Briole, P., Froger, J.-L. (2000). Volcano-wide fringes in ERS synthetic aperture radar interferograms of Etna (1992–1998): Deformation or tropospheric effect? *Journal of Geophysical Research: Solid Earth*, 105(B7), 16391–16402.

Berardino, P., Fornaro, G., Lanari, R., Sansosti, E. (2002). A new algorithm for surface deformation monitoring based on small baseline differential SAR interferograms. *IEEE Transactions on Geoscience and Remote Sensing*, 40(11), 2375–2383.

Biggs, J., Wright, T., Lu, Z., Parsons, B. (2007). Multi-interferogram method for measuring interseismic deformation: Denali Fault, Alaska. *Geophysical Journal International*, 170(3), 1165–1179.

Cao, N., Lee, H., Jung, H.C. (2015). Mathematical framework for phase-triangulation algorithms in distributed-scatterer interferometry. *IEEE Geoscience and Remote Sensing Letters*, 12(9), 1838–1842.

Cao, N., Lee, H., Jung, H.C. (2016). A phase-decomposition-based PSInSAR processing method. *IEEE Transactions on Geoscience and Remote Sensing*, 54(2), 1074–1090.

Casu, F., Elefante, S., Imperatore, P., Zinno, I., Manunta, M., De Luca, C., Lanari, R. (2014). SBAS-DInSAR parallel processing for deformation time-series computation. *IEEE Journal of Selected Topics in Applied Earth Observations and Remote Sensing*, 7(8), 3285–3296.

Chaussard, E., Bürgmann, R., Shirzaei, M., Fielding, E.J., Baker, B. (2014). Predictability of hydraulic head changes and characterization of aquifer-system and fault properties from InSAR-derived ground deformation. *Journal of Geophysical Research: Solid Earth*, 119(8), 6572–6590.

Chen, C.W. and Zebker, H.A. (2002). Phase unwrapping for large SAR interferograms: Statistical segmentation and generalized network models. *IEEE Transactions on Geoscience and Remote Sensing*, 40(8), 1709–1719.

Cohen-Waeber, J., Bürgmann, R., Chaussard, E., Giannico, C., Ferretti, A. (2018). Spatiotemporal patterns of precipitation-modulated landslide deformation from independent component analysis of InSAR time series. *Geophysical Research Letters*, 45(4), 1878–1887.

Colesanti, C., Ferretti, A., Novali, F., Prati, C., Rocca, F. (2003). SAR monitoring of progressive and seasonal ground deformation using the permanent scatterers technique. *IEEE Transactions on Geoscience and Remote Sensing*, 41(7), 1685–1701.

Costantini, M., Falco, S., Malvarosa, F., Minati, F. (2008). A new method for identification and analysis of persistent scatterers in series of SAR images. *IEEE International Geoscience and Remote Sensing Symposium*, 7–11 July.

Dalaison, M. and Jolivet, R. (2020). A Kalman filter time series analysis method for InSAR. *Journal of Geophysical Research: Solid Earth*, 125(7).

Daout, S., Doin, M.-P., Peltzer, G., Socquet, A., Lasserre, C. (2017). Large-scale InSAR monitoring of permafrost freeze-thaw cycles on the Tibetan Plateau. *Geophysical Research Letters*, 44(2), 901–909.

Daout, S., Doin, M.-P., Peltzer, G., Lasserre, C., Socquet, A., Volat, M., Sudhaus, H. (2018). Strain partitioning and present-day fault kinematics in NW Tibet from ENVISAT SAR interferometry. *Journal of Geophysical Research: Solid Earth*, 123(3), 2462–2483.

Davidson, G. and Bamler, R. (1999). Multiresolution phase unwrapping for SAR interferometry. *IEEE Transactions on Geoscience and Remote Sensing*, 37(1), 163–174.

De Zan, F., Zonno, M., Lopez-Dekker, P. (2015). Phase inconsistencies and multiple scattering in SAR interferometry. *IEEE Transactions on Geoscience and Remote Sensing*, 53(12), 6608–6616.

Doin, M.-P., Lasserre, C., Peltzer, G., Cavalié, O., Doubre, C. (2009). Corrections of stratified tropospheric delays in SAR interferometry: Validation with global atmospheric models. *Journal of Applied Geophysics*, 69(1), 35–50.

Doin, M.-P., Lodge, F., Guillaso, S., Jolivet, R., Lasserre, C., Ducret, G., Grandin, R., Pathier, E., Pinel, V. (2011). Presentation of the small baseline NSBAS processing chain on a case example: The Etna deformation monitoring from 2003 to 2010 using Envisat data. *Fringe Workshop*, Frascati.

Doin, M.-P., Twardzik, C., Ducret, G., Lasserre, C., Guillaso, S., Jianbao, S. (2015). InSAR measurement of the deformation around siling co lake: Inferences on the lower crust viscosity in central Tibet. *Journal of Geophysical Research: Solid Earth*, 2014JB011768.

Ducret, G., Doin, M.-P., Grandin, R., Lasserre, C., Guillaso, S. (2014). Dem corrections before unwrapping in a small baseline strategy for InSAR time series analysis. *Geoscience and Remote Sensing Letters*, 11(3), 696–700.

Ebmeier, S.K. (2016). Application of independent component analysis to multitemporal InSAR data with volcanic case studies. *Journal of Geophysical Research: Solid Earth*, 121(12), 8970–8986.

Elliott, J.R., Biggs, J., Parsons, B., Wright, T.J. (2008). InSAR slip rate determination on the Altyn Tagh Fault, northern Tibet, in the presence of topographically correlated atmospheric delays. *Geophysical Research Letters*, 35(12), L12309.

Emardson, T.R., Simons, M., Webb, F.H. (2003). Neutral atmospheric delay in interferometric synthetic aperture radar applications: Statistical description and mitigation. *Journal of Geophysical Research*, 108(B5), 2231.

Engelbrecht, J. and Inggs, M.R. (2016). Coherence optimization and its limitations for deformation monitoring in dynamic agricultural environments. *IEEE Journal of Selected Topics in Applied Earth Observations and Remote Sensing*, 9(12), 5647–5654.

Ferrero, A., Novali, F., Prati, C., Rocca, F. (2004). Advances in permanent scatterers analysis. Semi and temporary PS. *European Conference on Synthetic Aperture Radar*.

Ferretti, A., Prati, C., Rocca, F. (2000). Nonlinear subsidence rate estimation using permanent scatterers in differential SAR interferometry. *IEEE Transactions on Geoscience and Remote Sensing*, 38(5), 2202–2212.

Ferretti, A., Prati, C., Rocca, F. (2001). Permanent scatterers in SAR interferometry. *IEEE Transactions on Geoscience and Remote Sensing*, 39(1), 8–20.

Ferretti, A., Novali, F., De Zan, F., Prati, C., Rocca, F. (2008). Moving from PS to slowly decorrelating targets: A prospective view. *European Conference on Synthetic Aperture Radar*, 2–5 June.

Ferretti, A., Fumagalli, A., Novali, F., Prati, C., Rocca, F., Rucci, A. (2011). A new algorithm for processing interferometric data-stacks: SqueeSAR. *IEEE Transactions on Geoscience and Remote Sensing*, 49(9), 3460–3470.

Fialko, Y. (2006). Interseismic strain accumulation and the earthquake potential on the southern San Andreas fault system. *Nature*, 441, 968–971.

Fornaro, G., Monti Guarnieri, A., Pauciullo, A., De Zan, F. (2006). Maximum liklehood multi-baseline SAR interferometry. *IEE Proceedings – Radar, Sonar and Navigation*, 153(3), 279–288.

Fornaro, G., Pauciullo, A., Serafino, F. (2009). Deformation monitoring over large areas with multipass differential SAR interferometry: A new approach based on the use of spatial differences. *International Journal of Remote Sensing*, 30(6), 1455–1478.

Fornaro, G., Verde, S., Reale, D., Pauciullo, A. (2015). CAESAR: An approach based on covariance matrix decomposition to improve multibaseline–multitemporal interferometric SAR processing. *IEEE Transactions on Geoscience and Remote Sensing*, 53(4), 2050–2065.

Gatelli, F., Guamieri, A.M., Parizzi, F., Pasquali, P., Prati, C., Rocca, F. (1994). The wavenumber shift in SAR interferometry. *IEEE Transactions on Geoscience and Remote Sensing*, 32(4), 855–865.

Ghiglia, D.C. and Pritt, M.D. (1998). *Two-Dimensional Phase Unwrapping: Theory, Algorithms, and Software*. John Wiley & Sons, Inc., New York.

Gini, F., Lombardini, F., Montanari, M. (2002). Layover solution in multibaseline SAR interferometry. *IEEE Transactions on Aerospace and Electronic Systems*, 38(4), 1344–1356.

Goel, K. and Adam, N. (2014). A distributed scatterer interferometry approach for precision monitoring of known surface deformation phenomena. *IEEE Transactions on Geoscience and Remote Sensing*, 52(9), 5454–5468.

Goldstein, R.M. and Werner, C.L. (1998). Radar interferogram filtering for geophysical applications. *Geophysical Research Letters*, 25(21), 2517–2520.

Grandin, R., Socquet, A., Doin, M.-P., Jacques, E., de Chabalier, J.-B., King, G.C.P. (2010). Transient rift opening in response to multiple dike injections in the Manda Hararo rift (Afar, Ethiopia) imaged by time-dependent elastic inversion of interferometric synthetic aperture radar data. *Journal of Geophysical Research: Solid Earth*, 115(B9), B09403.

Grandin, R., Doin, M.-P., Bollinger, L., Pinel-Puysségur, B., Ducret, G., Jolivet, R., Sapkota, S.N. (2012). Long-term growth of the Himalaya inferred from interseismic InSAR measurement. *Geology*, 40(12), 1059–1062.

Guarnieri, A.M. and Tebaldini, S. (2008). On the exploitation of target statistics for SAR interferometry applications. *IEEE Transactions on Geoscience and Remote Sensing*, 46(11), 3436–3443.

Hanssen, R.F. (2001). *Radar Interferometry: Data Interpretation and Error Analysis*. Kluwer Academic Publishers, Dordrecht.

Ho Tong Minh, D. and Ngo, Y.-N. (2017). Tomosar platform supports for Sentinel-1 tops persistent scatterers interferometry. *IEEE International Geoscience and Remote Sensing Symposium (IGARSS)*, 1680–1683.

Ho Tong Minh, D. and Ngo, Y.-N. (2021). ComSAR: A new algorithm for processing big data SAR interferometry. *IEEE International Geoscience and Remote Sensing Symposium (IGARSS)*, 1–4.

Ho Tong Minh, D., Van Trung, L., Toan, T.L. (2015). Mapping ground subsidence phenomena in Ho Chi Minh City through the radar interferometry technique using ALOS PALSAR data. *Remote Sensing*, 7(7), 8543–8562.

Ho Tong Minh, D., Tran, Q.C., Pham, Q.N., Dang, T.T., Nguyen, D.A., El-Moussawi, I., Le Toan, T. (2019). Measuring ground subsidence in Ha Noi through the radar interferometry technique using TerraSAR-X and Cosmos SkyMed data. *IEEE Journal of Selected Topics in Applied Earth Observations and Remote Sensing*, 12(10), 3874–3884.

Ho Tong Minh, D., Hanssen, R., Rocca, F. (2020). Radar interferometry: 20 years of development in time series techniques and future perspectives. *Remote Sensing*, 12(9), 1364.

Hooper, A. (2008). A multi-temporal InSAR method incorporating both persistent scatterer and small baseline approaches. *Geophysical Research Letters*, 35(L16302), 1–5.

Hooper, A. and Zebker, H.A. (2007). Phase unwrapping in three dimensions with application to InSAR time series. *JOSA A*, 24(9), 2737–2747.

Hooper, A., Zebker, H.A., Segall, P., Kampes, B. (2004). A new method for measuring deformation on volcanoes and other natural terrains using InSAR persistent scatterers. *Geophysical Research Letters*, 31(23), L23611.

Hooper, A., Ofeigsson, B., Sigmundsson, F. (2011). Increased capture of magma in the crust promoted by ice-cap retreat in Iceland. *Nature Geoscience*, 4, 783–786.

Jiang, M., Ding, X., Hanssen, R.F., Malhotra, R., Chang, L. (2015). Fast statistically homogeneous pixel selection for covariance matrix estimation for multitemporal InSAR. *IEEE Transactions on Geoscience and Remote Sensing*, 53(3), 1213–1224.

Jolivet, R. and Simons, M. (2018). A multipixel time series analysis method accounting for ground motion, atmospheric noise, and orbital errors. *Geophysical Research Letters*, 45(4), 1814–1824.

Jolivet, R., Grandin, R., Lasserre, C., Doin, M.-P., Peltzer, G. (2011). Systematic InSAR tropospheric phase delay corrections from global meteorological reanalysis data. *Geophysical Research Letters*, 38(17), L17311.

Jolivet, R., Lasserre, C., Doin, M.-P., Peltzer, G., Avouac, J.-P., Sun, J., Dailu, R. (2013). Spatio-temporal evolution of aseismic slip along the Haiyuan Fault, China: Implications for fault frictional properties. *Earth and Planetary Science Letters*, 377–378, 23–33.

Kampes, B. (2006). *Radar Interferometry: Persistent Scatterer Technique*. Springer Publishing Company, New York.

der Kooij, V. (2003). Coherent target analysis. *Fringe Workshop*, Frascati.

Lanari, R., Mora, O., Manunta, M., Mallorqui, J.J., Berardino, P., Sansosti, E. (2004). A small-baseline approach for investigating deformations on full-resolution differential SAR interferograms. *IEEE Transactions on Geoscience and Remote Sensing*, 42(7), 1377–1386.

Lanari, R., Bonano, M., Casu, F., Luca, C.D., Manunta, M., Manzo, M., Onorato, G., Zinno, I. (2020). Automatic generation of Sentinel-1 continental scale DInSAR deformation time series through an extended P-SBAS processing pipeline in a cloud computing environment. *Remote Sensing*, 12(18), 2961.

López-Quiroz, P., Doin, M.-P., Tupin, F., Briole, P., Nicolas, J.-M. (2009). Time series analysis of Mexico City subsidence constrained by radar interferometry. *Journal of Applied Geophysics*, 69(1), 1–15.

Manunta, M., De Luca, C., Zinno, I., Casu, F., Manzo, M., Bonano, M., Fusco, A., Pepe, A., Onorato, G., Berardino, P., De Martino, P., Lanari, R. (2019). The parallel SBAS approach for Sentinel-1 interferometric wide swath deformation time-series generation: Algorithm description and products quality assessment. *IEEE Transactions on Geoscience and Remote Sensing*, 57(9), 6259–6281.

Maubant, L., Pathier, E., Daout, S., Radiguet, M., Doin, M.-P., Kazachkina, E., Kostoglodov, V., Cotte, N., Walpersdorf, A. (2020). Independent component analysis and parametric approach for source separation in InSAR time series at regional scale: Application to the 2017–2018 slow slip event in Guerrero (Mexico). *Journal of Geophysical Research: Solid Earth*, 125(3), e2019JB018187.

Monserrat, O., Crosetto, M., Cuevas, M., Crippa, B. (2011). The thermal expansion component of persistent scatterer interferometry observations. *IEEE Geoscience and Remote Sensing Letters*, 8(5), 864–868.

Mora, O., Mallorqui, J.J., Broquetas, A. (2003). Linear and nonlinear terrain deformation maps from a reduced set of interferometric SAR images. *IEEE Transactions on Geoscience and Remote Sensing*, 41(10), 2243–2253.

Morishita, Y., Lazecky, M., Wright, T.J., Weiss, J.R., Elliott, J.R., Hooper, A. (2020). LiCSBAS: An open-source InSAR time series analysis package integrated with the LiCSAR automated Sentinel-1 InSAR processor. *Remote Sensing*, 12(3), 424.

Narayan, A.B., Tiwari, A., Dwivedi, R., Dikshit, O. (2018). A novel measure for categorization and optimal phase history retrieval of distributed scatterers for InSAR applications. *IEEE Transactions on Geoscience and Remote Sensing*, 56(10), 5843–5849.

Perissin, D. and Wang, T. (2012). Repeat-pass SAR interferometry with partially coherent targets. *IEEE Transactions on Geoscience and Remote Sensing*, 50(1), 271–280.

Pinel-Puyssegur, B., Michel, R., Avouac, J.-P. (2012). Multi-link InSAR time series: Enhancement of a wrapped interferometric database. *IEEE Journal of Selected Topics in Applied Earth Observations and Remote Sensing*, 5(3), 784–794.

Rocca, F. (2007). Modeling interferogram stacks. *IEEE Transactions on Geoscience and Remote Sensing*, 45(10), 3289–3299.

Samiei-Esfahany, S., Martins, J.E., van Leijen, F., Hanssen, R.F. (2016). Phase estimation for distributed scatterers in InSAR stacks using integer least squares estimation. *IEEE Transactions on Geoscience and Remote Sensing*, 54(10), 5671–5687.

Schmidt, D.A. and Bürgmann, R. (2003). Time-dependent land uplift and subsidence in the Santa Clara valley, California, from a large interferometric synthetic aperture radar data set. *Journal of Geophysical Research*, 108(B9), 2416.

Shamshiri, R., Nahavandchi, H., Motagh, M., Hooper, A. (2018). Efficient ground surface displacement monitoring using Sentinel-1 data: Integrating distributed scatterers (DS) identified using two-sample t-test with persistent scatterers (PS). *Remote Sensing*, 10(5), 794.

Shanker, P. and Zebker, H.A. (2007). Persistent scatterer selection using maximum likelihood estimation. *Geophysical Research Letters*, 34(22), L22301.

Siddique, M.A., Wegmüller, U., Hajnsek, I., Frey, O. (2016). Single-look SAR tomography as an add-on to PSI for improved deformation analysis in urban areas. *IEEE Transactions on Geoscience and Remote Sensing*, 54(10), 6119–6137.

Spaans, K. and Hooper, A. (2016). InSAR processing for volcano monitoring and other near-real time applications. *Journal of Geophysical Research: Solid Earth*, 121(4), 2947–2960.

Usai, S., Del Gaudio, C., Borgstrom, S., Achilli, V. (1999). Monitoring terrain deformations at Phlegrean Fields with SAR interferometry. *Fringe 99 Workshop*, Liège.

Werner, C., Wegmuller, U., Strozzi, T., Wiesmann, A. (2003). Interferometric point target analysis for deformation mapping. *IEEE International Geoscience and Remote Sensing Symposium*, 7, 4362–4364

Wright, T.J., Parsons, B., Fielding, E. (2001). Measurement of interseismic strain accumulation across the North Anatolian Fault by satellite radar interferometry. *Geophysical Research Letters*, 28(10), 2117–2120.

Yu, C., Li, Z., Penna, N.T., Crippa, P. (2018). Generic atmospheric correction model for interferometric synthetic aperture radar observations. *Journal of Geophysical Research: Solid Earth*, 123(10), 9202–9222.

Yunjun, Z., Fattahi, H., Amelung, F. (2019). Small baseline InSAR time series analysis: Unwrapping error correction and noise reduction. *Computers and Geosciences*, 133, 104331.

Zebker, H.A. and Lu, Y. (1998). Phase unwrapping algorithms for radar interferometry: Residue-cut, least-squares, and synthesis algorithms. *Journal of the Optical Society of America A*, 15(3), 586–598.

Zebker, H.A. and Villasenor, J. (1992). Decorrelation in interferometric radar echoes. *IEEE Transactions on Geoscience and Remote Sensing*, 30(5), 950–959.

6

The Interferometric Phase: Unwrapping and Closure Phase

Béatrice PINEL-PUYSSÉGUR[1]**, Francesco DE ZAN**[2] **and Johann CHAMPENOIS**[1]

[1] *CEA, DAM, DIF, Arpajon, France*

[2] *Remote Sensing Technology Institute, German Aerospace Center (DLR), Oberpfaffenhofen, Germany*

6.1. Introduction

This chapter describes some properties of interferometric phase and the associated processing. Interferometric phase is naturally given as a wrapped number and needs to be unwrapped to obtain a valuable measurement. Besides its use in InSAR, phase unwrapping is a mathematical problem arising in several fields of optics, such as laser interferometry, sonar and nuclear magnetic resonance. In this chapter, only methods derived from the literature on InSAR are described. Historically, phase unwrapping algorithms were applied to single interferograms, as satellite acquisition rates were low. A fundamental property of the unwrapped phase in the spatial domain is that the closure of the phase around a spatial loop is null. This property, called irrotationality, is used by all unwrapping algorithms. The last 20 years have exhibited an unprecedented increase in the amount of data, which has led to the development of 3D phase unwrapping algorithms, the third dimension being time: instead of unwrapping each interferogram separately, the whole time series of interferograms is unwrapped at once. Naturally, the principles used in 2D algorithms were extended in 3D, especially irrotationality. However, in the last 10 years, several authors have noticed that the nullity of closure phase is

Surface Displacement Measurement from Remote Sensing Images,
coordinated by Olivier CAVALIÉ and Emmanuel TROUVÉ.
© ISTE Ltd 2022.

only approximate in the time dimension when the interferograms are multi-looked[1]. This chapter first describes the unwrapping problem then phase closure in the time domain.

As seen in the preceeding chapters, the unwrapping process is a crucial and difficult step that has given rise to an abundant literature. Section 6.2 gives an non-extensive overview of the phase unwrapping algorithms used in InSAR. Some common prerequisites are given in section 6.2.1, followed by the description of several algorithms. Despite the diversity of the algorithms, phase unwrapping errors still occur. Thus, various algorithms have been developed to automatically correct such errors and several are described in section 6.2.6, with some of them based on the analysis of closure phase.

Section 6.3 describes the closure phase and its mathematical properties and provides examples of physical phenomena that can explain it, such as volumetric effects associated with significant normal baselines or changes in the dielectric constant of the radar targets. The section shows also how the presence of non-zero closure phases is linked to a risk of a bias in the retrieved deformation parameters, depending on the algorithm and configuration.

6.2. Phase unwrapping algorithms and limitation of phase unwrapping errors

6.2.1. *The problem of phase unwrapping and some prerequisites*

6.2.1.1. *What is phase unwrapping?*

The phase of an interferogram is wrapped, i.e. its value is only known modulo 2π. The notations here follow those of Costantini (1998). Let $\Phi(i,j)$ be the unwrapped phase at indices (i,j), where i is the line index, $i = 0, 1, ..., N-1$ and j is the column index, $j = 0, 1, ..., M-1$. The rectangular definition domain of (i,j) is noted S. The wrapped interferogram phase is noted $\Psi(i,j)$ and lies in the interval $[-\pi, \pi[$[2]. The unwrapping process reconstructs $\Phi(i,j)$ from $\Psi(i,j)$. Precisely, for each $(i,j) \in S$, as $\Psi(i,j)$ is equal to $\Phi(i,j)$ modulo 2π, we have:

$$\Psi(i,j) = \Phi(i,j) + 2\pi k(i,j), k(i,j) \in \mathbb{Z} \qquad [6.1]$$

Thus, the unwrapping process can be equivalently seen as reconstructing the $k(i,j)$ from the $\Psi(i,j)$, for $(i,j) \in S$. In the absence of any noise and if the phase gradient absolute value between any two neighboring pixels is less than π, then phase

1. Strictly speaking, "closure phase" refers to a measurement whereas "phase closure" refers to the principle of phase triangularity, i.e. to the nullity of closure phase. However, some authors use the expression "phase closure" instead of "closure phase" (e.g. Benoit *et al.* 2020).

2. Depending on the author, the wrapped phase may lie in $[0, 2\pi[$ instead.

unwrapping is unambiguous. Problems arise if any of the two former conditions are not fulfilled, which is almost always the case in practice. In the following, some notions common to most algorithms as well as some notations are defined.

6.2.1.2. *Wrapped and unwrapped phase gradients*

In practice, most unwrapping algorithms first evaluate the wrapped phase gradients in the two directions (i.e. line and column) in order to estimate the unwrapped phase gradients. Once the phase gradients have been correctly estimated, the unwrapped phase is simply evaluated by integration. However, why begin with gradient estimation? In fact, the knowledge of Ψ required to reconstruct Φ without further assumptions is impossible: for each (i, j), any $\Phi(i, j)$ equal to $\Psi(i, j) - 2\pi k(i, j)$ with $k(i, j) \in \mathbb{Z}$ would be possible, leading to an infinity of solutions for each pixel. However, if Φ is continuous almost everywhere, as it should be intuitively, then its gradient should be "small" with respect to 2π almost everywhere. This essential and heuristic assumption is the basis of most unwrapping algorithms. However, it also creates unwrapping errors wherever it does not apply.

Still following the notations of Costantini (1998), let us define the subsets of the definition domain where the gradient is defined: $S_1 = (i, j), i \in 0, 1, ..., N-2, j \in 0, 1, ...M-1$ and $S_2 = (i, j), i \in 0, 1, ..., N-1, j \in 0, 1, ...M-2$. The discrete partial derivatives in the line and column directions of a function $F(i, j)$ defined on S are noted:

$$\Delta_1 F(i, j) = F(i+1, j) - F(i, j), (i, j) \in S_1 \quad [6.2]$$

$$\Delta_2 F(i, j) = F(i, j+1) - F(i, j), (i, j) \in S_2 \quad [6.3]$$

The gradient of Φ is a vector field:

$$\nabla\Phi(i, j) = \begin{pmatrix} \Delta_1\Phi(i, j) \\ \Delta_2\Phi(i, j) \end{pmatrix}, (i, j) \in S_0, \quad [6.4]$$

where $S_0 = S_1 \cap S_2$. Figure 6.1 illustrates the different notations. From the measurements Ψ, we aim to reconstruct $\nabla\Phi$. First, the gradient of Ψ is computed from equations [6.2] and [6.3]. Let us define the wrapping operator W such that[3]:

$$W(x) = x + 2k\pi, W(x) \in [-\pi, \pi[, k \in \mathbb{Z} \quad [6.5]$$

In order to fulfill the "smooth gradient" assumption, the gradient of Φ is heuristically estimated by the "wrapped-differences-of-wrapped-phases estimator", as named by Bamler and Hartl (1998):

$$\hat{\nabla}\Psi(i, j) = \begin{pmatrix} W(\Delta_1\Psi(i, j)) \\ W(\Delta_2\Psi(i, j)) \end{pmatrix} = \begin{pmatrix} \hat{\Delta}_1\Psi(i, j) \\ \hat{\Delta}_2\Psi(i, j) \end{pmatrix}, (i, j) \in S_0, \quad [6.6]$$

3. If the wrapped phase lies in $[0, 2\pi]$, then $W(x)$ is defined as: $W(x) = x[2\pi]$.

In practice, this last equation is the application of the "smooth gradient" assumption, as it constrains each component of $\hat{\nabla}\Psi$ to be smaller than π in absolute value. As $\hat{\nabla}\Psi$ is an estimator of $\nabla\Phi$, this estimate will be correct as soon as both components of $\nabla\Phi$ are smaller than π in absolute value, which is the case for most pixels in practice. However, whenever any of the components of $\nabla\Phi$ is greater than π in absolute value, $\hat{\nabla}\Psi$ underestimates the gradient in absolute value. This arises for noisy or aliased interferograms. As explained in the next section, the rotational is a tool used to detect such cases. Some authors have suggested robust approaches to estimate the fringe rate from noisy wrapped interferograms, which notably reduce the occurrence of such underestimation (Trouvé *et al.* 1998). Unfortunately, such methods are not systematically used.

6.2.1.3. *Notions of the rotational, residue and congruency*

The rotational (or curl) of the vector field $\nabla\Phi$ can be expressed in its discrete version (e.g. Bamler and Hartl 1998):

$$rot(\nabla\Phi)(i,j) = \\ \Delta_1\Phi(i,j) - \Delta_1\Phi(i,j+1) + \Delta_2\Phi(i+1,j) - \Delta_2\Phi(i,j), (i,j) \in S_0 \quad [6.7]$$

Figure 6.1a illustrates how the rotational is computed. Although the rotational is relative to the square delimited by indices (i,j), $(i+1,j)$, $(i+1,j+1)$ and $(i,j+1)$, it is attributed to the lower-left index (i,j). Following many authors, fractions of phase cycles are represented in Figure 6.1b and c, i.e. phases are divided by 2π. Thus, if the phase lies in $[-\pi,\pi[$, the fraction of cycles lies in $[-0.5,0.5[$. Figure 6.1b shows an example of unwrapped phase Φ, its gradient and its rotational. Due to the central pixel valuing 0.6, three gradients of Φ are greater than 0.5 in absolute value. However, the rotational of $\nabla\Phi$ is 0 for every square, as these high gradients are correctly compensated for by the three other terms of the rotational. Figure 6.1c represents the wrapped phase Ψ, the gradient components of $\hat{\nabla}\Psi$ and its rotational. Due to wrapping, the central value of Φ valuing 0.6 is replaced by -0.4 and the three highest gradients are underestimated. The rotational of $\hat{\nabla}\Psi$ values +1 and -1 on the two right-side squares as they each contain one erroneous gradient component. These two erroneous rotational values are called *residues*. More generally, a residue is a pixel for which the rotational computed as in equation [6.7] is non-zero. Due to wrapping, the rotational of a residue values $2k\pi, k \in \mathbb{Z}^{*}$[4]. If k is positive (respectively, negative), the residue is called positive (respectively, negative). Let us note that on the left-side squares, there are two erroneous estimates of the gradient components with opposite sign so that the rotational of $\hat{\nabla}\Psi$ is 0.

4. If the rotational is computed in terms of the number of cycles as in Figure 6.1, then it values $k, k \in \mathbb{Z}^{*}$.

This toy example illustrates the fundamental problem of unwrapping. An unwrapped phase field $\Phi(i,j), (i,j) \in S$ exists if and only if (from equation 5 in Costantini (1998)) the gradients in the line and column directions fulfill the following condition:

$$rot(\nabla\Phi)(i,j) = 0, \forall (i,j) \in S_0 \quad [6.8]$$

This condition means that $\nabla\Phi$, which is defined on a discrete space, should be a *conservative* or *irrotational* vector field. If this condition holds, then it is possible to find a real-valued function Φ by integration, up to an additive constant (Costantini 1998), which will be independent of the integration path. Thus, the aim of several unwrapping algorithms is to estimate the gradient field $\nabla\Phi$ such that condition [6.8] holds. In practice, only Ψ is known and $\hat{\nabla}\Psi$ may be non-conservative at some places. Some algorithms (minimum cost flow (MCF), branch cuts) identify where the rotational of $\hat{\nabla}\Psi$ is non-zero, i.e. where $\hat{\nabla}\Psi$ does not correctly estimate $\nabla\Phi$.

Finally, for some unwrapping algorithms, the reconstructed field $\hat{\Phi}$ has the following property, called *congruency*:

$$\forall (i,j) \in S, \hat{\Phi}(i,j) = \Phi(i,j) + 2k\pi, k \in \mathbb{Z} \quad [6.9]$$

Although most algorithms estimate a congruent unwrapped phase, it will be seen that some others do not.

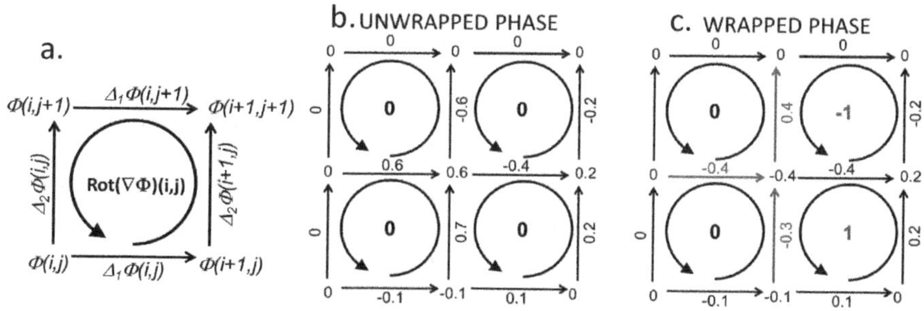

Figure 6.1. *(a) Notations for the phase gradient (represented by the straight arrows) and the rotational (represented by the curly arrows). (b) Example of an unwrapped phase and associated gradient and rotational. (c) Same example as in b but after wrapping. Values where the phase gradient is greater than a 0.5 cycle in absolute value are wrapped (red arrows) and residues (–1 and +1) arise even if the rotational is zero for unwrapped phase. For a color version of this figure, see www.iste.co.uk/cavalie/images.zip*

6.2.1.4. *Other useful prerequisites*

Different norms and pseudo-norms are used by unwrapping algorithms. The phase unwrapping problem is often stated as a minimization problem in order to evaluate the unknowns $\nabla\Phi$ from the measurements $\hat{\nabla}\Psi$. Weights $w_1(i,j)$ and $w_2(i,j)$ are attributed to each pixel (i,j) depending on the confidence on the gradient measure in the two directions. If the minimization function F is assumed to be separable in each direction, then it can be written as (e.g. Chen and Zebker (2000)):

$$F = \sum_{i=0}^{N-2}\sum_{j=0}^{M-1} w_1(i,j) \mid \Delta_1\Phi(i,j) - \hat{\Delta}_1\Psi(i,j) \mid^p + \sum_{i=0}^{N-1}\sum_{j=0}^{M-2} w_2(i,j) \mid \Delta_2\Phi(i,j) - \hat{\Delta}_2\Psi(i,j) \mid^p \quad [6.10]$$

The factor p determines the metric chosen to compare $\hat{\nabla}\Psi(i,j)$ with $\nabla\Phi(i,j)$. Depending on the author, F is minimized on Φ or on $\nabla\Phi$ and p is set to 0, 1 or 2, leading, respectively, to minimization with an L^0 pseudo-norm, L^1 norm or L^2 norm. The L^0 pseudo-norm of a vector is simply the number of non-zero elements of this vector. Strictly speaking, it is not a norm because it is not homogeneous. However, it is used by several authors (e.g. Chen and Zebker 2000). The MCF algorithm uses the L^1 norm (Costantini 1998), whereas the L^2 norm is used by the least-squares unwrapping algorithm (Ghiglia and Romero 1994). Chen and Zebker (2000) show that the L^0 pseudo-norm problem is NP-hard, which means that it cannot be solved exactly by any algorithm in a polynomial time (unless P and NP problem classes are equivalent). Thus, the approaches aiming to solve this problem are in fact heuristics to approximate the solution.

Finally, several authors use graph theory vocabulary. Indeed, a 2D interferogram can be viewed as a set of data points that form the nodes or vertices of a graph. When only a subset of pixels is selected (sparse data, for example, in the permanent scatterers technique), pixels are often linked by a Delaunay triangulation, forming the edges (or arcs) of the graph. In the temporal domain as well, the set of all acquisition dates can be viewed as a graph where each date is a node and each interferogram between two acquisitions is an edge. In a 3D domain (space and time), one pixel at a given date defines a node and the differential between two nodes (even if in different locations and different times) can be considered as an edge.

6.2.2. *A short history of unwrapping methods in InSAR*

One of the first unwrapping algorithms was the residue-cut tree algorithm (Goldstein *et al.* 1988), which identifies where the integration process is forbidden. As this algorithm is time-consuming, Ghiglia and Romero (1994) proposed a least-squares approach, which is very fast but has some drawbacks, as the unwrapped phase gradient is underestimated. Then, a family of algorithms using network-based approaches was developed, beginning with the minimum cost flow algorithm

(Costantini 1998). Since the 2000s, data increases have lead to the generalization of 2D unwrapping techniques to 3D, the third dimension being time. Another novelty of this period is that the sampling of data may not be regular anymore, following the development of InSAR time-series processing on sparse data points, such as the persistent scatterers technique (Ferretti *et al.* 2001). Finally, the redundancy of the InSAR time series also allowed the development of algorithms correcting unwrapping errors during the unwrapping process or *a posteriori*. The different unwrapping algorithms can be classified in several ways. Here, they are divided into the three large families originally developed for 2D problems and presented in three sections: the residue cut-type methods, the least-squares approach and the network-based algorithms. For the first and last families, 3D methods are also described.

6.2.3. *Residue-cut tree algorithm, variants and 3D generalization*

The *residue-cut tree algorithm* (Goldstein *et al.* 1988) is a still very widely used algorithm, despite being one of the first described in the InSAR literature. The principle of this algorithm is to identify every residue on the image by computing the rotational of $\hat{\nabla}\Psi$, following equations [6.6] and [6.7]. Then, every point where the rotational is non-zero is called either a positive or negative residue. Goldstein *et al.* (1988) state that an integration path enclosing a single residue produces an inconsistency in the unwrapped phase, whereas an integration path enclosing as many positive as negative residues does not. Thus, cuts are placed between couples of positive and negative residues to forbid any integration path across these cuts. The unwrapped phase difference between two pixels on opposite sides of a cut exceeds one half-cycle. In order to minimize the total discontinuity of the unwrapped phase, the aim is to find the overall shortest set of cuts pairing a positive and a negative residue. When a residue is found, another residue is searched for in a 3x3 box centered on the first one. If the second residue has the opposite sign, then a cut is placed in between and marked as "uncharged". Otherwise, the search begins again from the second residue. If no residue is found in the boxes, the size of the boxes is multiplied by 2. This algorithm works well in areas where residues are sparse but can create isolated regions in areas with a high density of residues. In such a case, only one isolated region can be unwrapped as there is no integration path between two regions that do not cross a cut. Thus, large parts on an interferogram may remain wrapped. Figure 6.2, adapted from Chen and Zebker (2000, 2001), shows an example of an interferogram affected by topographic fringes. This interferogram has been unwrapped by several unwrapping algorithms with adapted weights and the residual of the unwrapped interferogram with respect to a reference synthetic interferogram generated by an external DEM (see Chen and Zebker (2000) for details) is shown from b to g: the closer to zero the residual is, the better the unwrapping. The residual of the residue-cut tree algorithm solution is shown in Figure 6.2b. Although this algorithm performs well on the unwrapped area, a large area shown in black remains wrapped.

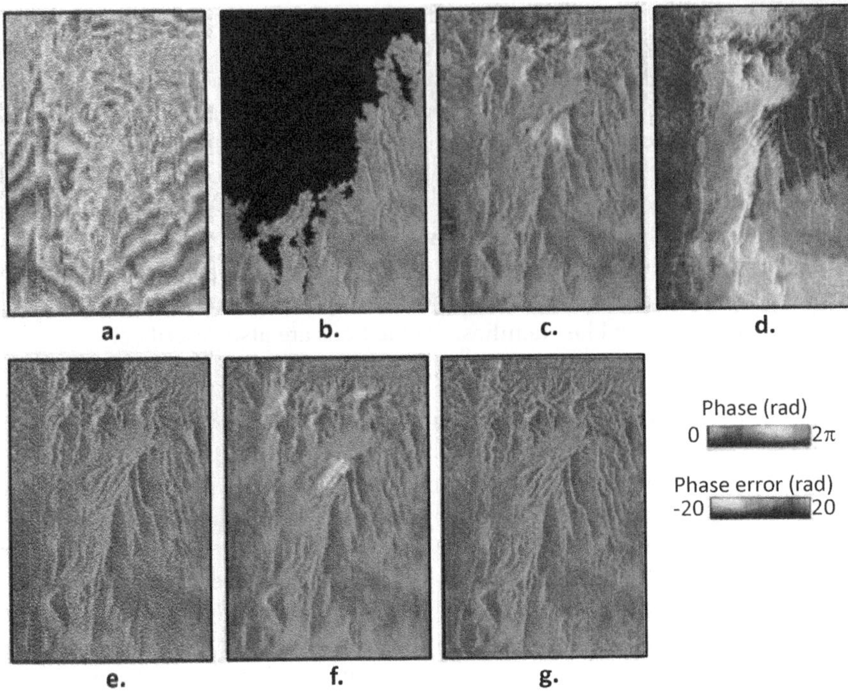

Figure 6.2. *(a) Interferogram affected by topography. (b–g) Residuals of interferograms unwrapped by different methods with respect to a reference synthetic interferogram: (b) residue-cut tree (c) minimum-spanning tree; (d) least squares; (e) minimum cost flow; (f) dynamic cost canceling; (g) the statistical-cost network-flow algorithm for phase unwrapping (SNAPHU). This figure is adapted with permission from the Journal of the Optical Society of America (JOSA). For a color version of this figure, see www.iste.co.uk/cavalie/images.zip*

Chen and Zebker (2000) enhanced the residue-cut tree algorithm to solve the problem of disconnected regions. This problem arises from the strategy of iteratively linking residues in the neighborhood of the current residue, which can lead to closed branch cuts for a high residue density. Instead, they proposed a *minimum-spanning tree algorithm*. As in the residue-cut approach, the tree forbids integration paths. The advantage of this approach is that, by definition, a tree never contains any branch closed on itself. Although performing well for the example shown and effectively unwrapping the whole interferogram (see Figure 6.2c), the authors mention that this approach suffers from a lack of accuracy when residue density is high.

Hooper and Zebker (2007) proposed two algorithms for 3D phase unwrapping. We focus here on the first algorithm, which highlights the differences between 2D and 3D unwrapping problems. It is called the *quasi-*L^{∞}*-norm 3D algorithm* and works well whenever no multiple-cycle phase discontinuity arises (i.e. phase discontinuity is always less than 4π in absolute value). Differences between 2D and 3D unwrapping are well-explained by Huntley (2001): in 3D, 2D pixels are replaced by voxels. For the sake of simplicity, the following explanations are given with regard to cube-shaped voxels. Each interferogram is in a 2D space domain and the third dimension is the time domain. Figure 6.3 shows an example of six unwrapped interferograms (in phase cycles) with six spatial indices in i and two in j. Residues are indicated by red curved arrows, and phase differences exceeding a 0.5 cycle are indicated by red lines. Each voxel has two faces between four pixels in a 2D interferogram, while the four other faces lie in a space-time dimension. As in space, the faces are between two successive dates in the time dimension. Then, it is possible to compute the rotational for each face as in 2D. Huntley (2001) explains that any residue on one face of the cube is "compensated" by a residue on another face of the cube, so that the singularity never terminates within a cube. The direction (entering or leaving the cube) should be taken into account: in Figure 6.3, a –1 residue enters and leaves the voxel between dates t_5 and t_6 and for i between 3 and 4. Then, it is possible to "follow" the residues from one face of a cube to another face, forming a 3D curve (dashed blue curve in Figure 6.3). This curve should be either a closed loop (which is the general case, as in Figure 6.3) or a partially closed loop if the curve enters the boundary of the space-time domain. In Figure 6.3, if there were only four indices in time (from t_1 and t_4), then the loop would be truncated of its upper part between residues A and B, as this part would be beyond the definition domain. Depending on the author, these loops are referred to as "phase singularity loops" (Huntley 2001) or as "residue loops" (Hooper and Zebker 2007). The intersection between a loop and an interferogram is a pair of residues of opposite signs. For example, in Figure 6.3, the negative residue A and the positive residue B in interferogram t_4 are unambiguously paired, as they belong to the same loop. However, considering only an interferogram (e.g. t_4), it is not possible to pair corresponding residues without further assumption. The heuristic algorithm developed by Goldstein *et al.* (1988) pairs the most probable residues, assuming their proximity. Thus, even if the pairing of residues is ambiguous in the 2D case, it becomes unambiguous in 3D insofar as far as the loop containing the residues is correctly identified.

Once a loop is identified, a surface limited by this loop should be defined: it should not be intersected during the integration process. The knowledge of the loop is not sufficient to define such a surface, as there is an infinity of them. The two teams mentioned above define the "surface cut" as the minimal surface whose limit is the loop, which is equivalent to the equilibrium soap film in zero gravity (in light blue in Figure 6.3). The 2D branch cuts can then be interpreted as the intersection of a surface cut with an interferogram in the space domain: in Figure 6.3, the segment $[AB]$ (in green) is the branch cut which is the intersection of plane t_4 and the surface cut. Note that although this latter is a plane in the example for visual clarity, a surface cut may be twisted in the time-domain dimension. The aim of a 3D unwrapping algorithm is thus to compute the residues on each face of the voxels, construct the closed or partial loop and define the surface cuts.

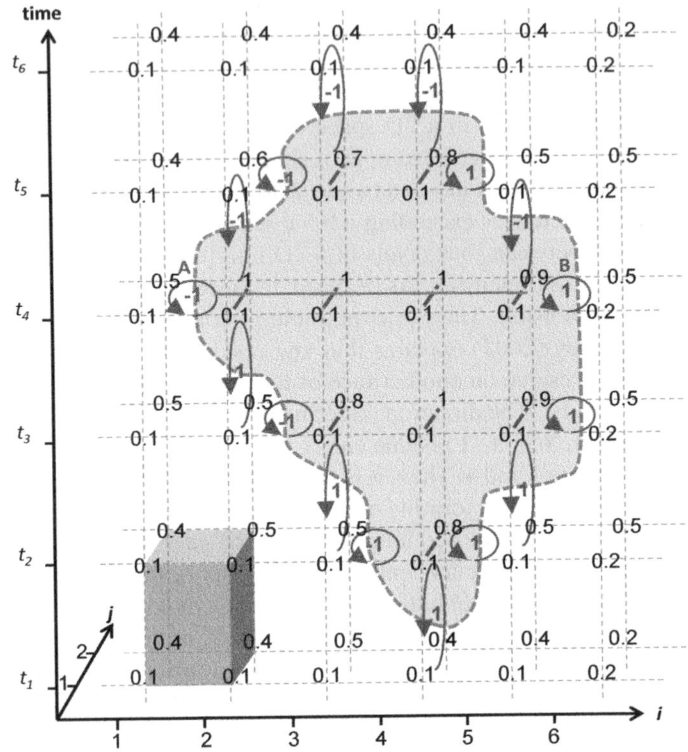

Figure 6.3. *Series of six unwrapped interferograms in phase cycles. A time-space voxel is represented in gray at the bottom left. Between dates t_2 and t_5, there are phase differences exceeding a 0.5 cycle where residues arise. In 3D, these "spatial" residues are linked to neighboring "space-time" residues and form a closed loop (the dashed blue curve). Here, the surface cut is represented in light blue. It is contained in a plane and bounded by the loop. For a color version of this figure, see www.iste.co.uk/cavalie/images.zip*

The quasi-L^{∞}-norm 3D algorithm (Hooper and Zebker 2007) processes interferometric time series, and the spatial sampling can be either regular or irregular (e.g. due to persistent scatterer processing). In the latter case, the voxels are wedge-shaped 3D elements instead of cubes. This algorithm identifies all residues, then links the residues to form a partial or closed residue loop. If the loop is closed, the surface cut can be defined as explained earlier. If it is not, a process tries to pair the partial loops in order to minimize the total cut surface. Once all the surfaces are defined, the integration is realized by a flood-fill algorithm that does not cross any surface cut. Hooper and Zebker show that this algorithm works well on simulated and real data.

6.2.4. *Least-squares method*

Ghiglia and Romero (1994) developed a fast unwrapping algorithm based on least-squares minimization, i.e. the L^2 norm, the first version being unweighted and the second being weighted. First, the wrapped phase gradient is computed as in equation [6.6]. Following the notations defined in section 6.2.1, the least-squares solution is the field Φ that minimizes the minimization function F of equation [6.10] with $p = 2$ and uniform weights. Hunt (1979) has shown that the least-squares solution can be written as:

$$\begin{aligned}\Phi(i+1,j) + \Phi(i-1,j) + \Phi(i,j+1) + \Phi(i,j-1) - 4\Phi(i,j)\\ = \hat{\Delta}_1\Psi(i,j) - \hat{\Delta}_1\Psi(i-1,j) + \hat{\Delta}_2\Psi(i,j) - \hat{\Delta}_2\Psi(i,j-1)\end{aligned} \quad [6.11]$$

If the known right-hand side of the equation is noted $\rho(i,j)$, equation [6.11] can be rewritten as a discrete equivalent of Poisson's equation:

$$\begin{aligned}\Phi(i+1,j) - 2\Phi(i,j) + \Phi(i-1,j) + \Phi(i,j+1) - 2\Phi(i,j)\\ + \Phi(i,j-1) = \rho(i,j)\end{aligned} \quad [6.12]$$

The left-hand side of the equation is the discrete equivalent of the Laplacian operator. It should be noted that the unwrapped phase is estimated only using $\rho(i,j)$, which can be considered as an estimate of the Laplacian of Φ from the measures Ψ. Then, the discrete cosine transform (DCT) leads to a very simple formulation of the problem, as the computation of the DCT of the exact solution $DCT(\Phi(i,j))$ is simply rewritten:

$$DCT(\Phi(i,j)) = \frac{DCT(\rho(i,j))}{2 \times (\cos\frac{\pi i}{N} + \cos\frac{\pi j}{M} - 2)} \quad [6.13]$$

Then, the solution can be easily computed with the inverse DCT. It can be noted that this algorithm directly gives an estimate of Φ, whereas most unwrapping algorithms first evaluate $\nabla\Phi$. The weighted version of this algorithm is an iterative process also based on the DCT.

Although very fast and simple to implement, this algorithm has two drawbacks. First, the L^2-norm is not the most appropriate because it has the property of distributing the errors, causing error spreading. Second, when the interferogram has a non-zero fringe rate, the slope of the phase is underestimated, as detailed in Bamler *et al.* (1998) and summarized hereafter. Let us first assume that there is no slope on the interferogram; it is then reasonable to assume that the unwrapped phase gradient $\nabla\Phi$ is centered on zero. Due to the wrapping operation to $[-\pi, \pi]$, if $\Delta\phi_i > \pi$, then $\hat{\Delta}\psi_i$ underestimates $\Delta\phi_i$ by $-2k\pi, k \in \mathbb{N}^*$ (here, i is the index corresponding to the direction of the gradient and is valued 1 or 2). In contrast, if $\Delta\phi_i < -\pi$, then $\hat{\Delta}\psi_i$ overestimates $\Delta\phi_i$ by $2k\pi, k \in \mathbb{N}^*$. Statistically, $\hat{\Delta}\Psi$ underestimates $\Delta\Phi$ as often as it overestimates it with a significant error of $\pm 2k\pi$. Let us now assume that there is a global positive slope α in the line direction on an interferogram (or a part of it); $\Delta_1\Phi$ is then centered on $\alpha > 0$. Consequently, there are more pixels where $\Delta_1\Phi > \pi$ than where $\Delta_1\Phi < -\pi$. Finally, $\hat{\Delta}_1\Psi$ underestimates $\Delta_1\Phi$ more often that it overestimates it which introduces a bias. In summary, whatever the sign of α, the phase gradient error has a non-zero expectation if $\alpha \neq 0$:

$$E[\hat{\Delta}_i\Psi - \Delta_i\Phi] \neq 0 \qquad [6.14]$$

Globally, the underestimation of $\mid \Delta\phi_i \mid$ leads to a reduced fringe rate for $\hat{\Phi}$ compared to Φ. Let us note that this underestimation also implies that the solution $\hat{\Phi}$ is non-congruent with Φ. Although proper weighting can lessen this effect, it would be impossible to tailor perfect weighting such that this underestimation effect would never occur. The least-squares approach thus leads to an imperfect estimate of the unwrapped phase. Figure 6.2d confirms that the least-squares algorithm does not perform as well as other algorithms.

6.2.5. *Network-based approaches*

6.2.5.1. *The MCF algorithm and its 3D generalization*

The *minimum cost flow* algorithm was first described in its original 2D version by Costantini (1998). It was then generalized to take into account the temporal dimension, as well as sparse data. The aim is to find the gradient fields of Φ from those of Ψ. First, two integer fields n_1 and n_2 are usually chosen such that:

$$\Delta_1\Psi(i,j) + 2\pi n_1(i,j) \in [-\pi, \pi[, (i,j) \in S_1 \qquad [6.15]$$

$$\Delta_2\Psi(i,j) + 2\pi n_2(i,j) \in [-\pi, \pi[, (i,j) \in S_2. \qquad [6.16]$$

Then, the discrete partial derivative residuals k_1 and k_2 are defined as:

$$k_1(i,j) = \frac{1}{2\pi}[\Delta_1\Phi(i,j) - \Delta_1\Psi(i,j) - 2\pi n_1(i,j)], (i,j) \in S_1 \qquad [6.17]$$

$$k_2(i,j) = \frac{1}{2\pi}[\Delta_2\Phi(i,j) - \Delta_2\Psi(i,j) - 2\pi n_2(i,j)], (i,j) \in S_2 \qquad [6.18]$$

k_1 (respectively, k_2) should be zero almost everywhere when $\Delta_1\Phi(i,j) \in [-\pi, \pi[$ (respectively, $\Delta_2\Phi(i,j)$) and if n_1 fulfills equation [6.15] (respectively, when n_2 fulfills equation [6.16]). The points where k_1 or k_2 are non-zero correspond to branch cuts (Goldstein *et al.* 1988). Then, the idea is to estimate these residuals so that the gradient fields of Φ fulfill the irrotational property. This is done through a constrained minimization of a function weighted by confidence maps $c_1(i,j), (i,j) \in S_1$ and

$c_2(i,j),(i,j) \in S_2$, expressing the confidence on k_1 (respectively, k_2). The function to be minimized is written:

$$\min_{k_1,k_2}\{\sum_{i=0}^{N-2}\sum_{j=0}^{M-1} c_1(i,j) \mid k_1(i,j) \mid + \sum_{i=0}^{N-1}\sum_{j=0}^{M-2} c_2(i,j) \mid k_2(i,j) \mid\} \qquad [6.19]$$

subject to the constraints:

$$\begin{aligned} & k_1(i,j+1) - k_1(i,j) - k_2(i+1,j) + k_2(i,j) \\ & = \frac{-1}{2\pi}[\Delta_1\Psi(i,j+1) - \Delta_1\Psi(i,j) - \Delta_2\Psi(i+1,j) + \Delta_2\Psi(i,j)] \\ & - [n_1(i,j+1) - n_1(i,j) - n_2(i+1,j) + n_2(i,j)] \\ & \qquad k_1(i,j) \in \mathbb{Z}, (i,j) \in S_1, k_2(l,m) \in \mathbb{Z}, (l,m) \in S_2 \qquad [6.20] \end{aligned}$$

Condition [6.20] expresses the constraint that $\nabla\Phi$ must fulfil the irrotational property. However, the function to minimize in equation [6.19] is nonlinear due to the absolute values. Thus, the problem is rewritten to get rid of the absolute values. It should be noted that the variables $k_1(i,j),(i,j) \in S_1$ and $k_2(i,j),(i,j) \in S_2$ can be considered as flows on non-oriented edges between any pair of neighboring pixels. As these flows can be positive or negative, the function to minimize should be applied on the absolute values of these flows. The variable $k_1(i,j)$ is in fact "replaced" by two positive variables: $x_1^+(i,j)$ if $k_1(i,j) \geq 0$ and by $-x_1^-(i,j)$ if $k_1(i,j) \leq 0$. Thus, a single edge between two neighboring pixels is replaced by two oriented edges with opposite directions. For example, $x_1^+(i,j)$ is the positive flow from node (i,j) to node $(i+1,j)$, and $x_1^-(i,j)$ is the positive flow from node $(i+1,j)$ to node (i,j). With these new variables, the problem becomes a minimum cost network flow problem that can be solved by dedicated algorithms. In order to limit processing time, the algorithm is applied on sub-blocks of the image with sufficient overlap. While processing a new sub-block, additional constraints from already processed blocks are applied on the overlapping parts. This algorithm generally gives good performance (see Figure 6.2e), even if some parts of the interferogram are incorrectly unwrapped in this example.

This algorithm has been generalized for sparse data where pixels are selected if their coherence is greater than a given threshold (Costantini and Rosen 1999). A Delaunay triangulation is then performed on the selected pixels. The generalization of MCF to sparse data then consists of imposing the irrotational property of the gradient of Φ to each elementary cycle of the Delaunay triangulation, instead of the classical square cycle shown in Figure 6.1.

Pepe and Lanari (2006) generalized the MCF algorithm in order to simultaneously unwrap a whole stack of multi-temporal interferograms: this algorithm is called *extended minimum cost flow* (EMCF). In the temporal dimension, the acquisitions are represented in the classical $T \times B_\perp$ plane, where T is time and $B_\perp$ is the perpendicular baseline. Edges on this plane represent the interferograms of the database. The unwrapping is performed in two steps. In the first step, each arc of the Delaunay triangulation in the spatial domain is unwrapped in the $T \times B_\perp$ dimension. More precisely, let us consider an arc between two neighboring pixels A and B in the spatial subset. The aim is to retrieve the unwrapped phase difference between these pixels in each interferogram. If there are M interferograms, the vector of the unwrapped phase differences between A and B is noted $\delta\Phi = (\delta\Phi_0, \delta\Phi_1, .., \delta\Phi_{M-1})$[5] and the vector of wrapped phase differences $\delta\Psi = (\delta\Psi_0, \delta\Psi_1, .., \delta\Psi_{M-1})$. Then, a parametric model m for the vector $\delta\Phi$ is introduced in order to help the unwrapping procedure. This model depends on two parameters (for each arc), namely Δz and Δv, which are the differential of elevation errors and the differential velocity between A and B. Then, the unwrapped vector is:

$$\delta\Phi = m + (\delta\Phi - m)[2\pi] + 2\pi K \qquad [6.21]$$

where K is a vector of integers. The aim is then to minimize:

$$\min_{k_j}\{\sum_{j=0}^{M-1} \mid k_j \mid\} \qquad [6.22]$$

However, instead of constraining the irrotational property on the azimuth/range plane, this constraint is imposed on the $T \times B_\perp$ plane. The elementary cycles are then the triangles formed in this plane by the database graph. If the three arcs of such a triangle are noted α, β and γ[6], then equation [6.22] is minimized, subject to constraints on $k_\alpha + k_\beta + k_\gamma$. It should be noted that the introduction of the parametric model aims at decreasing the value of the k_j by removing the predictable part of $\delta\Phi$. Finally, the first step gives a first estimate of the time series of the phase differences for each arc of the spatial triangulation. In the second step, the classical MCF spatial unwrapping algorithm is performed for each interferogram, beginning with the estimate given by the first step. The weights of equation [6.19] (in its irregular sampling version) are computed according to the quality of the temporal unwrapping, where greater weights are given to arcs whose temporal unwrapping is reliable. Pepe and Lanari show with simulated and real data that the EMCF performs better than classical MCF applied separately to each interferogram. This approach has been generalized to full-resolution interferograms (Pepe *et al.* 2011).

5. In contrast to the Pepe and Lanari's notation (2006), Φ refers to the unwrapped phase and Ψ refers to the wrapped one to ensure this chapter's internal coherence.

6. These three arcs are three interferograms derived from three acquisition dates.

6.2.5.2. *Redundant finite difference integration and phase unwrapping algorithm*

Costantini *et al.* (2012) proposed a generalized formulation that can be applied to data regularly or sparsely sampled in the spatial and temporal domains; this elegant formulation is called the *redundant finite difference integration and phase unwrapping algorithm*. In contrast to the EMCF and the 3D stepwise algorithm alternating 1D and 2D (see section 6.2.5.3), this approach unwraps the data in the full 3D domain at once. The authors consider two graphs: the first one, (S_1, A_1), where nodes are in S_1 and arcs are in A_1, represents the space domain; the second one, (S_2, A_2), is the time domain. For each graph, arcs not only link closest neighbors like in a Delaunay triangulation but some supplementary links between close nodes are also considered. Then, instead of working on phase differences between neighboring pixels, the authors consider a double difference of phase in space and time, namely $\Phi''_{ijkl} = \Phi_{ik} - \Phi_{il} - \Phi_{jk} + \Phi_{jl}$, where $(i, j) \in A_1$ and $(k, l) \in A_2$[7]. This double difference is considered because it is reasonable to assume that its absolute value is smaller than π in most cases, especially if an adequate model can be subtracted. Let (S_1, T_1) (respectively, (S_2, T_2)) be a spanning tree of (S_1, A_1) and B_1 (respectively, B_2) its strictly fundamental cycle basis[8]. If Ψ''_{ijkl} are preliminary estimates of Φ''_{ijkl}, then the aim is to minimize:

$$\min_{\delta_{ijkl}} \sum_{(i,j)\in A_1,(k,l)\in A_2} c_{ijkl} \mid \delta_{ijkl} \mid^p \tag{6.23}$$

subject to the constraints:

$$\Phi''_{ijkl} - \delta_{ijkl} = \Psi''_{ijkl}, (i, j) \in T_1, (k, l) \in T_2 \tag{6.24}$$

$$- \sum_{(k,l)\in\gamma_2} \delta_{ijkl} = \sum_{(k,l)\in\gamma_2} \Psi''_{ijkl}, \gamma_2 \in B_2, (i, j) \in T_1 \tag{6.25}$$

$$- \sum_{(k,l)\in\gamma_1} \delta_{ijkl} = \sum_{(k,l)\in\gamma_1} \Psi''_{ijkl}, \gamma_1 \in B_1, (k, l) \in T_2 \tag{6.26}$$

The last two constraints are irrotationality constraints in time and space. This formulation has the notable advantage of treating the time and space dimensions

7. The notations introduced here are different from the original ones (which are more general) to ensure this chapter's internal coherence. Φ_{ik} corresponds to f_{ik}, Φ''_{ijkl} to $f_{ik} - f_{il} - f_{jk} + f_{jl}$ and Ψ''_{ijkl} to f''_{ijkl}.

8. A unique cycle $(S_1, T_1 \cup (i, j))$ is associated with each arc $(i, j) \in A_1 \backslash T_1$. By definition, the set B_1 of all these cycles has a strictly fundamental cycle basis.

similarly. The originality of this method is that it makes it possible to consider the cycles in the temporal domain for each arc in the space domain and the cycles in the space domain for each arc in the time domain. Costantini *et al.* then study the influence of the connectivity of the graphs on phase unwrapping performance. The higher the connectivity is, the greater the redundancy. Redundancy either in the spatial domain or the time domain is shown to improve phase unwrapping performance.

6.2.5.3. *Stepwise 3D algorithm and its adaptation in StaMPS software*

Hooper and Zebker (2007) proposed an algorithm called the *stepwise 3D algorithm* that first unwraps in the temporal domain and then in the spatial domain, as in the EMCF algorithm. In the temporal domain, the differential phase between two neighboring pixels on an edge is considered and temporally unwrapped by temporal low-pass filtering, with the assumption that the double difference does not exceed π in absolute value. These first estimates are then the initial solution of the spatial unwrapping of each interferogram where any 2D unwrapping algorithm may be used. Let us note that the main steps of this algorithm are very similar to those of EMCF. Hooper (2010) implemented a variant of this last algorithm in the StaMPS software. The first step remains unchanged. The second step uses the statistical-cost network-flow algorithm for phase unwrapping (SNAPHU) for 2D unwrapping (see section 6.2.5.4). However, the statistical costs are derived from the first step.

6.2.5.4. *SNAPHU*

SNAPHU is a widely used, freely released 2D algorithm (Chen and Zebker 2001). Preliminary developments and explanations are given by Chen and Zebker (2000). First, the authors argue that the MCF algorithm is an L^1 algorithm for which multiple-cycle discontinuities (which may arise in layover areas) are difficult to estimate, and thus that L^0 pseudo-norm should be preferred as it is more adapted to the experimental content of many SAR interferograms. Indeed, multiple-cycle discontinuities are penalized by the L^1 norm compared to the L^0 pseudo-norm. The classical cost cycle algorithm (Ahuja *et al.* 1993) solves the minimum cost flow problem in the L^1 norm framework with an iterative implementation. The *dynamic cost cycle canceling* (DCC) algorithm is a variant of the former in the L^0 pseudo-norm framework. In both algorithms, costs per unit of flow are associated with any arc. In the original algorithm, these costs remain constant throughout the iterations, whereas in the DCC algorithm, the unit costs vary during the iterations, hence the "dynamic" update of the arcs costs. The DCC algorithm performs well (see Figure 6.2f), but the authors then generalize the L^0 framework by replacing the weighted distance between $\hat{\nabla}\Psi(i,j)$ and $\nabla\Phi(i,j)$ in the minimization function with a function of these variables in the SNAPHU method (see Figure 6.2g). Precisely, the minimization function of equation [6.10] is rewritten:

$$\min_{\Delta_1\Phi(i,j),\Delta_2\Phi(i,j)} \{\sum_{i=0}^{N-2}\sum_{j=0}^{M-1} g_{i,j}^{(1)}(\Delta_1\Phi(i,j),\hat{\Delta}_1\Psi(i,j))+$$

$$\sum_{i=0}^{N-1}\sum_{j=0}^{M-2} g_{i,j}^{(2)}(\Delta_2\Phi(i,j),\hat{\Delta}_2\Psi(i,j))\} \quad [6.27]$$

where subscript or superscript 1 (respectively, 2) corresponds to the range (respectively, azimuth) direction. SNAPHU's aim is to estimate the maximum *a posteriori* (MAP) solution $\hat{\Phi}$ given the measurements $\hat{\nabla}\Psi$. As the measurements are assumed to be independent, the MAP is equivalent to the maximization of the sum of the conditional log probabilities of $\nabla\Phi$ given $\hat{\nabla}\Psi$. The coherence of the interferogram is noted ρ, and the mean intensity of the two images forming the interferogram is noted I. If $f(\Phi|\Psi)$ denotes the conditional probability of Φ given Ψ, then the cost functions $g_{i,j}^{(1)}$ and $g_{i,j}^{(2)}$ are:

$$g_{i,j}^{(1)}(\Delta_1\Phi(i,j),\hat{\Delta}_1\Psi(i,j)) = -log[f(\Delta_1\Phi(i,j)|\hat{\Delta}_1\Psi(i,j),I,\rho)] \quad [6.28]$$

$$g_{i,j}^{(2)}(\Delta_2\Phi(i,j),\hat{\Delta}_2\Psi(i,j)) = -log[f(\Delta_2\Phi(i,j)|\hat{\Delta}_2\Psi(i,j),I,\rho)] \quad [6.29]$$

Several remarks should be made about these last equations. First, the functions $g_{i,j}^{(1)}$ and $g_{i,j}^{(2)}$ vary for each pixel (i,j), as the conditional probabilities depend on the *a priori* distribution of Φ on each pixel, as detailed later on. Second, for a given pixel, $g_{i,j}^{(1)}$ and $g_{i,j}^{(2)}$ differ, as specific phase distributions in range and in azimuth are taken into account. The core of SNAPHU is its automatic definition of the cost functions (given a few user-tuned parameters) from ρ and I.

The prior distribution of $\Delta\Phi$ takes into account the signal component and the noise component. The noise distribution depends on the coherence of the interferogram, whereas the signal distribution depends on the content of the interferogram. In their original paper, Chen and Zebker (2001) propose two cases, although the work may be adapted to other cases. The first is a "topographic interferogram", i.e. an interferogram from which a DEM would be estimated. The second aims at measuring deformation on a differential interferogram due to an earthquake, with a possible high fringe rate. For the topographic case, the model takes into account I, as high-intensity pixels often correspond to layover areas where $\Delta\Phi$ may take high values. It also takes into account the coherence, as low-coherence areas may also be prone to high values of $\Delta\Phi$. For the deformation case, low coherence may correspond to a sharp fringe rate and thus high values of $\Delta\Phi$. For both cases, if there is no hint that $\Delta\Phi$ may be high, the cost function is a parabola centered around the expected value of $\Delta\Phi$, which is similar to the L^1 norm. However, if I or ρ indicate that $\Delta\Phi$ may attain high values due to layover or a high deformation rate, then the cost function has two shelves extending the range of possible values for $\Delta\Phi$ "without extra cost", which is similar to the L^0 pseudo-norm. Moreover, SNAPHU only allows congruent solutions.

This approach has been generalized for large interferograms (Chen and Zebker 2002), which are cut into smaller individually unwrapped tiles. Then, phase offsets between regions are computed in order to obtain the whole interferogram.

6.2.5.5. *Edgelist phase unwrapping*

Shanker and Zebker (2010) proposed edgelist phase unwrapping, an algorithm to unwrap data in 3D that is similar to the EMCF algorithm but with another formulation. They consider N data points in the 2D or 3D domain (in 3D, N is the total number of pixels of the whole time series). For data points i, n_i is the integer such that $\Phi_i = \Psi_i + 2 \times \pi \times n_i$, with Ψ_i (respectively, Φ_i) being the wrapped (respectively, unwrapped) phase. Considering an edge (i, j) between data points i and j, the integer K_{ij} is the flow on this directed edge. Similar to the MCF formulation, K_{ij} is rewritten as the difference between two positive numbers (depending on the direction of the flow on the edge) as: $K_{ij} = P_{ij} - Q_{ij}$. The MCF algorithm searches for the optimal P_{ij} and Q_{ij}, whereas edgelist phase unwrapping also searches for the optimal n_i. If C_{ij} is the weight relative to the edge (i, j), then the function to minimize is:

$$\min_{P_{ij},Q_{ij}} \{\sum_{(i,j)} C_{ij} \times (P_{ij} + Q_{ij})\} \qquad [6.30]$$

subject to the constraints:

$$n_j - n_i + P_{ij} - Q_{ij} = round(\frac{\Psi_i - \Psi_j}{2\pi}), P_{i,j}, Q_{i,j} \geq 0, \forall(i,j) \qquad [6.31]$$

This minimization is solved by linear programming. As pointed out by the authors, the constraint is now on edges instead of loops, hence the name "edgelist". Although the number of unknowns is greater than that in the EMCF formulation, this formulation has some advantages: it is equivalent in 2D or 3D as the data points may belong to the same interferogram or not. Moreover, additional constraints from external data such as GPS can be added without any effect on the algorithm. The authors show with a time series for San Andreas Fault that the algorithm performs well despite the sharp discontinuity across the fault.

6.2.6. *Methods for correcting unwrapping errors*

Regardless of the algorithm used, unwrapping errors can occur; for example, where topography is sharp, producing high fringe rates. There are also several unwrapping errors that can arise in the borders of interferograms, as there are fewer paths to link pixels on the borders to other pixels in the center of the interferogram. Such problems have been identified by several authors; for example, Chen and Zebker

(2000). Although some techniques for correcting unwrapping errors can be applied to single interferograms, most of them are applied to temporal series of interferograms. Next, we first present a method applicable to a single interferogram and then methods exploiting the redundancy of the network.

6.2.6.1. *Bridging scheme applicable to single interferograms*

Unwrapping errors usually produce whole regions where the error spreads: each region is unwrapped well, but there is a $2k\pi$ offset between them, $k \in \mathbb{Z}^*$. Visual inspection of interferograms usually allows identification of such regions. One possibility is to then manually segment these regions and add the proper number of cycles to correct the unwrapping error by creating a "bridge" between neighboring regions and measuring the phase difference across the bridge. However, this process is highly time-consuming, especially for long time series. Another possibility is to discard interferograms where unwrapping errors are visible, at the expense of a poorer database. Yunjun *et al.* (2019) presented an approach to automatize the bridging scheme that can be applied on a single interferogram: the regions are automatically segmented, linked by bridges and their offsets are corrected. The authors showed good results with a synthetic interferogram. This approach is interesting when several parts of an interferogram cannot be linked; for example, when distinct islands are within the frame of the study area.

6.2.6.2. *Error correction applied to InSAR time series*

Biggs *et al.* (2007) observed that the closure phase on triplets of interferograms help identify unwrapping errors. A triplet is in fact the shortest closed loop in the temporal domain. The closure phase is defined in equation [6.32] and extensively described in section 6.3. If there is an unwrapping error in one of the interferograms of the triplet, then the closure is close to $2k\pi, k \in \mathbb{Z}^*$. The closure then indicates that one of the interferograms is incorrectly unwrapped. Based on this property, several unwrapping error correction methods have been developed. It should be noted that knowledge of the closure alone is insufficient to determine which one of the three interferograms is badly unwrapped. Biggs *et al.* (2007) proposed the visual inspection of each of the three interferograms to determine which one is incorrectly unwrapped. Following this, other methods were developed to automatize the correction process. It should be noted that 3D unwrapping algorithms such as Costantini *et al.*'s (2012) fulfill the phase closure condition by definition, which is optimal. In contrast, unwrapping error-correction algorithms *a posteriori* correct errors due to 2D unwrapping that may have been overlooked by 3D approaches.

Hussain *et al.* (2016) adapted the StaMPS unwrapping algorithm described in section 6.2.5.3 in order to automatically correct unwrapping errors on a time series of interferograms. This last algorithm is used iteratively, but between each iteration, a cost is attributed to each edge in the following way. The closure phase for a closed loop of interferograms is computed for every pixel in every interferogram. The pixel is labeled as correctly (respectively, incorrectly) unwrapped if its closure is smaller (respectively, greater) than a given threshold. The labels of the two pixels of an edge determine the cost attributed to the edge: the higher the cost (i.e. the more reliable the unwrapping), the more difficult the change of the phase difference on this edge will be at the next iteration. The authors showed that this technique decreased the number of incorrectly unwrapped pixels.

Yunjun *et al.* (2019) presented another approach based on the analysis of closure phase for triplets. Given a time series of interferograms, a vector of integers representing the number of cycles of the error is associated with each pixel. This vector should minimize the L^2 norm of the closure for all triplets of the time series. However, as the solution is not unique, a regularization term constrains the variations of this vector of integers. The authors show with a simulated database that the number of unwrapping errors significantly decreases, especially when the redundancy of the network increases, as in the approach developed by Costantini *et al.* (2012). As it excludes the regularization term, this approach is pixel-based.

Still based on triplet closure phase, Benoit *et al.* (2020) proposed an algorithm called the correction of phase unwrapping errors (CorPhU) algorithm. In contrast to the preceeding methods, which are pixel-based, this method identifies incorrectly unwrapped regions. Then, for each region, two indices are computed to determine which of the three interferograms is badly unwrapped: first, the flux between the borders of the region and a reference region is determined, which is *a priori* correctly unwrapped[9]; second, each interferogram may belong to several triplets and it is then possible to compute the mean closure for all these triplets. For a badly unwrapped interferogram, both the flux and the mean closure should be anomalous. The comparison of these indices for the three interferograms helps in identifying which interferogram is incorrectly unwrapped. The authors show with real and simulated data that the number of unwrapping errors significantly decreases.

Finally, in the NSBAS chain presented in Chapter 5, an unwrapping error correction process is performed during the inversion of the time series, which is pixel based. If the residual between inverted and original interferograms is large, then the interferogram may be badly unwrapped. The time-series inversion is computed again with a low confidence weight for such interferograms and the process is repeated iteratively until convergence of the inversion.

9. Although similar to the bridging scheme of Yunjun *et al.* (2019), the segmentation of the interferograms into regions unwrapped with an offset is derived from the closure phase.

6.2.7. *Summary and comparison*

As can be seen in the preceeding sections, InSAR phase unwrapping has given rise to a great number of methods. An interesting comparison of some 2D algorithms is given by Chen and Zebker (2000), especially the difference between L^0 and L^1 norms. Figure 6.4 shows an overview of the principal methods cited in this chapter. Globally, three families appear. First, the residue-cut tree algorithm has been enhanced and generalized in 3D. Second, the network-based methods deriving from the MCF algorithm are preponderant, as different teams worked on these approaches and generalized them to 3D. Third, the weighted least-squares algorithm has not been generalized to our knowledge, probably due to the problem of fringe rate underestimation described in section 6.2.4.

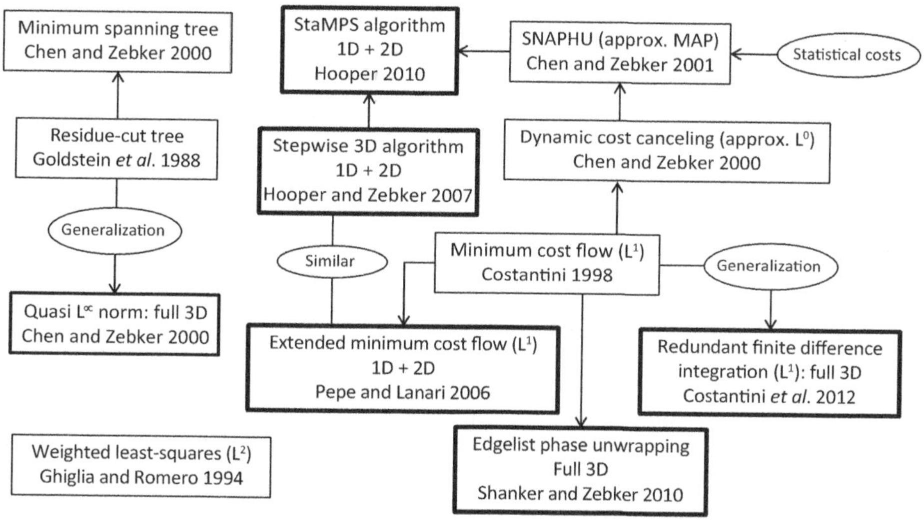

Figure 6.4. *Overview of the principal phase unwrapping methods described in this section and their relationships. Two-dimensional algorithms are in thin rectangles and 3D ones in bold rectangles*

With regard to 3D approaches, some are alternating 1D and 2D approaches, whereas others really consider the whole 3D problem at once, noted as "full 3D" in Figure 6.4. Thus, the quasi-L^∞ 3D algorithm is the strict 3D generalization of the residue-cut tree algorithm and the redundant finite difference integration algorithm is the strict generalization of the original MCF algorithm. By "strict generalization", we mean that all dimensions in 3D are equivalent, which is not the case for 1D + 2D methods. Edgelist phase unwrapping is part of the network-based family but differs from MCF, as constraints apply on edges rather than on loops. Finally, SNAPHU introduced a new way of weighting the minimization function by introducing statistical costs, which changed the problem from L^p-norm formulation to MAP formulation. A striking feature of the evolution of unwrapping methods is the growth of the network-based family. In our opinion, this is due to data increases and the capacity of this family to fully exploit data redundancy. Although the phase unwrapping problem is difficult and still a subject of research, redundancy should be one of the keys to solving this problem with well-adapted heuristics.

6.3. The (re)discovery of closure phases and their implications for SAR interferometry

The presence of statistical non-zero closure phases has been recognized in SAR interferometry since the development of triangulation methods (Monti Guarnieri and Tebaldini 2008). However, the discovery of *physical* non-zero closure phases is more recent (De Zan *et al.* 2012). This is somewhat surprising since the phenomenon is well-known in other fields, such as astronomy and, moreover, fundamentally invalidates well-established models for phase interpretation in multi-looked SAR interferograms and can strongly bias the deformation results. The basic equations we have been using for a long time (e.g. equations [4.5] and [5.1]) may look like crude approximations in the presence of closure phases, especially as we strive towards increased geodetic precision. At the same time, closure phases indicate a new world of physical phenomena that we will be able to observe.

6.3.1. *Introduction to closure phases*

With three SAR images, it is possible to generate three interferograms. The closure phase is the circular combination of the three interferometric phases:

$$\Psi_{1,2,3} = \varphi_{1,2} + \varphi_{2,3} + \varphi_{3,1} \mod 2\pi. \qquad [6.32]$$

For single pixels, the closure phase is always zero, as we can easily check. However, for multi-looked interferograms, there is no reason for the closure phase to be zero, contrary to common expectations. Consequently, to explore an interesting field, from now on we will assume that the interferometric phase is taken after a multi-look operation "$< \cdot >$" on the complex interferogram, for example:

$$\varphi^W_{1,2} = \angle\{< y_1 y_2^* >\}. \qquad [6.33]$$

The superscript W on the phase indicates that the phase is wrapped in this equation. With this notation, we could write the closure phase directly as the phase of a triple product:

$$\Psi_{1,2,3} = \angle\{< y_1 y_2^* >< y_2 y_3^* >< y_3 y_1^* >\}. \qquad [6.34]$$

Non-zero closure phases can result from very diverse mechanisms: first, we should clearly distinguish purely *statistical* explanations from *physical* explanations.

Statistical misclosure occurs every time the coherence magnitude is smaller than 1, i.e. all the time in real situations. Averaging more pixels, if possible, will reduce the statistical misclosure. Different triangulation algorithms have dealt with this kind of closure phase: Chapter 5 lists a number of such algorithms.

The second kind of misclosure is more fundamental and resists stronger multi-looking, i.e. larger spatial averaging. It is an essential type of inconsistency that requires some physical interpretation. From this perspective, the closure phase is actually a novel and interesting physical observable, especially considering that it is not sensitive to common interferometric signals, such as atmospheric delays or surface motion, as should become clear in the following.

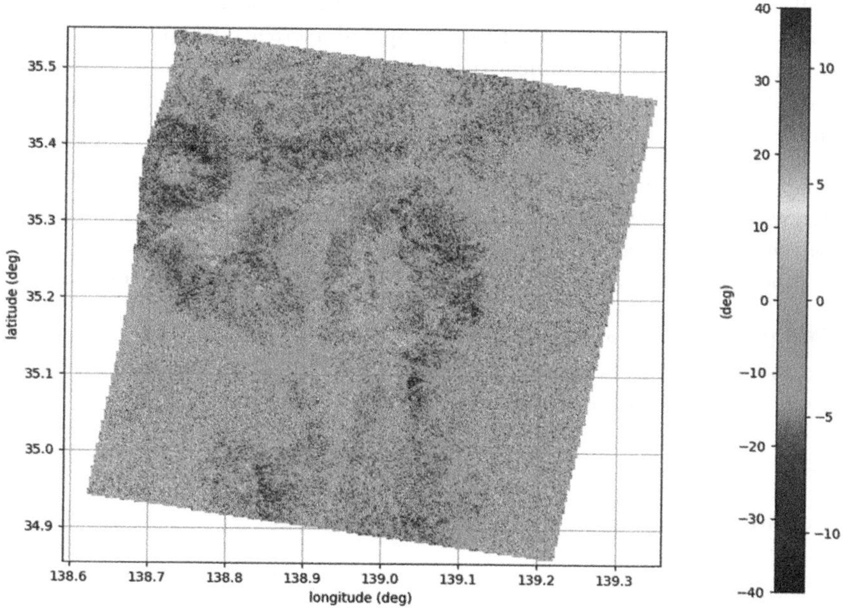

Figure 6.5. *The closure phase corresponding to three ALOS-2/PALSAR-2 acquisitions over Japan, acquired on May 7, 2015, July 16, 2015, and August 27, 2015. Mt. Fuji is recognizable in the northwest corner of the image. For a color version of this figure, see www.iste.co.uk/cavalie/images.zip*

Figure 6.5 shows an example of a full-scene closure phase. It has been derived from ALOS-2/PALSAR-2 data[10] (L-band) acquired over Japan. The Izu peninsula

10. The ALOS-2/PALSAR-2 data were kindly provided by the Japanese Aerospace Exploration Agency (JAXA) under the sixth Research Announcement (RA) Program.

(north–south) occupies the center of the image. To the east and west, it is possible to recognize the water areas, displaying zero-mean, noisy closure-phase.

Finally, the presence of closure phases forces us to reconsider the phase estimation in an interferometric stack of SAR images. The choice and weight given to the various interferograms can potentially lead to very different results in the reconstruction of the phase history. The phase triangulation algorithms were meant to deal with statistical misclosure, not with physical misclosure. Surprisingly, at least to a point, the triangulation algorithms display some robustness to physical misclosures too.

6.3.2. *Mathematical properties of closure phases*

Initially, the reader could be surprised in hearing that phases are always perfectly closed for single pixels and almost never for spatial averages of interferograms. The clean explanation lies in the nonlinearity of the arc-tangent operation: there is actually no reason why a linear operation in the real and imaginary part of the interferogram – like multi-looking – should commute with a nonlinear function.

The false zero-closure-phase expectation is also supported by the usual models for the phase, which describe the phase as a pure distance to a single target (equation [4.1]). Many physical signals indeed obey the piston-like paradigm (Zwieback *et al.* 2016), such as tropospheric delays, and might easily suggest a hasty generalization.

NOTE.– If the interferometric phase can be written as the difference between two terms, depending separately on each of the two interfering images, then the closure phase will always be zero. In other words, if

$$\varphi_{n,k} = \phi_n - \phi_k \qquad [6.35]$$

then, necessarily

$$\Psi_{1,2,3} = \phi_1 - \phi_2 + \phi_2 - \phi_3 + \phi_3 - \phi_1 = 0 \mod 2\pi. \qquad [6.36]$$

Zero closure phase is the immediate consequence of postulating separable phases, as in equation [6.35]. This is a correct assumption for many physical models, but not for all. Non-zero closure phases exist because equation [6.35] (i.e. the separability) does not hold in general.

Closure phases are invariant to phase calibration[11], i.e. closure phases do not change if we multiply each image by a phasor $\exp[j\phi_n]$. Checking this is trivial, as the effect will be opposite signs on two interferograms of the triplet.

11. In radio astronomy, closure phases are considered a robust estimator not affected by the local, unknown atmospheric delay.

Closure phases as a robust invariant are not unique. Astronomers also exploit closure amplitudes, which are a particular combination of what we would call correlation magnitudes (Chael *et al.* 2018).

A couple of simple properties define a circular permutation symmetry:

$$\Psi_{n,k,h} = \Psi_{h,n,k} = \Psi_{k,h,n}, \qquad [6.37]$$

and an odd symmetry for anti-circular permutations:

$$\Psi_{n,k,h} = -\Psi_{n,h,k}. \qquad [6.38]$$

If the total number of closure phases in a stack with M images is $\binom{M}{3}$, the number of *independent* closure phases is only quadratic in M (De Zan *et al.* 2015; Zwieback *et al.* 2016):

$$N_c = \binom{M-1}{2}. \qquad [6.39]$$

It is possible to see this by noting that the N_c closure phases of the form $\Phi_{1,n,k}$ with $k > n > 1$ constitute a basis. They are enough to express any closure phase:

$$\Psi_{n,k,h} = \Psi_{1,n,k} + \Psi_{1,k,h} + \Psi_{1,h,n} \mod 2\pi, \qquad [6.40]$$

whereas each $\Psi_{1,n,k}$ is necessary as it is the only element containing the interferometric phase $\phi_{n,k}$.

N_c is therefore the dimension of the inconsistent (non-zero closure) subspace; adding the $M-1$ consistent phases (e.g. all relative to a reference), we get:

$$N_c + (M-1) = \binom{M-1}{2} + (M-1) = \binom{M}{2} \qquad [6.41]$$

which is the total number of independent phases in the (temporal) covariance matrix of M acquisitions, considering the Hermitian symmetry.

6.3.3. *Physical misclosures*

The simplest scattering model that allows for non-zero closure phases is a two scatter model with independent phase variations:

$$y(t) = a + b \cdot e^{j\phi(t)}. \qquad [6.42]$$

Let us assume that the two scatterers a and b are zero-mean uncorrelated random variables so that $\mathrm{E}[ab^*] = 0$. The expected value of the interferogram, thanks to this assumption, is:

$$\mathrm{E}[y(t_1)y^*(t_2)] = \sigma_a^2 + \sigma_b^2 \cdot e^{j\phi(t_1) - j\phi(t_2)}. \qquad [6.43]$$

Provided that $\sigma_a^2 \neq \sigma_b^2$, any trajectory comprising three different values for $\phi(t)$ will generate a non-zero closure phase. The two scatterers represent different populations in the same resolution cell, with different phase histories. Sometimes, the multi-looking operation can be responsible for mixing two populations that are well separated in the image. In this special case, the closure phase would be an artifact.

A more general model would include many (possibly infinite) scatterers, each modeled by a zero-mean uncorrelated stochastic process with a peculiar phase history:

$$y(t) = \sum a_n(t)e^{j\phi_n(t)} \quad [6.44]$$

and real correlation $\mathrm{E}[a_n(t_1)a_k^*(t_2)] = \gamma_n(t_1, t_2)\,\sigma_n^2\,\delta_{nk}$, where δ_{nk} is the Kronecker symbol.

The corresponding interferograms will be:

$$I_{1,2} = \mathrm{E}[y(t_1)y^*(t_2)] = \sum \mathrm{E}[a_n(t_1)a_n^*(t_2)]e^{j\phi_n(t_1)-j\phi_n(t_2)} \quad [6.45]$$

$$= \sum \gamma_n(t_1, t_2)\sigma_n^2 e^{j\phi_n(t_1)-j\phi_n(t_2)}. \quad [6.46]$$

This is simply the sum of the individual interferograms corresponding to the elemental scatterers.

How can we distinguish non-zero closures that are physical from statistical non-closures? We can build a statistical hypothesis test. Knowing the coherence magnitudes and the number of independent samples L, the closure-phase variance for the zero hypothesis with enough samples is (De Zan *et al.* 2015):

$$\mathrm{Var}[\Psi_{1,2,3}] = \mathrm{E}[\Psi_{1,2,3}^2] = \mathrm{E}[(\varphi_{12} + \varphi_{23} + \varphi_{31})^2] \quad [6.47]$$

$$\simeq \frac{1}{2L\gamma_{12}^2\gamma_{23}^2\gamma_{31}^2}\left[3\,\gamma_{12}^2\gamma_{23}^2\gamma_{31}^2 + \gamma_{12}^2\gamma_{23}^2 + \gamma_{23}^2\gamma_{31}^2 + \gamma_{31}^2\gamma_{12}^2\right.$$

$$\left. -2\,\gamma_{12}\gamma_{23}\gamma_{31}(\gamma_{12}^2 + \gamma_{23}^2 + \gamma_{31}^2)\right] \quad [6.48]$$

As we would expect, this variance tends to zero for increasing L. We must say that this formula is difficult to apply in practice; for instance, if there are not enough samples or the coherence estimation is biased. In many cases, however, closure phases display such an evident physical structure that a proof of their physical significance is not needed.

Zwieback *et al.* (2016) have developed a statistical test for the multi-image case.

It is possible to find examples of physical closure phases in all frequency bands, for P-band to K-band. The mere mathematical possibility of their existence is in fact defined by a rich variety of cases. There is no single physical phenomenon that explains all observed misclosures and, in reality, different physical mechanisms will be at work simultaneously, which might complicate the retrieval of physical parameters.

This is not a new situation in SAR, where even in the ordinary interferometric phase, it is possible to recognize a plethora of contributions (deformation, topography, atmospheric propagation, etc.). It is an error to expect closure phases to derive from a single physical origin. The following sections present different physical models that can yield closure phases to clarify this concept.

6.3.3.1. *A volumetric scatterer observed at different angles*

This example of physical non-closure is really equivalent to the model used in astrophysics.

Let us assume that we have a volumetric scatterer ξ that statistically depends only on the vertical coordinate such that $\mathrm{E}[\xi(z)\xi^*(z')] = f(z)\delta(z-z')$. The function $f(z) \geq 0$ then describes the expected intensity along the vertical direction.

If the volume is observed under different incidence angles, the resulting images can be modeled as integrals:

$$y(k_n) = \int \xi(z)e^{jk_n z}dz. \tag{6.49}$$

as a function of the vertical wavenumber k_n, for the n-th acquisition. The interferograms will be:

$$I_{n,h} = \mathrm{E}[y(k_n)y^*(k_h)] = \int f(z)e^{j(k_n-k_h)z}dz, \tag{6.50}$$

i.e. a function of the differential wavenumber $k_n - k_h$. Normalizing $f(z)$ such that $\int f(z)dz = 1$, which makes it equivalent to a probability distribution, it is possible to show (De Zan *et al.* 2015) that for small wavenumber variations:

$$\Psi_{1,2,3} = \angle\{< y_1 y_2^* >< y_2 y_3^* >< y_3 y_1^* >\} \tag{6.51}$$

$$\approx -\frac{1}{2}\mathrm{E}[(z-\mathrm{E}[z])^3](k_1-k_2)(k_2-k_3)(k_3-k_1). \tag{6.52}$$

This gives a simple interpretation of the closure phase as being proportional to the skewness $\mathrm{E}[(z-\mathrm{E}[z])^3]$ of the scattering intensity profile $f(z)$. This interpretation is of course shared in astronomical imaging with closure phases.

It is possible to estimate the maximum closure phase for a given volume height $z_0 > 0$ in the small angle approximation. Let us consider the vertical distribution

$$f(z) = \delta(z)(1 - a) + \delta(z - z_0)a, \qquad [6.53]$$

where δ is the Dirac distribution, which represents one scatterer at the ground level and one scatterer at height z_0. The parameter a controls the relative intensity of the two scatterers. The skewness is

$$\mathrm{E}[(z - \mathrm{E}[z])^3] = (2a^3 - 3a^2 + a)\, z_0^3, \qquad [6.54]$$

and its maximum absolute value is attained for $a = (3 \pm \sqrt{3})/6$, where

$$\left|\mathrm{E}[(z - \mathrm{E}[z])^3]\right| = \frac{1}{6\sqrt{3}}\, z_0^3. \qquad [6.55]$$

Therefore, the maximum closure phase will be:

$$\max|\Psi_{1,2,3}| \approx \frac{1}{12\sqrt{3}}|(k_1 - k_2)(k_2 - k_3)(k_3 - k_1)|\, z_0^3. \qquad [6.56]$$

This formula can help in discriminating between different physical mechanisms, given the magnitude of the observed closure phase.

This type of closure phase mechanism can impact deformation estimation in time series where the distribution of the baselines has some structure in time, for example, if the baselines are left to drift for periods longer than the repeat pass and then are suddenly corrected. Indeed, orbital drift is not random, for example, because of atmospheric drag. In other cases, the mission plan may contemplate a certain systematic baseline variation, as in the case of the BIOMASS mission.

6.3.3.2. *A simple interferometric model for soil moisture*

A possible interferometric model for a soil under different moisture conditions is to think of the soil as a dielectric volume, whose permittivity ϵ varies as a function of the moisture state and soil composition (Hallikainen *et al.* 1985; Bircher *et al.* 2016). Following the Born approximation, little scatterers dispersed inside the volume will generate a backscattered field, each scatterer with amplitude and phase proportional to the incident background field (De Zan *et al.* 2012). The resulting model for the backscatter is very similar to the one for a volume observed at different angles:

$$y(k_n) = \int \xi(z) e^{-j2k_n z} dz. \qquad [6.57]$$

One first difference is that the volume here is likely only a few centimeters thick. A second difference is that the vertical wavenumber k_n is typically complex. It is a function of permittivity, approximately $k_n = \sqrt{\omega^2 \mu \epsilon(n)}$, where ω is the angular frequency, μ is the dielectric permeability and $\epsilon(n)$ the dielectric permittivity at the time of the nth acquisition.

To derive the expected value of the interferograms, we adopt the usual assumption that $\mathrm{E}[\xi(z)\xi^*(z')] = f(z)\delta(z - z')$ and obtain:

$$I_{n,h} = \mathrm{E}[y(k_n)y^*(k_h)] = \int f(z)e^{-j2(k_n - k_h^*)z}dz. \quad [6.58]$$

The scattering profile does not have to follow any particular dependence on depth. If it is exponential, $f(z) = \exp(-2\alpha z)$, with $\alpha \geq 0$, the integral has a simple analytical solution:

$$I_{n,h} = \frac{1/2}{j(k_n - k_h^*) + \alpha}. \quad [6.59]$$

For $\alpha = 0$, the formula covers also the case of a constant scattering profile.

This model seems to describe reasonably well the closure phases in L-band over bare soil, at least in the example given by De Zan *et al.* (2014). Inverting the soil moisture model from closure phases is not trivial; as Zwieback *et al.* (2017) point out, there is an ambiguity in the relation between moisture histories and the corresponding closure phases. However, with the additional help of coherence magnitudes or intensities, the inverse problem seems feasible at least for a limited sequence of acquisitions (De Zan and Gomba 2018).

In theory, moisture variations can produce a velocity bias in the time series. The bias can arise from the moisture cycle, which normally sees a sudden wetting and a slow drying. If the phase depends in a nonlinear way on moisture differences, then a fast wetting will be more than compensated by a slow drying. In the long term, by integrating short-term pairs, we should observe an apparent decrease in the distance, i.e. a motion towards the satellite. This might have been observed, but conclusive proof is still needed.

6.3.3.3. *Other reasons for differential motion in a resolution cell*

There are likely many other sources of differential motion in a resolution cell that can ultimately yield non-zero phase closure. For example, there could be differential motion in ice (as a function of depth) when dry ice is observed with sufficiently long wavelengths. There could be thermal expansion displacing different targets of different amounts.

Another possibility is vegetation superimposing a time-variable phase screen. If not all targets are covered by the same amount of vegetation, then there could be differences that produce non-zero closure phases. This tentative explanation is given by Ansari *et al.* (2021) to explain a phase bias that did no display behavior that the moisture model would predict. In this case, the prevailing small steps were biomass accumulation, i.e. delay increase.

Different sources of phase inconsistency might be present at once, and if they are generated by independent (orthogonal) mechanisms, then it is likely safe to assume that they contribute to the closure phase independently. For example, if one differential mechanism is in the time domain and the other in the Doppler domain, then the resulting closure phase would be the sum of the closure phases for each independent mechanism.

6.3.4. *Implications for interferometric phase estimation*

6.3.4.1. *Statistical misclosure and phase triangulation algorithms*

Different from physical misclosures, statistical misclosures have been recognized for a long time. The phase triangulation algorithms (e.g. Monti Guarnieri and Tebaldini 2008; Ansari *et al.* 2018; Michaelides *et al.* 2019) exploit the hypothesis of underlying phase closure to improve the phase estimates in a stack. Even if these were mostly not designed to deal with physical closure phases, they happen to usually do a good job in retrieving the phase related to long-term coherent scatterers (Ansari *et al.* 2021).

This is possible thanks to redundancy. The available interferograms are $M(M-1)/2$, and each of them describes a time difference between only M independent acquisition times. We realize that there are many alternative paths to reconstruct the phase difference between a given pair of acquisition times. The problem is finding which paths are good and how to weight them properly.

There is a common pitfall in choosing the paths: letting coherence drive the choice. Indeed, we can show that many high-coherence steps might be worse than a single or a few low-coherence steps, depending on the coherence scenario. In theory, the phase triangulation algorithms deal with this complexity in an optimal way. In practice, they have limitations related to limited or biased knowledge of the temporal coherence structure.

6.3.4.2. *Biases everywhere*

A particularly dangerous situation arises when the physical closure phases present some particular structure in time, for example, when they tend to be always biased in one direction. This can happen with any of the physical phenomena analyzed above. For instance, the closure phases with three images in a successive acquisition order could tend to have the same sign in the moisture model, if moisture increase is a rare event, and moisture decrease can be observed over many repeat cycles. Similar reasoning can be applied for any of the mentioned possible physical mechanisms.

Another, more abstract, perspective is given by considering the following complex coherence model:

$$\gamma(\Delta t) = \gamma_a + \gamma_b \cdot e^{-|\Delta t|/\tau_b} e^{-j2\pi f_b \Delta t} + (1 - \gamma_a - \gamma_b)\delta(\Delta t). \qquad [6.60]$$

This is a model with two scatterers, one stable and the other decorrelating (with temporal characteristic τ_b) and simultaneously moving (with "velocity" f_b). This kind of behavior was observed by Ansari *et al.* (2021). The model should be understood as if we had compensated a phase so that the first scatterer appears not to have moved at all; the second scatterer shows its relative or residual motion with respect to the first. The Kronecker δ is just there to guarantee a coherence of 1 for the auto-interferogram.

We might find it disturbing that the second scatterer appears to move indefinitely, but this is accommodated by the coherence loss, under which a steady state might be continuously recovered under the invisibility cloak of decorrelation.

According to this model, long-term interferograms will see the "correct" velocity of the first scatterer, since the second scatterer is completely decorrelated. However, short-term interferograms will see another apparent motion, partially biased by the motion of the second scatterer. Considering interferograms of different time lags, the estimated velocity might be very different. Figure 6.6 displays the velocity bias as a function of the temporal baseline for a certain choice of the parameters (based roughly on C-band observations over Sicily reported by Ansari *et al.* (2021)). In this example, the bias can reach about 10 mm/year for a short temporal baseline and is less than 1 mm/year only when the separation is greater than 30 days.

The reality is likely more complex that the simple model presented here. There might be different contributions with different time constants and different apparent velocities. If a long-term coherent scatterer is present, it can act as an anchor to keep the estimates from drifting. This is done implicitly by using a full-covariance estimator or explicitly by avoiding short-term interferograms, as in the study by Daout *et al.* (2020).

It might be possible to quantify the bias for short-term interferograms by estimating closure phases with very asymmetric temporal lags, i.e. with two long arms and one short arm. In the long arms, the bias will probably be very small so that the closure phase would be representative of the short-arm bias. Of course, this requires some care, since in the long arms, the decorrelation will be strong.

6.3.4.3. *Outlook: multi-frequency and polarimetry*

Closure phases remind us that the phase of the scattered signal does not reflect a simple delay, and this can have significant consequences on the deformation products and complex implications for the processing algorithms. A deeper understanding of the scattering mechanisms requires more physical modeling, which will have to explain backscatter and phase behavior across different frequencies and polarizations.

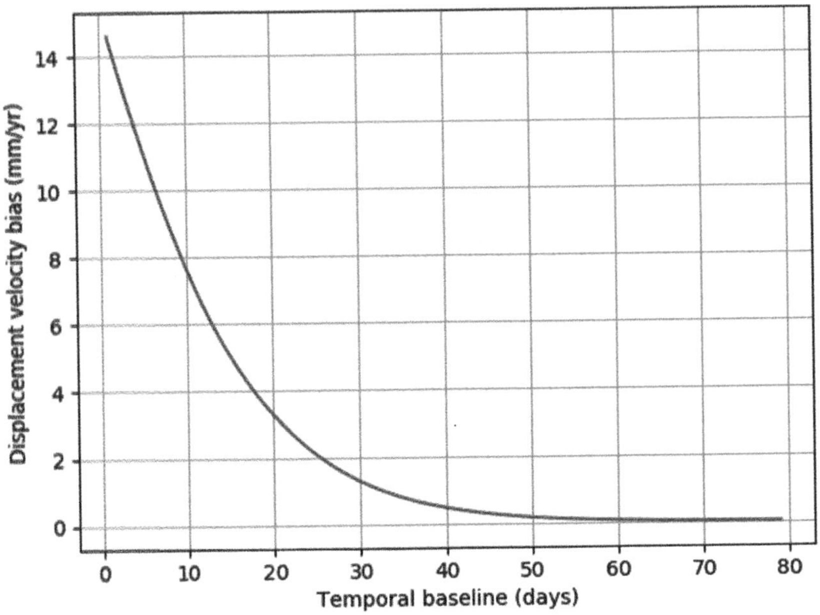

Figure 6.6. *Velocity bias as a function of temporal baseline for the model in equation [6.60] for* $\gamma_a = 0.38$, $\gamma_b = 0.18$, $\tau_b = 11$ days, $f_b = 0.03$ rad/day

6.4. Acknowledgments

We thank Professor Mario Costantini, Professor Paco López-Dekker and Mrs Yan Yuan for their careful reviews and comments that helped to improve this chapter.

6.5. References

Ahuja, R.K., Magnanti, T.L., Orlin, J.B. (1993). *Network Flows: Theory, Algorithms, and Applications*. Prentice Hall, Englewood Cliffs.

Ansari, H., De Zan, F., Bamler, R. (2018). Efficient phase estimation for interferogram stacks. *IEEE Transactions on Geoscience and Remote Sensing*, 56(7), 4109–4125.

Ansari, H., De Zan, F., Parizzi, A. (2021). Study of systematic bias in measuring surface deformation with SAR interferometry. *IEEE Transactions on Geoscience and Remote Sensing*, 59(2), 1285–1301.

Bamler, R. and Hartl, P. (1998). Synthetic aperture radar interferometry. *Inverse Problems*, 14.

Bamler, R., Adam, N., Davidson, G.W., Just, D. (1998). Noise-induced slope distortion in 2-D phase unwrapping by linear estimators with application to SAR interferometry. *IEEE Transactions on Geoscience and Remote Sensing*, 36(3), 913–921.

Benoit, A., Pinel-Puysségur, B., Jolivet, R., Lasserre, C. (2020). CorPhU: An algorithm based on phase closure for the correction of unwrapping errors in SAR interferometry. *Geophysical Journal International*, 221(3), 1959–1970.

Biggs, J., Wright, T., Lu, Z., Parsons, B. (2007). Multi-interferogram method for measuring interseismic deformation: Denali Fault, Alaska. *Geophysical Journal International*, 170(3), 1165–1179.

Bircher, S., Demontoux, F., Razafindratsima, S., Zakharova, E., Drusch, M., Wigneron, J.-P., Kerr, Y.H. (2016). L-band relative permittivity of organic soil surface layers – A new dataset of resonant cavity measurements and model evaluation. *Remote Sensing*, 8(12), 1–17.

Chael, A.A., Johnson, M.D., Bouman, K.L., Blackburn, L.L., Akiyama, K., Narayan, R. (2018). Interferometric imaging directly with closure phases and closure amplitudes. *The Astrophysical Journal*, 857(1), 23.

Chen, C.W. and Zebker, H.A. (2000). Network approaches to two-dimensional phase unwrapping: Intractability and two new algorithms. *Journal of the Optical Society of America A*, 17(3), 401–414.

Chen, C.W. and Zebker, H.A. (2001). Two-dimensional phase unwrapping with use of statistical models for cost functions in nonlinear optimization. *Journal of the Optical Society of America A*, 18(2), 338–351.

Chen, C.W. and Zebker, H.A. (2002). Phase unwrapping for large SAR interferograms: Statistical segmentation and generalized network models. *IEEE Transactions on Geoscience and Remote Sensing*, 40(8), 1709–1719.

Costantini, M. (1998). A novel phase unwrapping method based on network programming. *IEEE Transactions on Geoscience and Remote Sensing*, 36(3), 813–821.

Costantini, M. and Rosen, P. (1999). A generalized phase unwrapping approach for sparse data. *IGARSS'99*, 1, 267–269.

Costantini, M., Malvarosa, F., Minati, F. (2012). A general formulation for redundant integration of finite differences and phase unwrapping on a sparse multidimensional domain. *IEEE Transactions on Geoscience and Remote Sensing*, 50(3), 758–768.

Daout, S., Dini, B., Haeberli, W., Doin, M.-P., Parsons, B. (2020). Ice loss in the northeastern Tibetan Plateau permafrost as seen by 16 yr of ESA SAR missions. *Earth and Planetary Science Letters*, 545, 116404.

De Zan, F. and Gomba, G. (2018). Vegetation and soil moisture inversion from SAR closure phases: First experiments and results. *Remote Sensing of Environment*, 217, 562–572.

De Zan, F., Parizzi, A., Prats, P. (2012). A proposal for a SAR interferometric model of soil moisture. *IEEE International Geoscience and Remote Sensing Symposium*, 630–633.

De Zan, F., Parizzi, A., Prats-Iraola, P., López-Dekker, P. (2014). A SAR interferometric model for soil moisture. *IEEE Transactions on Geoscience and Remote Sensing*, 52(1), 418–425.

De Zan, F., Zonno, M., Lopez-Dekker, P. (2015). Phase inconsistencies and multiple scattering in SAR interferometry. *IEEE Transactions on Geoscience and Remote Sensing*, 53(12), 6608–6616.

Ferretti, A., Prati, C., Rocca, F. (2001). Permanent scatterers in SAR interferometry. *IEEE Transactions on Geoscience and Remote Sensing*, 39(1), 8–20.

Ghiglia, D.C. and Romero, L.A. (1994). Robust two-dimensional weighted and unweighted phase unwrapping that uses fast transforms and iterative methods. *Journal of the Optical Society of America A*, 11(1), 107–117.

Goldstein, R., Zebker, H.A., Werner, C. (1988). Satellite radar interferometry: Two-dimensional phase unwrapping. *Radio Science*, 23, 713–720.

Hallikainen, M.T., Ulaby, F.T., Dobson, M.C., El-Rayes, M.A., Wu, L.-K. (1985). Microwave dielectric behavior of wet soil – Part 1: Empirical models and experimental observations. *IEEE Transactions on Geoscience and Remote Sensing*, 23(1), 25–34.

Hooper, A. (2010). A statistical-cost approach to unwrapping the phase of InSAR time series. *Fringe Workshop*, 30 November–4 December, Frascati.

Hooper, A. and Zebker, H.A. (2007). Phase unwrapping in three dimensions with application to InSAR time series. *Journal of the Optical Society of America A*, 24(9), 2737–2747.

Hunt, B.R. (1979). Matrix formulation of the reconstruction of phase values from phase differences. *Journal of the Optical Society of America A*, 69(3), 393–399.

Huntley, J.M. (2001). Three-dimensional noise-immune phase unwrapping algorithm. *Applied Optics*, 40(23), 3901–3908.

Hussain, E., Hooper, A., Wright, T.J., Walters, R.J., Bekaert, D.P.S. (2016). Interseismic strain accumulation across the Central North Anatolian Fault from iteratively unwrapped InSAR measurements. *Journal of Geophysical Research: Solid Earth*, 121(12), 9000–9019.

Michaelides, R.J., Zebker, H.A., Zheng, Y. (2019). An algorithm for estimating and correcting decorrelation phase from InSAR data using closure phase triplets. *IEEE Transactions on Geoscience and Remote Sensing*, 57(12), 10390–10397.

Monti Guarnieri, A. and Tebaldini, S. (2008). On the exploitation of target statistics for SAR interferometry applications. *IEEE Transactions on Geoscience and Remote Sensing*, 46, 3436–3443.

Pepe, A. and Lanari, R. (2006). On the extension of the minimum cost flow algorithm for phase unwrapping of multitemporal differential SAR interferograms. *IEEE Transactions on Geoscience and Remote Sensing*, 44(9), 2374–2383.

Pepe, A., Euillades, L.D., Manunta, M., Lanari, R. (2011). New advances of the extended minimum cost flow phase unwrapping algorithm for SBAS-DInSAR analysis at full spatial resolution. *IEEE Transactions on Geoscience and Remote Sensing*, 49(10), 4062–4079.

Shanker, A.P. and Zebker, H.A. (2010). Edgelist phase unwrapping algorithm for time series InSAR analysis. *Journal of the Optical Society of America A*, 27(3), 605–612.

Trouvé, E., Nicolas, J.M., Maître, H. (1998). Improving phase unwrapping techniques by the use of local frequency estimates. *IEEE Transactions on Geoscience and Remote Sensing*, 36(6), 1963–1972.

Yunjun, Z., Fattahi, H., Amelung, F. (2019). Small baseline InSAR time series analysis: Unwrapping error correction and noise reduction. *Computers and Geosciences*, 133, 104331.

Zwieback, S., Liu, X., Antonova, S., Heim, B., Bartsch, A., Boike, J., Hajnsek, I. (2016). A statistical test of phase closure to detect influences on DInSAR deformation estimates besides displacements and decorrelation noise: Two case studies in high-latitude regions. *IEEE Transactions on Geoscience and Remote Sensing*, 54(9), 5588–5601.

Zwieback, S., Hensley, S., Hajnsek, I. (2017). Soil moisture estimation using differential radar interferometry: Toward separating soil moisture and displacements. *IEEE Transactions on Geoscience and Remote Sensing*, 55(9), 5069–5083.

PART 2

Applications for Surface Displacements

7

Remote Sensing of the Earthquake Deformation Cycle

Mathilde MARCHANDON[1], Tim J. WRIGHT[2] and James HOLLINGSWORTH[1]

[1] *ISTerre, Grenoble, France*
[2] *COMET, University of Leeds, UK*

7.1. Introduction

Our most basic understanding of the earthquake deformation cycle comes from the theory of the elastic rebound elaborated by Reid (1910) after the 1906 San Francisco earthquake. As tectonic plates slowly move, elastic strain is accumulated on faults. When the accumulated strain is large enough to overcome the fault strength, all the strain is suddenly released by a sudden slip on the fault, producing an earthquake. Fault slip induces displacements at the ground surface that can be measured and used to better understand earthquake processes. Displacements can also be measured during the long "inter-seismic" buildup to earthquakes and the rapid "post-seismic" transients that follow many earthquakes.

Before the advent of space-based geodesy, surface displacement measurements could only be made at sparse locations and only in regions where a surveyed network was in place before the earthquake. Thirty years ago, the development of the interferometric synthetic aperture radar (InSAR) technique revolutionized our ability to measure the surface displacement fields of earthquakes. By comparing the phase

Surface Displacement Measurement from Remote Sensing Images,
coordinated by Olivier CAVALIÉ and Emmanuel TROUVÉ.
© ISTE Ltd 2022.

of RADAR satellite images, InSAR can precisely measure the surface displacements of earthquakes with nearly complete spatial continuity. Since its first application for measuring the coseismic displacements of the 1992 Landers earthquake (Massonnet *et al.* 1993), and thanks to the dramatic increase in data availability, InSAR has been applied to document the surface deformation of more than 100 earthquakes (Funning and Garcia 2019). InSAR also allows us to track the deformation produced by inter-seismic loading and aseismic transient processes, such as post-seismic deformation or inter-seismic creep events (e.g. Peltzer *et al.* 1998; Wright *et al.* 2001b; Rousset *et al.* 2016). InSAR data have been critical in enabling scientists to constrain the distribution of slip and fault geometry at depth, determine the rheology of the lithosphere and assess the frictional properties of fault zones. Furthermore, InSAR data can now be used to map inter-seismic loading over thousands of square kilometers, maps that might be used in the near future to improve models of seismic hazard assessment (Weiss *et al.* 2020).

More recently, another space-based geodetic technique, optical correlation, has been developed. This technique makes it possible to measure the horizontal displacements of an earthquake by measuring the pixel shifts between two pictures of the ground. The vertical component of the displacement can also be assessed if stereoscopic images are available. Although less precise than InSAR, this technique has the considerable advantage of being able to measure the near-field deformation of surface-rupturing earthquakes, where InSAR often cannot provide measurements due to the decorrelation of the interferometric phase. The geometry of the surface rupture and the surface slip can be readily observed at spatial resolutions not achievable by field investigations (or other geodetic techniques). This new type of data has allowed researchers to better characterize the surface rupture of large continental earthquakes (e.g. Michel and Avouac 2002), estimate the amount of deformation occurring in the volume around the fault (e.g. Zinke *et al.* 2014) and better constrain fault slip models (e.g. Xu *et al.* 2016).

In this chapter, we review the main advances brought by InSAR and optical correlation for our understanding of earthquake and fault mechanics. The chapter is divided into two parts. The first part provides an overview of what we have learned from 30 years of tectonic InSAR (section 7.2). Section 7.2.1 focuses on the coseismic phase of the seismic cycle, section 7.2.2 on the inter-seismic phase and section 7.2.3 on the post-seismic phase. Each section reviews the history of InSAR studies and what we have learned about faults and earthquakes from them. For a detailed review of the technical aspects of InSAR, we refer the reader to Chapters 4, 5 and 6 of this book. In the second part of the chapter, we focus on the study of surface-rupturing earthquakes using optical correlation (section 7.3). First, the advantages of the technique compared to more traditional methods are reviewed and a brief history of the technique is presented (section 7.3.1). Then, we briefly describe the general workflow used to obtain the 2D and 3D displacement field of an earthquake from optical images, as well as the main sources of error and limitations (section 7.3.2). For the theoretical aspect of optical correlation, we refer the reader to Chapter 2 of this book. In section 7.3.3, we show examples of surface displacement fields from recent and pre-modern geodesy (<1990s) for earthquakes measured by optical correlation. Finally, we review the insights on earthquake and fault mechanics brought by high-resolution optical correlation displacement fields (section 7.3.4).

7.2. What have we learned about faults from nearly three decades of tectonic InSAR?

Zebker and Goldstein (1986) were the first to demonstrate that an interferogram could be formed from two radar images. In this case, they were acquired simultaneously from an airborne platform, and the interference patterns were due to topography. Two years later, Goldstein *et al.* (1988) published an interferogram from two images acquired three days apart by the short-lived SeaSat mission (Barrick and Swift 1980). Shortly afterwards, Gabriel *et al.* (1989) proposed that differential synthetic aperture radar interferometry could be used to map small ground movements over large areas. However, it was not until the launch of the first European Remote Sensing Satellite (ERS-1) by the European Space Agency (ESA) in 1991 that there was a platform stable enough to perform differential InSAR reliably. The first coseismic interferogram by Massonnet *et al.* (1993), famously gracing the cover of *Nature*, showed the line-of-sight displacement field for the 1992 M_w 7.3 Landers earthquake in spectacular detail.

The Landers earthquake paper marked the beginning of nearly three decades of tectonic applications of InSAR, which we review below. We organize the review according to the phase of the earthquake cycle. For each phase, we describe the history of InSAR studies, how those studies have evolved and progressed over the last 30 years and what we have learned about faults from the research. The review is not intended to be an comprehensive overview of every tectonic study, and the selection of research highlighted below is certainly biased by the authors' own experiences and prejudices. We were motivated by being challenged with the question, "what have we learned from InSAR about tectonics and faulting?", and we apologize if what follows reads like an answer to Monty Python's famous question, "What have the Romans ever done for us?"[1] (Jones 1979).

7.2.1. *Coseismic deformation*

7.2.1.1. *History and recent progress in coseismic InSAR studies*

Before InSAR, it was extremely challenging to measure surface displacements caused by earthquakes. Earthquakes in a specific location are rare, so taking

1. "All right, but apart from the sanitation, the medicine, education, wine, public order, irrigation, roads, the fresh water system and public health…".

displacement measurements relied on serendipity – earthquakes could only be measured if precise survey data had been acquired prior to the event (e.g. Stein and Barrientos 1985) or if natural data, such as uplifted shorelines, could be interpreted (e.g. Plafker 1965). As such, displacements from around 20 earthquakes or fewer had been measured (Wright 2000). Data from these studies appeared to confirm the elastic rebound theory popularized by Reid (1910) but were in general too sparse to be used to determine details of earthquake mechanisms and slip distributions.

The widespread availability of SAR data acquired before and after earthquakes, along with the proliferation of measurements from global navigation satellite systems (such as the US Global Positioning System), means that spatially dense, precise observations of surface displacements are now available for almost all recent earthquakes where there is significant (1 cm or more) movement of the land surface. Earthquake deformation has been mapped for events ranging in size from small ($M_w \sim 5$), shallow (< 10 km) events on continental faults (e.g. Lohman and Simons 2005) to on-land deformation from great earthquakes on subduction megathrusts (e.g. Kobayashi *et al.* 2011).

The ability of InSAR to capture coseismic deformation relies on pre- and post-event data existing. Mission managers at the ESA recognized this opportunity and implemented systematic background acquisition plans for the ERS and Envisat missions (Attema *et al.* 2000; Miranda *et al.* 2013) that ensured data were acquired regularly over tectonically active areas, although the 35-day revisit time and relatively wide orbital tube for ERS and Envisat meant that the method did not work for many earthquakes. Nevertheless, by the end of the Envisat mission in 2010, studies on 60 earthquakes using InSAR data had been published (Weston *et al.* 2012).

The launches of the EU Copernicus Sentinel-1 C-band SAR satellites, Sentinel-1A (launched in 2014) and Sentinel-1B (launched in 2016), have radically increased the volume and quality of SAR data available for InSAR analysis. Together, the satellite constellation can acquire SAR data with a six-day revisit, and a small orbital tube means that interferograms can be formed between all acquisitions (Torres *et al.* 2012). Funning and Garcia (2019) examined Sentinel-1A data for 35 "potentially detectable" earthquakes that occurred between April 2015 and December 2016, finding that around 70% of these had coherent signals. However, they found that 31% of the events, mostly in vegetated low latitude regions, could not be detected with the available C-band data from Sentinel-1A. Since that study, Sentinel-1 has increased its duty cycle and acquires data globally every 6 or 12 days, depending on the location, including on both ascending and descending tracks in the tectonic belts. Several automatic systems have been built that provide rapid response interferometric data to the community for analysis (e.g. Barnhart *et al.* 2019; Lazecký *et al.* 2020).

C-band data ($\sim$6 cm wavelength) can suffer from poor coherence in vegetated areas. Further improvements can be made in these cases using data from longer-wavelength L-band SAR missions ($\sim$20 cm wavelength), such as from PALSAR on the Japanese ALOS-2 satellite (Rosenqvist *et al.* 2014). The launch of the joint US–India L-band SAR mission NISAR, currently scheduled for 2022 (Rosen *et al.* 2017), will ensure L-band SAR data are acquired globally for all ascending and descending passes with a 12-day revisit time. The EU Copernicus program is also planning an operational L-band SAR mission to complement the C-band data from Sentinel-1 (Elliott 2020).

7.2.1.2. *What have we learned about faults from InSAR studies of coseismic deformation*

For individual earthquakes on land, the detailed maps of surface deformation provided by InSAR typically provide more accurate locations and depths than are available from global teleseismic seismology, often revealing significant mislocations in the seismic catalogues (Weston *et al.* 2011). For shallow events larger than $M_w \sim 6$, it is usually possible to derive detailed slip distributions from InSAR, again with the geodetic solutions resolving finer spatial details than is possible from global seismology. The large number of dense geodetic earthquake deformation observations has led to a significant increase in our knowledge of earthquake behavior in several key areas. Here, we summarize some of the major advances in our understanding about faults and earthquakes that have been made possible through InSAR observations.

Fault complexity: a key area where our understanding of earthquakes has changed because of the availability of precise, dense geodetic data from InSAR is earthquake complexity. Earthquakes with M_w less than around 6 can typically be modeled with uniform slip on a single rectangular fault plane in an elastic half-space (Okada 1985), particularly if they do not rupture the surface. However, larger, shallow earthquakes usually require more complex models of variable slip on segmented faults or multiple fault strands (e.g. Elliott *et al.* 2016). Several recent earthquakes, including the 2010 M_w7.1 Darfield (New Zealand) earthquake (Elliott *et al.* 2012), the 2010 M_w7.2 El Mayor Cucapeh (Mexico) earthquake (Huang *et al.* 2017) and the 2019 M_w6.4/7.1 Ridgecrest (USA) sequence (Xu *et al.* 2020), have required significant complexity to match the geodetic observations, with slip occurring on multiple fault strands.

The type example of a complex, multi-fault rupture is the 2016 M_w7.8 Kaikoura (New Zealand) earthquake, which ruptured more than 170 km of faults in the northeast part of the South Island of New Zealand (Figure 7.1; Hamling *et al.* 2017). In total, more than 20 crustal faults had surface ruptures, and these include faults that were previously mapped as well as previously unknown structures (Litchfield *et al.* 2018). Both crustal faults and the deeper subduction interface may have failed simultaneously (Furlong and Herman 2017). The earthquake defied conventional wisdom about the ability of fault segmentation to arrest/control rupture; the rupture appeared to jump across fault stepovers as high as 15–20 km (Hamling *et al.* 2017). Without the detailed geodetic imaging work from InSAR and optical correlation, it would have been extremely challenging for field teams to map the complexity of this earthquake – scarps from surface ruptures erode very rapidly in New Zealand. The new understanding of the potential for complex, multi-fault ruptures is now being incorporated in the latest generation of seismic hazard models.

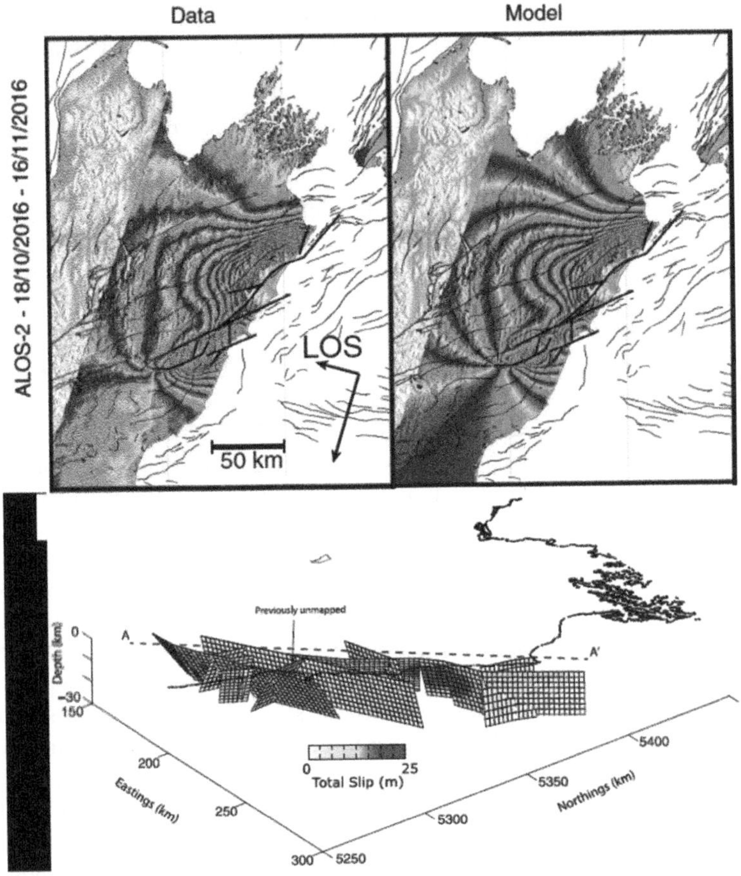

Figure 7.1. *Deformation and model for the 2016 Kaikoura earthquake, adapted from (Hamling* et al. *2017). Top left: coseismic interferogram from ALOS-2. Top right: predicted interferogram using dislocation models. Bottom: 3D view of the complex slip model for the earthquake, showing only the crustal faults that were modeled as slipping in the event. For a color version of this figure, see www.iste.co.uk/cavalie/images.zip*

Surface versus deep fault slip during earthquakes: another area where dense geodetic imaging data have changed our understanding of earthquakes is the relationship between fault slip seen at the surface in an earthquake and the slip that occurs at depth. Earthquakes on "blind" thrust faults, where minimal slip reaches the surface, were known before InSAR (e.g. Yielding *et al.* 1981; Ekström *et al.* 1992), but we now know that for all types of earthquake, the surface slip is in fact only a partial guide to the slip that occurs at depth. For example, modeling of the 1995 $M_w6.4$ Dinar earthquake, a normal-faulting earthquake in southwest Turkey, requires 1.2–1.4 m of slip as shallow as 2 km beneath the surface, but the maximum observed surface rupture magnitude was only 0.25 m (Fukahata and Wright 2008). Similarly, in the 2003 $M_w6.5$ Bam earthquake (strike-slip, Iran), models require 2–3 m of slip at $\sim$ 4 km depth, but the maximum surface slip was again 0.25 cm. This observation is not limited to moderate earthquakes – Kaneko and Fialko (2011) compared the slip profiles of the Bam earthquake and four other strike-slip faults with $M_w > 7.1$, showing that they all had a shallow slip deficit as high as 60%. Although more detailed analysis of near-field date by Xu *et al.* (2016) reduced the magnitude of the discrepancy for some of these earthquakes, it is clear that the surface slip often only provides a minimum estimate for the fault slip that occurred in the subsurface.

The observation of a discrepancy between surface observations and slip at depth has important implications for studies of paleoseismology – magnitudes estimated from paleo-offset data and empirical relationships between slip and magnitude (Wells and Coppersmith 1994) are likely to be lower bounds of the true magnitude of earthquakes. The causes and implications of the shallow slip deficit (SSD) have also been investigated. There are several plausible explanations. The first is that fault slip branches near the surface onto multiple strands, and without precise and detailed near-field data, this can map into an apparent deficit (e.g. Xu *et al.* 2016). A second possibility is that friction in the shallowest portion of the fault is velocity-strengthening – some shallow slip then occurs as afterslip or inter-seismic creep. Hussain *et al.* (2016) investigated this possibility for part of the North Anatolian Fault that ruptured in the 1999 Izmit/Kocaeli earthquake, finding that without additional periods of rapid shallow slip, this mechanism could only partially explain the observed shallow slip deficit. Finally, it is possible that inelastic distributed deformation, for example due to folding, accommodates some of the strain in the shallowest portion of the fault. It seems likely that these mechanisms may operate concurrently in some cases. High-resolution near-field displacement data from optical image correlation is invaluable for investigating these mechanisms (see discussion later in this chapter).

Dynamic triggering: although geodetic imaging techniques like InSAR provide a snapshot of the total deformation that occurs between two image acquisitions, the dense spatial coverage of the observations has revealed unexpected deformation patterns in a few instances that were initially missed by seismology. One of the best observations of dynamic triggering is from Pagli *et al.* (2003), who observed deformation on the Reykjanes Peninsula in Iceland that was consistent with an undetected $M_w5.8$ strike-slip earthquake. They showed that the earthquake was dynamically triggered by a $M_w6.6$ earthquake around 100 km away. The triggered earthquake was not initially detected by seismic networks because the seismic waveform was partially hidden by the waveforms from the larger event.

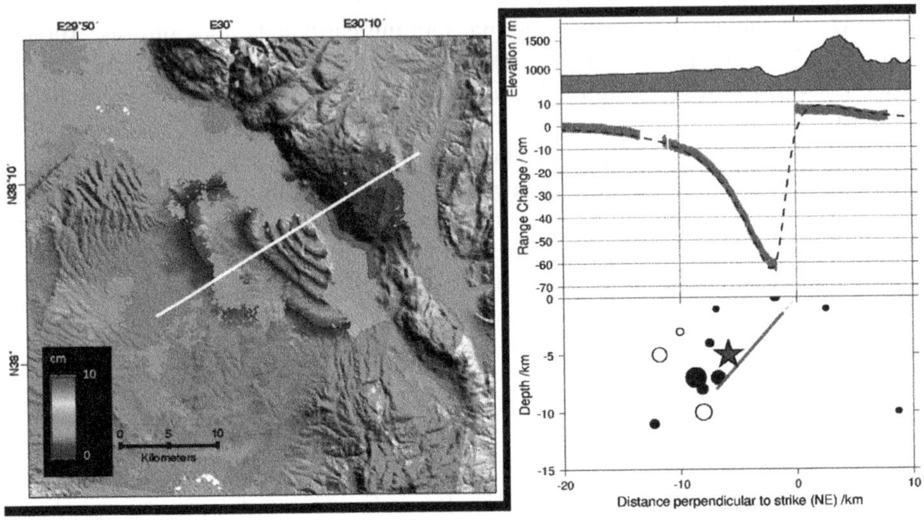

Figure 7.2. *Surface displacements in the 1995 M_w 6.4 Dinar (Turkey) earthquake. Left: Coseismic interferogram formed between an ERS-1 acquisition on August 13, 1995 and an ERS-2 acquisition on January 1, 1996, unwrapped and rewrapped with 10 cm fringes, draped on Shuttle Radar Topography Mission (SRTM) topography. The red line shows the surface rupture location. Right: Profiles along the white line in the map showing topography and range change (negative values are movement away from the satellite in this figure), and a cross-section showing fault geometry and seismicity within two weeks of the mainshock (from the Kandilli Observatory catalogue; foreshocks: open circles, aftershocks: filled, mainshock: star). The red line is the observed ground movement from InSAR and the black dashed line is a model with uniform slip. Elastic dislocation models require 1.2–1.4 m of slip at depth; however, the largest observed surface rupture in the field was 0.25 m (Wright* et al. *1999; Fukahata and Wright 2008). For a color version of this figure, see www.iste.co.uk/cavalie/images.zip*

Another beautiful example of dynamic triggering was presented by Nissen *et al.* (2016), who investigated an M_w7.1 reverse-faulting earthquake that occurred in Pakistan in 1997. They found that the earthquake was in fact not a single event but a doublet, with two clear areas of uplift associated with separate faults. Joint analysis of the InSAR with seismology showed that the initial rupture was followed just 19 s later by rupture of a second reverse fault, located $\sim$ 50 km away. Again they invoke dynamic triggering to explain the occurrence of this doublet. Dynamic triggering is also likely to have played a part in the Kaikoura (New Zealand) earthquake (Figure 7.1) discussed earlier. The potential for dynamic triggering is an additional factor that needs to be accounted for in future seismic hazard models and scenarios.

Structural control of earthquake rupture: although some earthquakes, such as Kaikoura, appear to propagate through structural barriers (Hamling *et al.* 2017), InSAR has documented other examples where pre-existing structure does appear to play an important role in controlling earthquake rupture. Several good examples of this have been observed in the India–Asia collision zone. The 2001 $M_w7.8$ Kokoxili earthquake ruptured more than 400 km of the Kunlun Fault in northern Tibet, with slip reaching as high as 8 m in places. Lasserre *et al.* 2005 showed that areas of significant moment release corresponded to the first-order along-strike structural segmentation of the fault, with very low slip observed and modeled on some segment boundaries. Elliott *et al.* (2011) showed that depth segmentation can also act as a barrier to rupture – they investigated two $M_w6.3$ reverse-faulting earthquakes that occurred in the North Qaidam Thrust system (Tibet) at almost the same location but separated in time by nearly 300 days. They showed that the first of the two earthquakes ruptured the lower part of the seismogenic crust, with the later event rupturing a near continuation of the same fault plane in the upper seismogenic crust. They speculated that the gap between the two earthquakes was due to a structural barrier caused by the intersection of a reverse fault dipping in the opposite direction.

Structural control also appears to have played an important role in the 2015 $M_w7.8$ Gorkha (Nepal) earthquake, which partially ruptured the locked shallow Himalayan megathrust (Grandin *et al.* 2015; Elliott *et al.* 2016a). Qiu *et al.* (2016) used a structural geology model to show that the coseismic slip was confined to a flat region of the megathrust surrounded by two steeper ramps. They also ran dynamic simulations that showed that their fault geometry was sufficient to confine some (but not all) ruptures to the lower or upper part of the seismogenic fault; some earthquakes in their simulations were larger and ruptured the entire locked megathrust. Importantly for Nepal, the shallow, locked portion of the megathrust remains locked and has not moved following the earthquake (Ingleby *et al.* 2020) – the hazard remains very high there.

7.2.2. Inter-seismic deformation

7.2.2.1. History and recent progress in inter-seismic InSAR studies

Mapping inter-seismic strain accumulation with geodetic methods has been a long-standing goal of the geodetic community as it has the potential to strongly inform seismic hazard models. If Reid's 1910 elastic rebound hypothesis is correct then measuring where strain is accumulating, and at what rate, should allow us to quantify the long-term rates of fault movement, and hence, the rate of seismicity. Elliott *et al.* (2016) showed that in the western US and in eastern Turkey, areas with dense GNSS coverage and high tectonic strain rates, there is a good correlation between seismicity rate and geodetic strain rate, and data from the GNSS-derived Global Strain Rate Model (Kreemer *et al.* 2014) have been used to make global seismicity forecasts that compare well with forecasts based on instrumental seismicity (Bird and Kreemer 2015). However, dense GNSS data are not available globally. Here, we review progress in using InSAR to measure inter-seismic strain and show that it has the potential to significantly improve strain rate models built from GNSS data alone.

Using InSAR to measure inter-seismic strain is challenging compared to using InSAR for earthquakes – the inter-seismic signal in any one short-interval interferogram is much smaller than the noise, particularly the noise arising from spatial and temporal variations in tropospheric delays. The first studies to clearly show the signature of inter-seismic strain in InSAR results were able to reduce the noise in individual interferograms by stacking a relatively small number of datasets from ERS-1 and ERS-2 with multi-year time spans (Peltzer *et al.* 2001; Wright *et al.* 2001b). With the ESA operating Envisat with a more systematic background acquisition strategy in tectonic areas, alongside increasing volumes of data from systems such as ALOS, increasing computer power, and improved algorithms (e.g. Doin *et al.* 2011; Agram *et al.* 2013; Morishita *et al.* 2020), an increasing number of studies have been able to map inter-seismic tectonic deformation over ever larger and/or more challenging areas (e.g. Grandin *et al.* 2012; Wang and Wright 2012; Tong *et al.* 2013; Cakir *et al.* 2014; Pagli *et al.* 2014; Hussain *et al.* 2018; Wang *et al.* 2019).

The launch of the Sentinel-1 constellation has again provided a radical increase in the volume of SAR data acquired over tectonic areas (Elliott *et al.* 2015). The shorter revisit time increases coherence and means that average crustal velocities can be obtained with improved accuracy. Several groups have developed tools capable of mass processing Sentinel-1 data (e.g. Bekaert *et al.* 2019; Foumelis *et al.* 2019; Lazecký *et al.* 2020), with the ESA's Geophysical Exploitation Platform (Foumelis *et al.* 2019) and COMET's LiCSAR system (Lazecký *et al.* 2020) focused on mass producing interferometric products for the continental seismic belts. Weiss *et al.* (2020) presented the largest regional analysis of Sentinel-1 to date, processing Sentinel-1 data using LiCSAR and LiCSBAS (Morishita *et al.* 2020) over 800,000 km^2 of Anatolia. They produced average velocities for both ascending and descending tracks and used these in combination with GNSS data to provide east–west and vertical velocities and to estimate strain (Figure 7.3). The results reveal both large-scale features, such as the North Anatolian Fault, as well as localized deformation, for example due to fault creep or water extraction. Even in Anatolia, where GNSS coverage is good, the inclusion of InSAR data in the inversion of strain rate sharpens the strain rate model considerably (compare the North Anatolian Fault in panels (c) and (d) of Figure 7.3). The work in Anatolia is being expanded at present to produce complete velocity fields with this approach for the entire Alpine–Himalayan Belt and East African Rift.

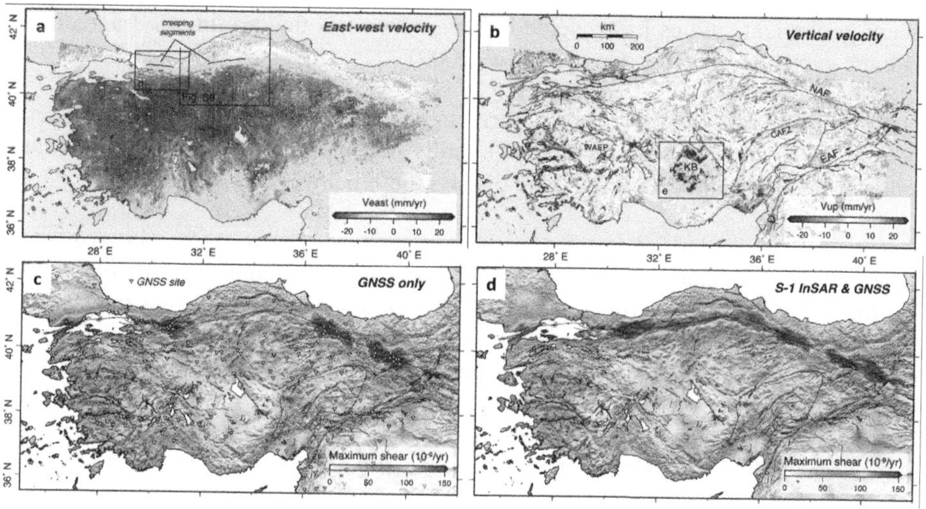

Figure 7.3. *Inter-seismic deformation across Anatolia derived from Sentinel-1 InSAR and GNSS data, adapted from Weiss* et al. *(2020). (a) East–west velocity field derived from ascending and descending Sentinel-1 InSAR data. Boxes highlight areas of the North Anatolian Fault that are creeping. (b) Vertical velocity field. Box shows a region of significant vertical deformation due to water extraction. (c) Maximum shear strain rate estimated from GNSS only using the VELMAP approach (Wang and Wright 2012). (d) Maximum shear strain rate estimated from a combination of InSAR and GNSS data. Faults are from Emre* et al. *(2018). For a color version of this figure, see www.iste.co.uk/cavalie/images.zip*

7.2.2.2. What have we learned about faults from InSAR studies of inter-seismic deformation

Geodetic determination of fault slip rates: for individual faults, InSAR has regularly been used to obtain estimates of their present-day slip rate (e.g. Peltzer *et al.* 2001; Wright *et al.* 2001b; Cavalié *et al.* 2008; Wang *et al.* 2009; Karimzadeh *et al.* 2013; Fattahi and Amelung 2016). To do so requires a few assumptions – the rheology of the seismogenic layer is typically assumed to be elastic such that strain observed in the crust away from the fault is eventually be released by slip on the fault, and the rate of strain accumulation observed geodetically is typically assumed to be representative of the long-term strain rate. If we make those assumptions, geodetic slip rates can be estimated, alongside parameters such as "locking depth" that are indicative of the seismogenic layer thickness (Wright *et al.* 2013). Many of the geodetic slip rates, for example on the North Anatolian and San Andreas Faults, agree with estimates from geology/paleoseismology, but in a few cases, the geodetic slip rates have disagreed with the prevailing wisdom. For example, Wright *et al.* (2004) used InSAR to show that the faults in western Tibet were generally accumulating strain at a slower rate than had been estimated in some geological models (e.g. Tapponnier *et al.* 2001). More recent geological studies have yielded Quaternary rates that are in good agreement with the geodetic rates (e.g. Cowgill *et al.* 2009).

Beyond the rates and styles of deformation at individual fault systems, geodetic measurements of inter-seismic strain at faults have led to new insights into fault behavior in a number of key areas. We discuss these below.

Focused strain throughout the earthquake cycle for most faults: inter-seismic deformation, particularly at strike-slip faults, is often modeled using a simple elastic rheology, in which the deep part of a fault slips at a steady rate whereas the shallow part is locked. For strike-slip faults, a screw dislocation formulation can be used (Savage and Burford 1973); other fault geometries can be modeled with a backslip formulation (Savage 1983). These elastic dislocation models give rise to relatively focused areas of inter-seismic strain around active faults; in the standard screw dislocation model, 70% of the motion for a strike-slip fault, for example, occurs within a distance of two locking depths from the fault.

Wright *et al.* (2013) noted that these simple elastic dislocation models were ubiquitous in studies of inter-seismic deformation; their review found 187 examples in the literature. The fact that focused strain is seen at most faults suggests that it occurs at all phases of the earthquake cycle. This argument is supported by observations of late-stage inter-seismic deformation at several major strike-slip faults that later failed in earthquakes, including the North Anatolian Fault at Izmit (Turkey), the San Andreas Fault at Parkfield (California), the Denali Fault in Alaska and the Manyi Fault in Tibet (Elliott *et al.* 2016). This is a somewhat surprising result – if the viscosity of the lower crust is as low as some studies of post-seismic deformation have suggested (see below), then we would not expect focused inter-seismic strain late in the earthquake cycle (e.g. Hetland and Hager 2006).

There are a few exceptions, however, where faults do not appear to have clear focused strain associated with them during the inter-seismic period. A clear recent example is from Wang *et al.* (2019), who used ERS and Envisat data to map inter-seismic tectonic strain across south-central Tibet. Of the five extensional grabens in their study region, only two had clear focused extensional strain in the inter-seismic geodetic velocity field.

Temporal invariance of inter-seismic strain rate: typical inter-event times for major earthquakes on large faults are hundreds to thousands of years, so modern geodetic measurements are inevitably only available for a small portion of this period. To assess whether strain rates vary throughout this time period, indirect methods are required. Meade *et al.* (2013) compared short-term geodetic estimates of slip rate for 15 continental strike slip faults with long-term geological estimates of slip rate, estimated over multiple earthquake cycles, finding very good agreement overall between these rates. They argued, persuasively, that because the geodetic rates sample random short intervals within the earthquake cycle, the geodetic rates must be fairly constant throughout the earthquake cycle, apart from during a relatively short post-seismic period.

Hussain *et al.* (2018) took a different approach to the same problem, examining present-day strain rates along the entire North Anatolian Fault from InSAR alongside estimates of strain from GPS measurements before and after the 1999 Izmit earthquakes. As major earthquakes have occurred at different times in the past, these spatial samples of strain rate can be thought of as sampling different time periods of the earthquake cycle, assuming the entire fault behaves in a similar way. They found that strain rates were approximately constant apart from an approximately 10-year period following the 1999 earthquakes (Figure 7.4). This observation requires that the relaxation time for the majority of the mid to lower crust be as long as the inter-event time (Figure 7.4; Yamasaki *et al.* 2014; Hussain *et al.* 2018).

The fact that, for large strike-slip faults at least, the inter-seismic strain rate outside the post-seismic period appears to be fairly time invariant is good news for hopes to use geodetic observations to help constrain seismic hazard models. The primary input for most modern probabilistic seismic hazard models are instrumental and historical records of previous earthquakes. These can be biased, particularly in areas of moderate to low strain, because of the relatively short observation period compared to the inter-event time (see e.g. Stein *et al.* 2012). If short-term geodetic strain rates really are largely representative of long-term rates of fault activity, then they have the potential to be an important dataset for constraining earthquake hazard (e.g. Bird and Kreemer 2015).

7.2.3. *Post-seismic deformation and aseismic deformation transients*

7.2.3.1. *History and recent progress in the study of aseismic deformation transients*

As geodetic data from InSAR and GNSS have shown improvements in accuracy and sampling period, it has become increasingly evident that aseismic deformation transients are an important phenomenon in fault zones. They can occur as periods of accelerated post-seismic deformation following earthquakes, when the rate of deformation is higher than the inter-seismic period (e.g. Nur and Mavko 1974), or as "slow earthquakes", where the aseismic deformation transients are not obviously associated with the earthquake cycle (e.g. Sacks *et al.* 1978).

The first post-seismic studies using InSAR came just a few years after the first coseismic studies by Massonnet *et al.* (1996) and Peltzer *et al.* (1996). Although these authors studied the same earthquake, their observations were limited to just a few post-seismic interferograms and they invoked different mechanisms to explain the data. A number of studies in the 2000s presented coarsely sampled time series showing how post-seismic deformation evolved through time (e.g. Fialko 2004; Ryder *et al.* 2007; Biggs *et al.* 2009). Prior to the launch of Sentinel-1 in 2014, there were around 23 earthquakes with geodetic observations of post-seismic deformation, of which more than half had been detected by InSAR. The latest generation of InSAR missions, including Sentinel-1, allow the spatial and temporal history of post-seismic deformation to be mapped with unprecedented detail (e.g. Figure 7.5; Floyd *et al.* 2016; Ingleby *et al.* 2020; Wang and Bürgmann 2020).

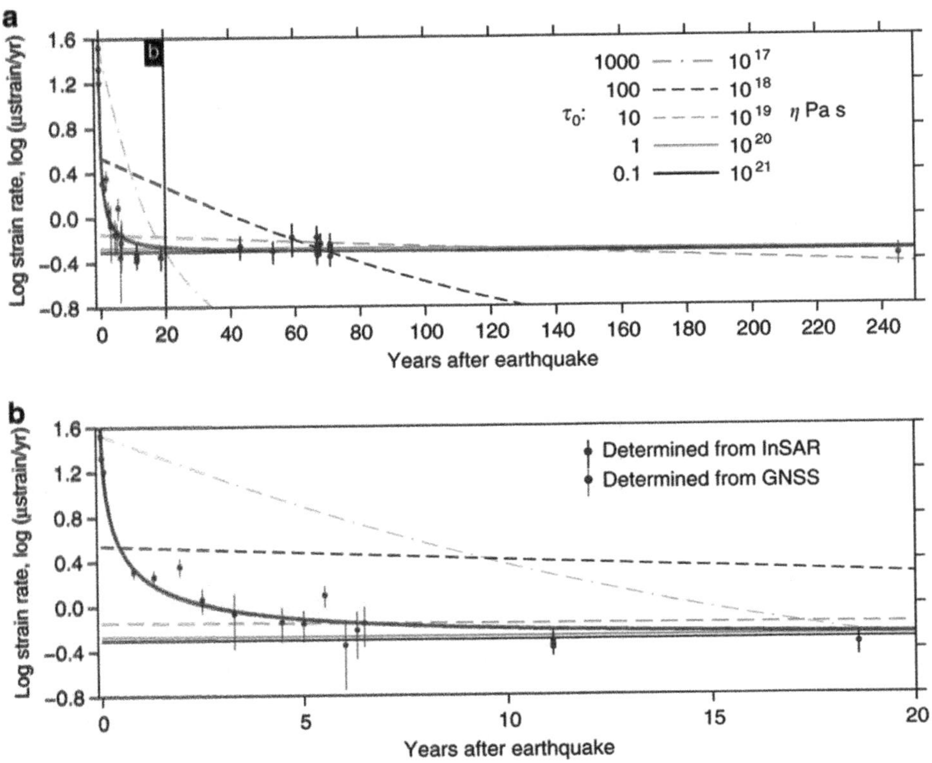

Figure 7.4. *Strain rate throughout the earthquake cycle for the North Anatolian Fault, reproduced under creative commons license from Hussain* et al. *(2018). (a) Strain rates derived from InSAR and GNSS plotted as time since the most recent earthquake in the location of the measurement, providing a pseudo-history of the strain rate throughout a single earthquake cycle. Black lines show predictions from a simple two-layer viscoelastic coupling model (Savage and Prescott 1978), with the viscosity of the substrate shown in the legend. The red line is a best-fit model that includes post-seismic afterslip and viscoelastic relaxation embedded within a stronger substrate. (b) Magnification of the post-seismic period. For a color version of this figure, see www.iste.co.uk/cavalie/images.zip*

Aseismic deformation has also been observed with InSAR. Some of the earliest InSAR observations of aseismic deformation were for creeping sections of major faults, including the Hayward and San Andreas faults in California (Bürgmann *et al.* 2000; Johanson and Bürgmann 2005) and the North Anatolian Fault in Turkey (Cakir *et al.* 2005). With more recent data it has been possible to map the spatial and temporal distribution of creep, showing that it often accrues in aseismic transient slip events ("slow earthquakes"). Furuya and Satyabala (2008) were the first to present InSAR observations of a slow earthquake, for the Chaman fault system in Afghanistan – lasting around a year and with an equivalent moment magnitude of 5.0. By using time-series analysis, several authors have now been able to map the spatial and temporal distribution of creep events at continental strike-slip faults (Rousset *et al.* 2016; Pousse Beltran *et al.* 2016; Khoshmanesh and Shirzaei 2018). Cavalié *et al.* (2013) presented the first observations from InSAR of a slow slip event on a subduction zone, at Guerrero, Mexico, but in general, the deformation caused by slow slip events at subduction zones has been too small to be reliably detected using InSAR.

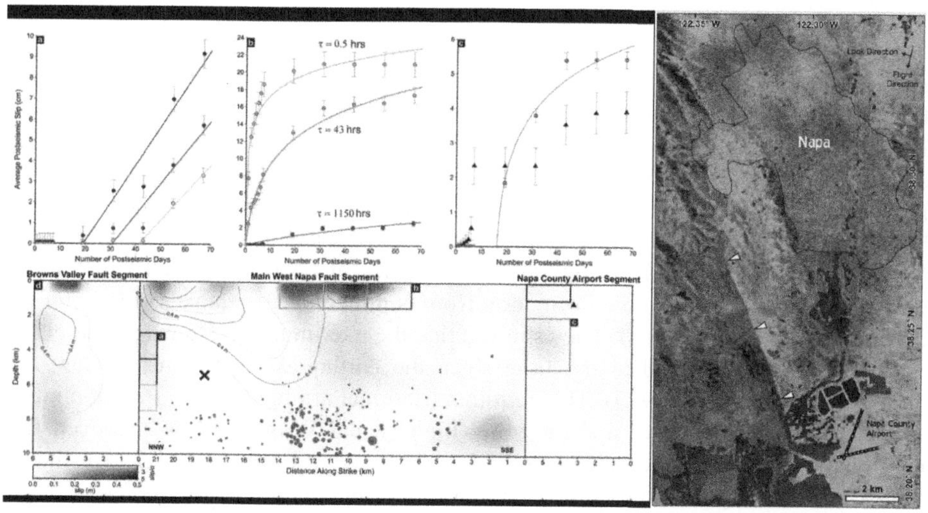

Figure 7.5. *Post-seismic deformation for the 2014 South Napa earthquake, adapted from Floyd* et al. *(2016) and Elliott* et al. *(2015) Right-hand side: phase for 12-day post-seismic interferogram spanning surface rupture of the South Napa earthquake (Elliott* et al. *2015). Left-hand side: (d) modeled distribution of post-seismic slip, in colors, determined from joint inversion of INSAR and GNSS time series, compared to coseismic slip contours (black contours) and aftershocks (pink circles). (a–c) Slip histories for the patches of the fault model – locations are shown in (d). For a color version of this figure, see www.iste.co.uk/cavalie/images.zip*

7.2.3.2. *What have we learned about faults from InSAR studies of aseismic deformation transients*

Post-seismic deformation is now commonly observed: for most moderate to large earthquakes, post-seismic deformation is now routinely observed – it is now recognized as an important part of the earthquake deformation cycle. However, multiple mechanisms have been invoked to explain the observations (Wright *et al.* 2013), notably frictional afterslip (e.g. Floyd *et al.* 2016), viscoelastic relaxation (e.g. Ryder *et al.* 2007) and poroelastic rebound (e.g. Jónsson *et al.* 2003). Often, different mechanisms are invoked for the same earthquake from geodetic datasets with different time or spatial scales (Wright *et al.* 2013). For the 1992 Landers earthquake, post-seismic deformation has been attributed to afterslip and fault zone contraction (Massonnet *et al.* 1996), poroelastic deformation (Peltzer *et al.* 1996), deep afterslip (Savage and Svarc 1997), a combination of poroelastic deformation and deep afterslip (Fialko 2004), linear Maxwell viscoelastic relaxation of the lower crust (Deng *et al.* 1998) and power-law viscoelastic relaxation of the mantle (Freed and Bürgmann 2004).

Multiple processes contribute to post-seismic deformation: when looking at individual earthquakes with detailed time series of geodetic data, such as was possible for the 2014 M_w 6.1 South Napa earthquake by combining GNSS with Sentinel-1 InSAR, the spatial and temporal distribution of deformation appears as somewhat complex (Figure 7.5; Floyd *et al.* 2016); afterslip appears to start at different times in different locations and have different temporal decay patterns (compare the slip histories in panels (a) to (c) of Figure 7.5). To try to make sense of the complexity, Ingleby and Wright (2017) compiled time histories of post-seismic velocity for all observations of post-seismic deformation from continental earthquakes (151 velocity observations from 34 events). They showed that the maximum post-seismic velocities are inversely proportional to the time since the earthquake, like Omori's law for aftershocks (Utsu *et al.* 1995). This requires frictional afterslip or a highly nonlinear viscoelastic deformation of a shear zone (Ingleby and Wright 2017). However, the maximum post-seismic velocities studied by Ingleby and Wright (2017) are likely to be dominated by shallow fault zone processes, such as afterslip. Studies of large earthquakes show that a combination of frictional afterslip and deeper viscoelastic deformation is required (e.g. Weiss *et al.* 2019). To separate the contributions of these mechanisms requires long time periods of precise observations over large spatial scales.

A wide range of complex aseismic phenomena occur: in addition to post-seismic deformation, InSAR has helped identify a range of complex and varied aseismic creep behavior, including triggered slip, fault creep and slow earthquakes. Triggered aseismic slip of nearby faults has been observed following a number of major earthquakes (e.g. Wright *et al.* 2001a; Fielding *et al.* 2004; Fujiwara *et al.* 2016). Xu *et al.* (2020) presented a compelling analysis of 169 fractures away from the main faults that deformed during the 2019 Ridgecrest earthquake sequence (M_w 6.4, 7.1,

California). They showed that fractures that experienced coseismic stress changes that were in the same sense as the regional stress field had narrow deformation zones and were likely to have failed due to permanent inelastic deformation, such as fault slip. Faults that were stressed coseismically in the opposite sense to the regional stress had wider deformation zones, with deformation likely to be the result of enhanced strain in a compliant fault zone.

Detailed studies of time series of aseismic creep are starting to reveal complexities in the temporal behavior of creep. For example, Rousset *et al.* (2016) used a dense time series of high-resolution InSAR data to document a pulse of fault creep that took place over around a month in 2013. The pulse had a magnitude of 15 mm and was the only significant fault motion to occur in their observation window (July 2013 to April 2014). Similarly, Khoshmanesh *et al.* (2015) showed that creep at the Parkfield section of the San Andreas Fault was not steady in time, instead occurring in pulses separated by one to three years. Khoshmanesh and Shirzaei (2018) suggest that the pulsing is caused by variations in normal stress that result from pore fluids trapped in hydraulically-isolated parts of the fault.

Aseismic behavior is related to lithology: spatial maps of aseismic fault creep have been powerful in illuminating the close relationship between aseismic creep and fault zone lithology. In Taiwan, Thomas *et al.* (2014) showed that creep occurred on the Longitudinal Valley Fault in Taiwan in areas where fault zone gouge was dominated by a particular lithological unit, derived from unlithified subduction zone mélange, that favored pressure solution creep. In Turkey, Cetin *et al.* (2014) showed that creep on the North Anatolian Fault is fastest in areas where the lithology has low frictional strength, including andesitic-basaltic, limestone and serpentine units caught up in the fault zone. The relationship between aseismic deformation and lithology has also been observed in post-seismic data – regions of shallow afterslip that occurred following the 2014 South Napa earthquake correlated with younger sediments and alluvium; Floyd *et al.* (2016) argued that the complex variations in spatial and temporal behavior (Figure 7.5) may be a function of different clay contents, mineralogy or pore pressures within these units.

As the quality of InSAR time series improve globally, we expect to see further advances in our understanding of the aseismic behavior of faults.

7.3. Investigating earthquake surface ruptures with optical image correlation

7.3.1. *A history of the optical correlation technique in the study of earthquakes*

Developed for the study of earthquakes in the early 2000s, the optical correlation technique allows the retrieval of 2D or 3D ground displacements produced by an earthquake from the measurement of pixel shifts between optical images acquired before and after the event (Van Puymbroeck *et al.* 2000). The main advantage of this technique is the efficiency in capturing the near-fault deformation of surface-rupturing earthquakes (Mw 6.5+). Optical correlation therefore addresses one of the main limitations of the various traditional methods typically used to characterize the coseismic displacements in continental earthquakes. On the one hand, field investigations undertaken following an earthquake allow the on-fault slip at the surface to be measured from the offset of linear features displaced across the fault (e.g. roads, fences, canals and geomorphic features). However, these measurements are made only at a limited number of locations where easily recognizable features can be reliably measured across the fault, and therefore do not give a continuous representation of the surface fault slip. Moreover, the deformation is not always localized on a structurally simple fault plane, but can be expressed as a broad zone of distributed deformation that can be hundreds of meters wide. In that case, quantifying the total displacement that occurs across the entire fault zone is challenging. Finally, field investigations can provide the along-strike and vertical component of the slip but the measurement of the fault-perpendicular component is usually not possible. On the other hand, the InSAR technique allows surface displacements to be measured with very high precision (mm-scale) and at high spatial resolution. However, when the displacement gradients are too large, the interferometric phase decorrelates, leading to an incomplete description of surface displacements near the fault rupture. Moreover, this technique only measures the displacement in one direction, along the satellite line-of-sight, which is approximately oriented obliquely east–west due to the near-polar orbit of the satellites. The sensitivity to any along-track (i.e. north–south) displacement is thus limited, while combination of ascending and descending data are needed to distinguish vertical from east–west horizontal ground displacement. Finally, the GPS technique gives very precise measurements of 3D ground displacements, albeit at a very limited number of locations.

The optical correlation technique was first applied to the measurement of earthquake-related displacements by Van Puymbroeck *et al.* (2000), who measured the horizontal surface displacement field produced by the 1992 Mw 7.3 Landers earthquake using Satellite Pour l'Observation de la Terre (SPOT) satellite images of 10 m resolution. The east–west (EW) and north–south (NS) displacement fields they obtained clearly show the 80 km long discontinuity corresponding to the surface rupture of the Landers earthquake, demonstrating the high potential of optical correlation for the study of earthquake ruptures in large continental earthquakes. In the following years, motivated by the successful demonstration for the Landers earthquake, the method used by Van Puymbroeck *et al.* (2000) was applied to other surface-rupturing earthquakes, such as the Izmit (Michel and Avouac 2002), Chi-Chi (Dominguez *et al.* 2003) and Kashmir earthquakes (Avouac *et al.* 2006), using satellite images. Michel and Avouac (2006) extended the application to aerial photos, from which they resolved the displacement field for the Kickapoo step-over (part of the

Landers rupture), demonstrating the potential of using historical aerial photo archives to study past surface-rupturing earthquakes. The development of easy-to-use software for optical correlation, such as COSI-Corr (Leprince *et al.* 2007b) and MicMac (Rosu *et al.* 2015), has greatly helped to establish this geodetic technique for earthquake studies. Nowadays, more and more optical data of increasing resolution are available and near-field displacements are commonly measured using optical images each time a major continental earthquake occurs. At the time of writing, the surface displacement field of about 30 earthquakes has been measured using optical correlation (e.g Zinke *et al.* 2014; Milliner *et al.* 2015; Vallage *et al.* 2015; Hollingsworth *et al.* 2017; Marchandon *et al.* 2018a; Barnhart *et al.* 2020; Delorme *et al.* 2020).

7.3.2. Measuring earthquake surface displacements from optical images: methodology

Optical correlation allows the measurement of the surface displacement by comparing two or more optical images spanning an earthquake. Contrary to InSAR, optical correlation is very flexible as one can correlate images acquired by different sensors (Hollingsworth *et al.* 2012; Marchandon *et al.* 2018a), with different look angles, different resolutions (although images must be resampled to a common resolution prior to correlation), and with very long temporal baselines (increasing time difference can nevertheless impact the correlation quality as surface changes accumulate). The detection threshold of the optical correlation technique is ~1/10th of the image pixel size (Leprince *et al.* 2007b), which is limited by noise in the images (e.g. illumination bias; see Lacroix *et al.* (2019)). The choice of the images to use depends on the amplitude of the displacement signal, as well as the spatial resolution needed for the final displacement field. For large continental earthquakes (Mw > 7), medium resolution images (10–15 m; for example, Landsat 8, SPOT-1–4 and Sentinel-2) are sufficient to describe the first-order features of the rupture. However, any displacement below the detection threshold (<1–1.5 m) will not be detected and fine details of the rupture, such as small geometrical complexities or small-scale slip variations, will not be resolvable. For smaller earthquakes, or when a detailed description of the rupture is required, high-resolution images must be used (0.3–5 m, for example, SPOT-5–7, Pléiades, WorldView, Planet and aerial photos). However, medium resolution images typically have very large footprints (e.g. Sentinel-2 across-track width is 290 km), thereby allowing an entire surface rupture (potentially extending up to several 100s of kilometers in length) with only a few pairs of images (e.g. Klinger *et al.* 2006). In contrast, high-resolution images cover much smaller areas (e.g. Pléiades across-track width is 20 km); consequently, a large number of images are necessary to characterize the entire rupture, especially for a rupture oriented east–west (i.e. ~perpendicular to the satellite track). Processing high-resolution images is therefore time-consuming and supplementary processing steps, such as the bundle adjustment[2] of the images, are needed in order to precisely mosaic the images together with subpixel precision.

7.3.2.1. *Optical image correlation: general workflow*

Measuring the horizontal surface displacement field from raw optical images requires three main steps: (1) a pre-processing step during which the pre- and post- earthquake images are co-registered and orthorectified, (2) the correlation of the pre- and post-orthoimages and (3) the post-processing of the output surface displacement maps. Some optical datasets are provided in a processed format (with radiometric and geometric corrections applied), e.g. Landsat 8, and Sentinel-2. In such cases, we may proceed directly to the correlation without any pre-processing; however, any artifacts arising from the pre-processing stages cannot be easily corrected from the processed imagery (for example, DEMs used by space agencies to orthorectify images are often lower resolution than the images, which can lead to related topographic noise in the final displacement map). If datasets are provided without geometric correction, the user must pre-process the data prior to correlation. For more details about the pre-processing and the correlation steps, we refer the reader to Chapter 2 of this book as well as to the review papers by Avouac and Leprince (2015) and Hollingsworth *et al.* (2020).

Basic post-processing to minimize the influence of noise on the output surface displacement fields typically includes outlier removal, based either on displacements with unrealistic values, low signal-to-noise ratio, or values which differ significantly from their neighbors (e.g. Zinke *et al.* 2019), and various spatial smoothing operations (e.g. low-pass, median or non-local mean filtering). Non-local mean filtering, which is a spatial averaging filter with weights applied based on the similarity with each target pixel, is particularly appropriate for denoising displacements containing sharp discontinuities, such as those produced by surface-rupturing earthquakes, because it preserves sharp edges, and thus preserves fine details of the surface rupture (Buades *et al.* 2005). More specific corrections can be applied depending on the case, such as the removal of ramps, stripes, or undulations due to misregistration errors, charged-couple device misalignment or unknown variations of the satellite attitude. More details about these errors and their corrections can be found in section 7.3.2.3.

Several software packages are available to process optical images. The most widely used by the earthquake science community include: COSI-Corr

2. Bundle adjustment is the process by which the camera position and orientation of a network of images are simultaneously optimized to ensure that the observations in multiple images of a single ground feature are self-consistent (i.e. a single feature has the same x,y and z coordinates in all the images in which this feature is present).

(Co-registration of Optically Sensed Images and Correlation (Leprince *et al.* 2007b)), MicMac (Multi Images Correspondances par Méthodes Automatiques de Corrélation (Rosu *et al.* 2015)) and ASP (Ames Stereo Pipeline (Beyer *et al.* 2018)).

7.3.2.2. *Measuring the 3D displacement field of earthquakes from stereo optical images*

With the increasing number of stereo optical images available, it is now possible to measure the vertical component of the displacement in addition to the horizontal component. Three different methods can be used; all of them require pre- and post-stereo pairs of optical images[3] (with enough stereoscopic information to isolate the pre- and post-earthquake topographic signal from the tectonic signal).

DEM differencing: the most straightforward technique is to (1) compute pre- and post-earthquake DEMs from pre- and post-stereo optical images, (2) orthorectify the pre- and post-images using the relevant DEMs, (3) correlate the orthorectified images to obtain the horizontal displacement field, (4) resample the pre-DEM to incorporate the horizontal displacement and (5) compute the vertical difference between the two DEMs. If the horizontal displacement is not accounted for when differencing DEMs, the resulting vertical displacement field will be overprinted by noise correlated with misregistered topography (e.g. Oskin *et al.* 2012). If the amount of horizontal displacement is small enough, or in areas of smooth topography, then the pre- and post-DEMs can be simply subtracted without accounting for the horizontal displacements (Kuo *et al.* 2019). This technique has been successfully applied to measure the 3D displacement field of several earthquakes, such as the Norcia, Hualien and Balochistan earthquakes (Barnhart *et al.* 2019a; Kuo *et al.* 2019; Delorme *et al.* 2020).

3D triangulation: in the second approach, the 3D location of each point before and after the earthquake is triangulated from displacement maps computed for the stereo pre- and post-event images (Avouac and Leprince 2015; Zinke *et al.* 2019). Let us consider two stereo pairs of optical images covering the same area: one taken before the earthquake and the second one after (Figure 7.6). One of the pre-earthquake images is defined as the primary (i.e. reference) image, and three disparity maps are computed for the primary image and the three secondary images (Figure 7.6). As stereo images are taken with off-nadir incidence angles, the pre-earthquake disparity map contains only topographic information, while the two pre-/post-earthquake disparity maps contain a combination of stereoscopic effects and ground displacements from the earthquake (Figure 7.6). Using a ray-tracing approach, the x,y and z positions at each grid location in the pre-event correlation map can be retrieved using the pre-event correlation map and the camera positions of the pre-stereo images. Similarly,

3. A stereo pair of images is a pair of images of the same area taken with different incidence angles.

the x,y and z positions of these same points after the event can be retrieved using the post-event camera positions, and the pre-/post-correlation maps. The 3D displacements are then obtained by simply differentiating the pre- and post-3D positions. An example of computing a 3D displacement field is given in section 7.3.3.2.

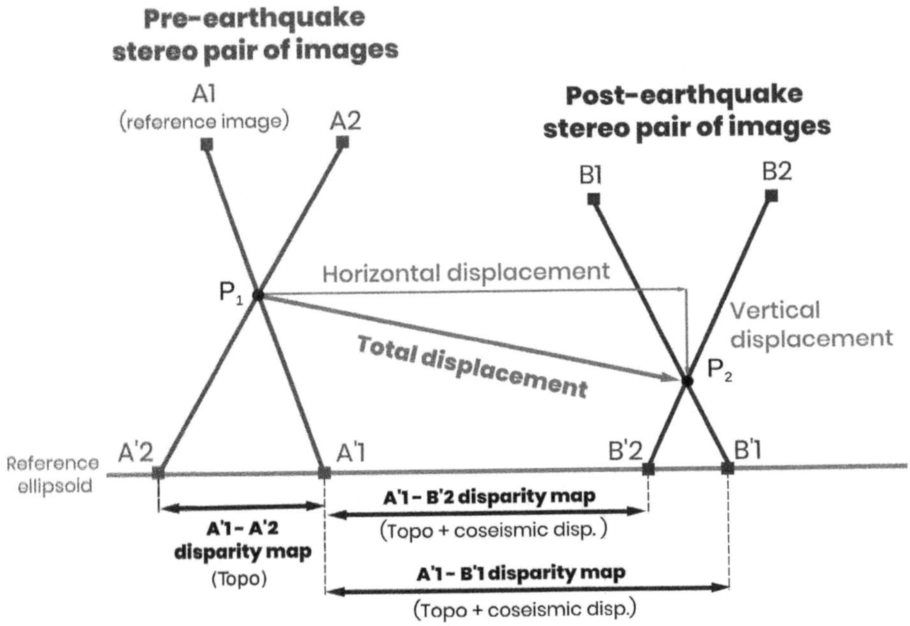

Figure 7.6. *Computing the 3D displacement field from correlation of pre- and post-stereo pairs of optical images. Figure modified from Avouac and Leprince (2015). For a color version of this figure, see www.iste.co.uk/cavalie/images.zip*

Point-cloud registration: finally, the third approach is to use point-cloud registration algorithms (e.g. the iterative closest point and fast global matching algorithms) on pre- and post-earthquake DEMs computed from stereo optical images. This method allows more expressive transforms to be retrieved, albeit with limited subpixel precision, thereby revealing, in the case of an affine transform, local changes in rotation, scale and shear, in addition to rigid 3D translations (e.g. Nissen *et al.* 2014). The advantage of using this technique instead of image correlation-based approaches is that the similarity metric is based on surface shape/morphology rather than surface reflectance. Therefore, optical images with vastly different reflectance properties (e.g. agricultural settings, regions with strong seasonal changes in vegetation and snow) are likely to fail with traditional correlation. However, if the pre- and post-event DEMs can be well-resolved for each epoch and precisely co-registered, point-cloud registration techniques can be used to retrieve the 3D ground deformation. Point-cloud registration is commonly used to measure the 3D surface deformation with LiDAR datasets (e.g. Nissen *et al.* 2014; Scott *et al.* 2018), which are generally high resolution (negating the need for subpixel performance), very high quality and well-resolved at high frequencies. This is not always the case with DEMs derived from stereo imagery, although with increasing resolution, image quality and availability of satellite and aerial datasets, and tools for processing them, this technique is finding increasing applications with optical datasets (e.g. Barnhart *et al.* 2019a).

7.3.2.3. *Limitations and sources of error*

Error and bias in surface displacement fields obtained with optical image correlation have a variety of sources. First, errors can arise from imperfect modeling of the imaging system. For example, systematic misregistration can produce a ramp over the entire displacement field. Moreover, unrecorded variations of the satellite attitude, such as pitch oscillations, can lead to across-track undulations in the displacement field (e.g. Dominguez *et al.* 2003; Leprince *et al.* 2007a), while misalignment of the charge-coupled devices (CCDs) of pushbroom sensors can cause along-track stripes (e.g. Leprince *et al.* 2007b). These artifacts are very common (Avouac and Leprince 2015) and can be mitigated using detrending and destriping tools (e.g. Ayoub *et al.* 2017). When using film-based images, unaccounted for distortions of the film (e.g. thermomechanical warping) and scanning artifacts can also bias the correlation maps (e.g. Michel and Avouac 2006; Ayoub *et al.* 2009). Bias affecting the entire displacement field (e.g. orbital error and film distortion) make the measurement of the absolute displacement difficult. Indeed when such errors affect the correlation, only the measured offsets across the fault are reliable whereas the absolute values of displacements are not. These errors can be mitigated by setting the far-field displacements to zero (e.g. Vallage 2015). However, when working with high-resolution images, the image coverage is too tight, so there are no zones of zero displacement. In that case, a possible solution is to work with images at different scales, anchoring high-resolution correlation with lower-resolution correlation but larger coverage. If GPS data are available, they can also be used to estimate a correction to apply to the whole optical displacement field (e.g. Dominguez *et al.* 2003). Topographic artifacts can appear when images are orthorectified with a lower resolution DEM, if the same DEM is used to orthorectify both pre- and post-images despite a change in topography during the earthquake, and if images are not well registered to the topography (e.g. Ayoub *et al.* 2009). Finally, any changes of the ground surface between the pre- and post-event images will impact the resulting displacement field. The correlation will fail in areas of drastic change, such as those due to urbanization, farming, snow, clouds or slope failure, while less drastic changes will likely be expressed as coherent noise in the correlation map. For example, if the pre- and post-images are not taken at the same time of the year, shifts in the positions of shadows between the pre- and post-acquisitions will be resolved by the correlator, leading to coherent noise in the surface displacements. This bias can be avoided by using pre- and post-images taken at the same epoch of the year (i.e. with the same illumination conditions). Alternatively, shadowing bias can also be characterized and corrected using time-series analysis (Lacroix *et al.* 2019; Hollingsworth *et al.* 2020). Indeed, because the position of the sun (i.e. azimuth and elevation) at the time of image acquisition changes throughout the year, the orientation and length of the shadows will change locally, depending on the topography. Therefore, the displacement bias induced by the shadows will be different depending on the time of the year at which the pre- and post-images were taken. By correlating all possible combinations of images taken throughout the year, the redundancy of the measurements can be used to invert a robust displacement time series for each pixel. As the time series obtained is the sum of various signals (e.g. tectonic signal + illumination bias + other noise), the isolation and removal of the illumination component can be achieved using, for example, blind source separation approaches (e.g. independent component analysis) or decomposition into known basis functions (if their functional form is known).

7.3.2.4. *Perspectives and future developments*

One of the main limitations of the optical correlation technique is the difficulty of measuring small displacements (cm to mm) produced by small earthquakes or by transient processes such as post-seismic deformation or inter-seismic creep (especially if the displacement signal at the surface has a long spatial wavelength). In the near future, optical images with increasing spatial and temporal resolutions will be increasingly available, along with improved correlation schemes. This large archive of optical data will thus allow measurement of displacements of increasingly smaller amplitude and with higher temporal resolution. In particular, we will be able to make use of the large archive of optical images to construct dense time series of optical displacements. Indeed, the correlation of every possible combination of optical image available between two dates will provide a set of displacement measurements with high redundancy that can be inverted for a robust time series. Time-series analysis is already used to measure small displacements using InSAR data and has started to be used with optical data to monitor slow-moving landslides and glaciers (which typically experience larger displacements compared to those produced throughout the earthquake cycle). With the increase in temporal and spatial resolution of optical imagery, we might hope that such processing methods will enable the measurements of transient deformation relating to the seismic cycle, such as post-seismic deformation or inter-seismic creep. Furthermore, machine-learning approaches applied to optical geodesy will help to improve correlator performance (faster, increased robustness and better accuracy) and may be used to detect and correct sources of noise coming from the imaging system (CCD and jitter), thereby increasing our ability to robustly measure small displacements.

7.3.3. *Some examples of near-fault displacement fields from recent and historical earthquakes*

In this section, we present several examples of displacement fields derived from optical data. Each example is different and is chosen for its particular methodological aspects. The first example shows how a large number of aerial images (>400) can be processed in a semi-automatic manner in order to derive a high-resolution and seamless displacement field. The second example illustrates the performance of the 3D triangulation method in retrieving the 3D displacement field of an earthquake. Finally, the third example shows how archives of historical aerial or satellite imagery can be used to study the surface rupture of past earthquakes.

7.3.3.1. *Horizontal surface displacement of the 1992 Mw 7.3 Landers earthquake from high-resolution aerial images*

As a first application example, we describe the horizontal displacement field of the 1992 Mw 7.2 Landers, California, earthquake measured from the correlation of aerial images. The pre-earthquake dataset was composed of ~40 National Aerial Photography Program (NAPP) overlapping photographs of 1 m resolution acquired in July 1989 (i.e. two years before the earthquake). The post-earthquake dataset was composed of nearly 400 photographs of 20 cm resolution taken two days after the earthquake. The post-earthquake dataset was acquired specifically to map the rupture trace of the Landers earthquake and was composed of several segments of overlapping photos that follow the surface rupture (Figure 7.7). Due to the very high resolution of the images, the spatial coverage is very narrow (image width of ~2 km, Figure 7.7). The aerial photos of each dataset were acquired with a frame camera, meaning that the incidence angle increases radially outwards from the optical center. Partially overlapping images therefore have different incidence angles and can be used to extract a DEM for the overlapping region. Therefore, for each dataset, DEMs were extracted and used to orthorectify the images. Each segment of the post-earthquake dataset was processed independently. The only metadata available for the post-images were the approximate latitude/longitude coordinates of the image center and the focal length of the camera. Therefore, using the scale of the images (estimated from several GCPs), the camera elevation was estimated and the camera information (position and orientation) refined using the bundle adjustment tool in ASP to ensure that the overlapping images were consistent with each other (i.e. to make sure that the x,y and z coordinates of features seen in several overlapping images were identical). The DEM was then extracted and registered to a reference DEM (used as ground truth) using the iterative closest point (ICP) alignment tool in ASP ("pc_align"). We used a 5 m DEM covering the entire study zone (extracted from stereo Worldview satellite images). The alignment step with the reference DEM ensured that each segment was well-registered to the same external reference and therefore consistent with neighboring segments. Once a satisfactory alignment was achieved (the alignment could be evaluated by simply differencing the aerial and reference DEMs), the aerial DEM was used to orthorectify the aerial photos. Finally, the orthoimages belonging to a given segment were mosaiced together.

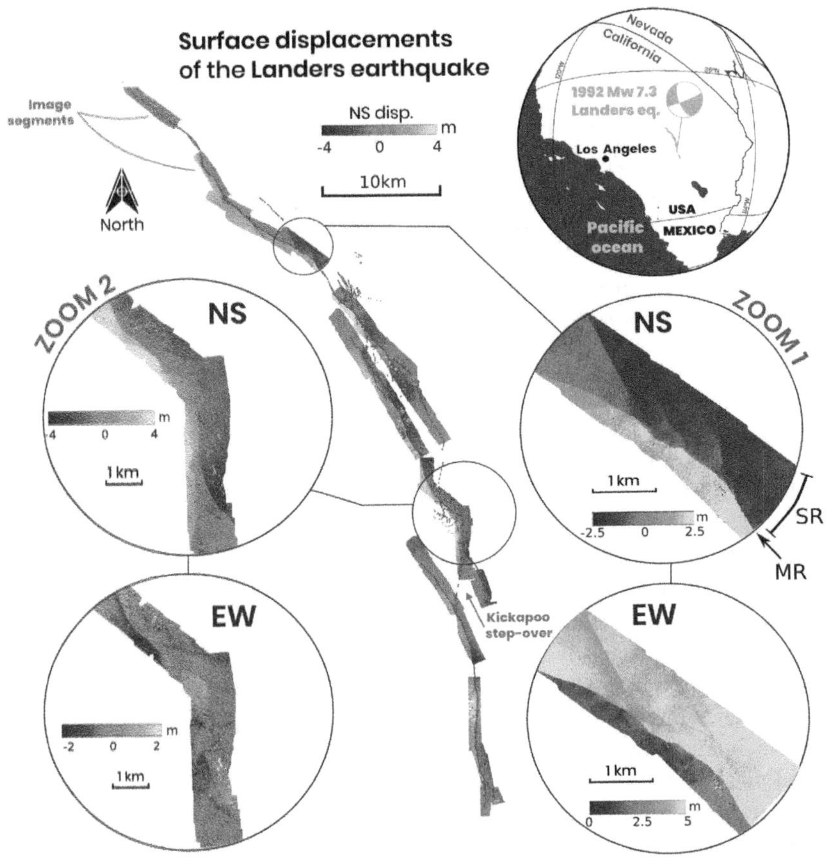

Figure 7.7. *Horizontal surface displacement field of the 1992 Mw 7.3 Landers, California, earthquake. Main rupture (MR) and secondary rupture (SR) are shown, respectively. For a color version of this figure, see www.iste.co.uk/cavalie/images.zip*

The same processing chain was used for the pre-earthquake images except that all images were processed simultaneously (they were all overlapping). Both the pre- and post-images were orthorectified at 1 m resolution to match the lower resolution of the pre-earthquake images. Then, the pre- and post-orthorectified images were correlated using the phase correlator of COSI-Corr with an initial window size of 64×64 pixels, a final window size of 32×32 pixels and a step size of 8 pixels. This resulted in a displacement field with ground sampling every 8 m and truly independent displacements every 32 m (although the windowing function used in the correlator resulted in nearly independent measurements every ~16 m). Finally, measurements with a low signal-to-noise ratio (< 0.97) and outliers with displacement values higher than ± 10 m were discarded and the EW and NS displacement fields were smoothed using the non-local-mean filter implemented in COSI-Corr (Buades *et al.* 2005).

The surface displacement field is shown in Figure 7.7. The processing chain applied (bundle adjustment and alignment with a reference DEM) allowed a seamless horizontal displacement map to be obtained (Figure 7.7). The high-resolution displacement field allowed detection of secondary ruptures, which are clearly visible as sharp discontinuities yet were only partially mapped in the field (Figure 7.7, zoom 1). From these measurements, we can also characterize a wide zone of complex deformation that occurred north of the Kickapoo step-over (Figure 7.7, zoom 2).

7.3.3.2. *Three-dimensional surface displacement of the 2016 Mw 7.8 Kaikoura earthquake from Worldview satellite images*

On November 13, 2016, an Mw 7.8 earthquake struck the southeast coast of the South Island of New Zealand, 60 km southwest of the city of Kaikoura. This earthquake produced the most complex surface rupture ever observed, while the source mechanism indicated about equal thrust slip and right lateral strike-slip motion. Field investigations measured surface strike-slip up to 12 m and vertical slip up to 6.5 m. Considering the significant role that thrust faulting played in this earthquake, computation of both horizontal and vertical surface displacements was essential to sufficiently characterize the kinematics of this earthquake. Using stereoscopic Worldview 1–3 images of approximately 0.5 m resolution, Zinke *et al.* (2019) measured the 3D displacement field of the Kaikoura earthquake using the method describe in section 7.3.2.2 (3D triangulation approach). Before proceeding to the correlation, they first corrected sub-pixel artifacts in the raw images (from CDD array mis-alignments) using ASP. Then, for each overlapping pair of pre- and post-stereo images, the pre-earthquake image with an incidence angle closest to nadir was selected as the primary (to maximize the correlation with the three other images). Then, the three secondary images were coregistered to the master pre-images, and the four images were orthorectified using a coarse DEM (8 m) to remove the long-wavelength topographic distortions (thus allowing the remaining high-frequency stereoscopic distortions to be retrieved with a relatively small 32×32 correlation window). Then, the primary image was correlated with the three secondary images to compute disparity maps that contained information on both the short-wavelength (<8 m) topographic distortions and the tectonic displacement. Finally, they used a ray tracing approach between the disparity maps and satellite position to obtain the pre- and post-event 3D positions of each point of the ground. The 3D displacement was then obtained by simply differentiating the pre- and post-3D positions of each point of the ground (Zinke *et al.* 2019). ICP alignment was then used to align each correlation map with its neighbor, and the full displacement map was then aligned to either an independent correlation map (e.g. derived from correlation of Landsat 8 images) or GPS data.

The 3D displacement field of the Kaikoura earthquake is shown in Figure 7.8 and highlights a very geometrically and kinematically complex rupture composed of 19 faults, including one fault previously unmapped structure, the Snowflake Spur fault (Zinke *et al.* 2019). The surface slip amplitude is highly variable, with significantly larger slip in the northern region, where the slip is subparallel to the plate motion direction, than in the southern part, where the slip is subperpendicular to the plate motion direction (Figure 7.8).

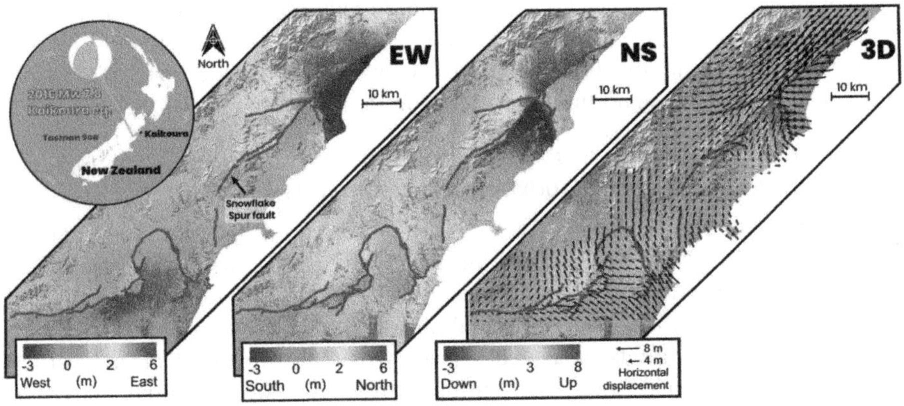

Figure 7.8. *Three-dimensional surface displacement field of the 2016 Mw 7.8 Kaikoura, New Zealand, earthquake processed by Zinke* et al. *(2019). Red lines highlight the complex surface rupture trace. For a color version of this figure, see www.iste.co.uk/cavalie/images.zip*

7.3.3.3. *Horizontal surface displacement field of the 1979 Mw 7.1 Khuli-Boniabad, Iran, earthquake from KH-9 spy satellite images*

The optical image correlation technique allows us to take advantage of the substantial archive of historical aerial and satellite images to document the surface displacement fields of past earthquakes. The oldest events that have currently been studied with optical data occurred in 1975–1984 (Krafla seismo-volcanic rifting crisis crisis (Hollingsworth *et al.* 2012, 2013)), 1978 (Tabas earthquake (Zhou *et al.* 2016)) and 1979 (Khuli-Boniabad earthquake (Marchandon *et al.* 2018a)); nevertheless, any earthquake can be studied for which pre-imagery exists, potentially extending as far back as the beginning of aerial photography in the early 20th century. Correlation of aerial photos therefore provides a powerful method to considerably increase the database of documented surface ruptures, which are important for fault mechanics and seismic hazard research (as in the SURE database). For this purpose, the declassified KH-9 US spy satellite images are particularly useful. The KH-9 spy satellite was active between 1973 and 1980 and acquired ~29,000 images globally with a footprint of 250×125 km and a resolution of 6–9 m. The KH-9 negatives are scanned by the USGS at 7 μm (3,600 dpi) using a high-performance photogrammetric film scanner, and the digital copies can be purchased from the USGS Earth Explorer website for USD 30. Although KH-9 image metadata have not been declassified, KH-9 images were acquired with a camera similar to the Large Format Camera (LFC) used by NASA in 1984, which featured a 23×46 cm frame camera with a focal length of 30.5 cm. For the purposes of orthorectification, they can be treated as very high altitude aerial photos (Surazakov and Aizen 2010; Hollingsworth *et al.* 2012). In the example presented in this section, the surface displacement field of the 1979 Mw 7.1 Khuli-Boniabad, Iran, earthquake was measured using a KH-9 pre-earthquake image and a 10 m Sentinel-2 post-earthquake image (Marchandon *et al.* 2018a; Figure 7.9). A Sentinel-2 image taken in the same season (i.e. January) as the KH-9 image was used to limit the illumination bias in the correlation (Hollingsworth *et al.* 2017; see also section 7.3.2.3). Using the COSI-Corr software package, the KH-9 image was coregistered and orthorectified using the post-earthquake Sentinel-2 image as reference and a 0.5 m resolution Pléiades-derived DEM of the studied area. A very long-wavelength trend was removed from both components of the EW and NS displacement fields by subtracting a second-order polynomial using the detrending tool of COSI-Corr. Such distortions are commonly seen when correlating film-based images and are usually due to thermo-mechanical warping of the photographic film (Michel and Avouac 2006). Nevertheless, the sharp discontinuity associated with the rupture was unaffected by long-wavelength denoising.

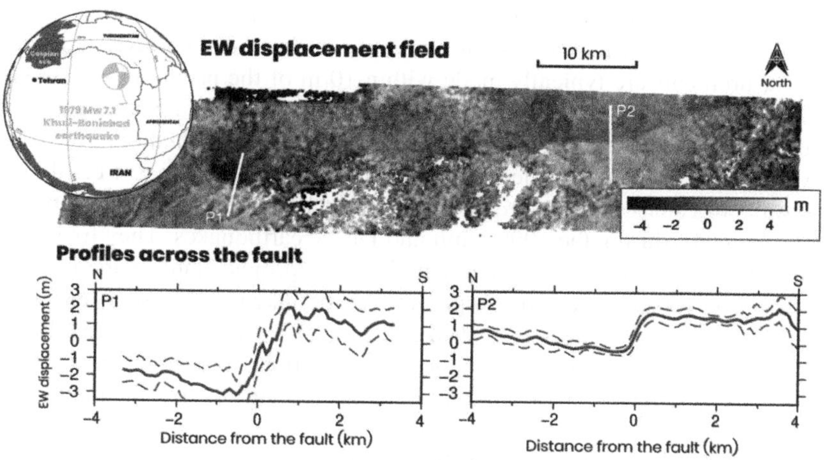

Figure 7.9. *EW surface displacement field and profiles across the fault of the 1979 Mw 7.1 Khuli-Boniabad, Iran, earthquake processed by Marchandon* et al. *(2018a). For a color version of this figure, see www.iste.co.uk/cavalie/images.zip*

The left-lateral surface rupture trace is clearly visible on the EW displacement field (Figure 7.9). The 50-km long rupture trace trends ~N70°, before bending abruptly N160° (west of point F) at its western termination. Surface displacements in the western part of the rupture (west of point F) are typically higher than in the eastern part of the rupture (between points B and C). This is shown by the two example profiles P1 and P2, which reveal fault offsets of ~4.92 ± 0.73 m and ~2.24 ± 0.26 m.

7.3.4. *New insights from high-resolution near-fault displacement maps*

Optical correlation has revolutionized the characterization of surface ruptures of large continental earthquakes. Using surface displacement maps derived from optical correlation, the geometry of the surface rupture and along-strike slip variability can be easily quantified, typically much more quickly than field-based approaches, particularly in remote regions. In addition to enabling an efficient characterization of the surface rupture, these data can be used in a variety of ways to increase our understanding of earthquakes. The principal ones are described in the following sections.

7.3.4.1. *Characterization of off-fault deformation*

During an earthquake, a fraction of the coseismic strain is released off the main fault, either as localized slip on minor secondary faults or as inelastic deformation expressed as warping, granular flow or rigid-block rotation (Figure 7.10; Nelson and Jones 1987; Shelef and Oskin 2010). Given its nature, coseismic distributed deformation is difficult to observe and quantify during field surveys. Indeed, field offset measurements are typically made within 10 m of the primary fault strand and thus capture only the on-fault component of the deformation. Some field studies have also attempted to capture the coseismic distributed deformation. For example, Rockwell *et al.* (2002) used linear features crossing the fault (e.g. tree alignment, fences, walls or canals) to measure deformation accommodated off the main fault by inelastic processes during the 1999 İzmit and Düzce earthquakes. They found that up to 40% of the total slip was accommodated off the primary fault plane over lateral distances up to 100 m. The same approach can be taken with visual inspection of high-resolution optical images (e.g. Zinke *et al.* 2014; Gold *et al.* 2015). However, estimation of off-fault deformation in this way can be time-consuming and interpretive and may only be achieved in areas where long linear features cross the fault.

By capturing the pattern of near-field deformation, optical correlation allows measurement of the total shear accommodated across the fault at high spatial resolution. Optical displacement fields can thus be used to characterize the off-fault deformation (e.g. magnitude, spatial distribution and width) in recent surface-rupturing earthquakes, and to evaluate which parameters control its characteristics (e.g. Zinke *et al.* 2014, 2019; Milliner *et al.* 2015; Vallage *et al.* 2015). The total surface slip accommodated across the fault zone, along with the width over which it is accommodated, are usually measured from fault-perpendicular stacked profiles spaced regularly along the fault strike (e.g. Zinke *et al.* 2014; Milliner *et al.* 2015). On each side of the displacement profile, a linear regression is fitted from the far-field to the point where the displacement profile shows an inflection near the fault (care must be taken to account for any near-fault artifacts introduced by smoothing operations, as well as correlated noise). Then, the linear regressions are extrapolated to the fault trace and the relative difference between the extrapolated regressions defines the offset across the entire fault zone, while the distance between the inflection points located each side of the fault defines the fault zone width (Figure 7.10). The amount of off-fault deformation is then quantified by subtracting the on-fault slip (e.g. offsets measured in the field) from the total slip measured from optical displacement maps (i.e. total slip across the entire fault zone), and expressed as a percentage of the total slip. The method relies on the assumption that the differences between optical and on-fault offsets reflect only the degree of optical frequency division (OFD) and not a reduction of localized slip on the primary fault plane in the uppermost tens of meters of the crust. A comparison between the along-strike offsets measured from optical displacement maps and those measured within a few meters of the main fault strand (by inspection of satellite imagery) of the 2013 Mw 7.7 Balochistan earthquake (Zinke *et al.* 2014) is shown in Figure 7.11.

The average amount of off-fault deformation inferred from high-resolution near-fault displacement fields for various earthquakes ranges from 20% to 60%, while the mean fault-zone width ranges from ~100 to ~400 m (Zinke *et al.* 2014; Gold *et al.* 2015; Milliner *et al.* 2015; Kuo *et al.* 2019; Scott *et al.* 2018; Zinke *et al.* 2019). It should be noted, however, that the data and method used to estimate OFD can provide significantly different results for a given earthquake (e.g. 45% vs. 28% for the Balochistan earthquake in Zinke *et al.* (2014) and Gold *et al.* (2015), respectively). The parameters controlling the amount of OFD from one rupture to another are not completely understood, although fault zone maturity is thought to play a significant role: with increasing fault maturity (e.g. increasing cumulative offset, see Manighetti *et al.* (2007)), the fault becomes geometrically smoother (Brodsky *et al.* 2011), while developing wider and more enhanced damage zones (Faulkner *et al.* 2011) and more established (and mechanically weak) gouge zones, ultimately leading to increased localization of slip during an earthquake (Dolan and Haravitch 2014; Socquet *et al.* 2019). Cumulative long-term OFD has also been estimated using long-term markers of deformation; for example, Salyards *et al.* (1992) measured the paleomagnetic rotation of sediments at a site on the San Andreas fault and found that as much as 78% of the deformation was distributed over a fault zone width of ~45 m. Shelef and Oskin (2010) used geologic features (i.e. dikes and inactive faults) offset by active strike-slip faults in the Eastern California Shear Zone to conclude that distributed displacement accommodates up to 25% of the total displacement within 100–200 m of the studied faults; such values are also consistent with coseismic estimates from a structurally mature section of the 2016 Kaikoura earthquake (Zinke *et al.* 2019), with ~20% OFD occurring within ~100 m of the fault zone.

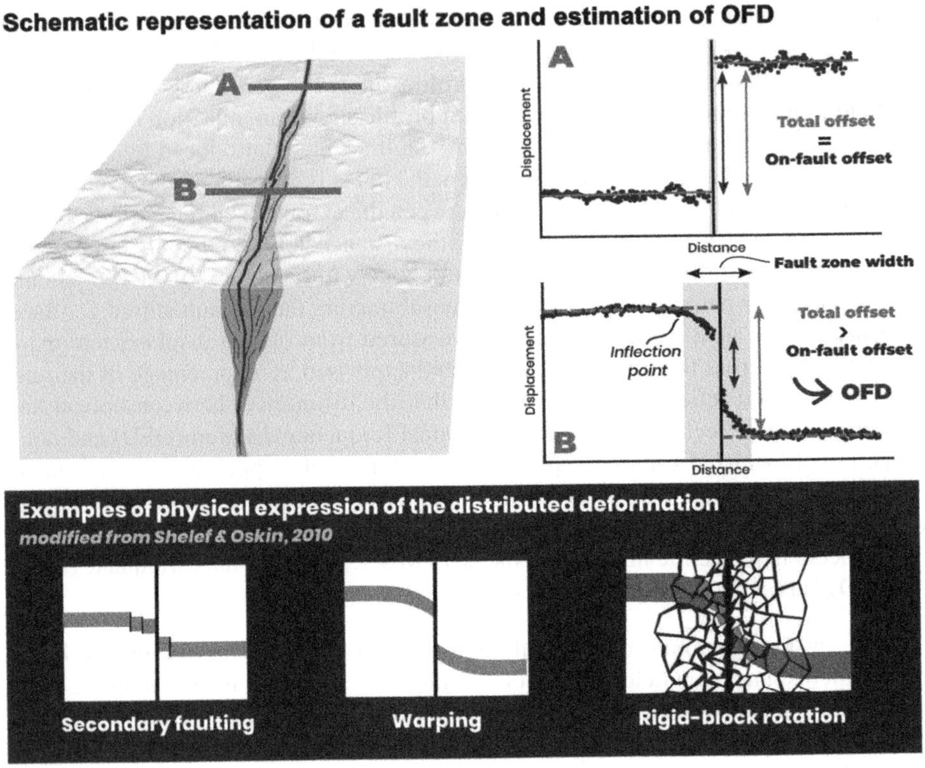

Figure 7.10. *Schematic representation of the distribution of deformation within a fault zone. For a color version of this figure, see www.iste.co.uk/cavalie/images.zip*

Along the length of a surface rupture, the amount of OFD can vary significantly, with OFD ranging from 0% (deformation entirely localized at the primary fault core) to 100% (completely distributed deformation). However, on average, the deformation is more localized at structurally simple sections of the fault and more distributed at structurally complex ones. For example, for the 2013 Mw 7.7 Balochistan earthquake, the percentage of OFD was 26% for structurally simple segments of the rupture, and 56% for structurally complex segments (Zinke *et al.* 2014). Similarly, the mean amount of OFD for the 1992 Mw 7.2 Landers earthquake was 15% in structurally simple parts of the fault and 50% in structurally complex parts of the fault (Milliner *et al.* 2015).

The type of near-fault materials also seems to control the amount of OFD. For the Balochistan earthquake, Zinke *et al.* (2014) found a general tendency for the deformation to be more distributed in younger and thicker sediments, although the trend was weak. Similarly, for the Hector-Mine earthquake, Milliner *et al.* (2016a) found larger OFD where the rupture propagated through sediments compared with bedrock (~40% vs. 20%). This observation is not systematic however. For the Landers earthquake, Milliner *et al.* (2015) found similar and large ranges of OFD in areas of sediment–sediment and basement–basement interfaces (mean of 49±24% and 52±25%, respectively) and lower OFD (mean of 20%) in areas with a strong bimaterial interface (i.e. where the fault juxtaposes bedrock against sediment). Lower amounts of OFD on bimaterial interfaces has also been observed for the 2018 Mw 7.5 Palu earthquake (Socquet *et al.* 2019) and likely results from enhanced localization along the strong contrast in material rigidity (Andrews and Ben-Zion 1997).

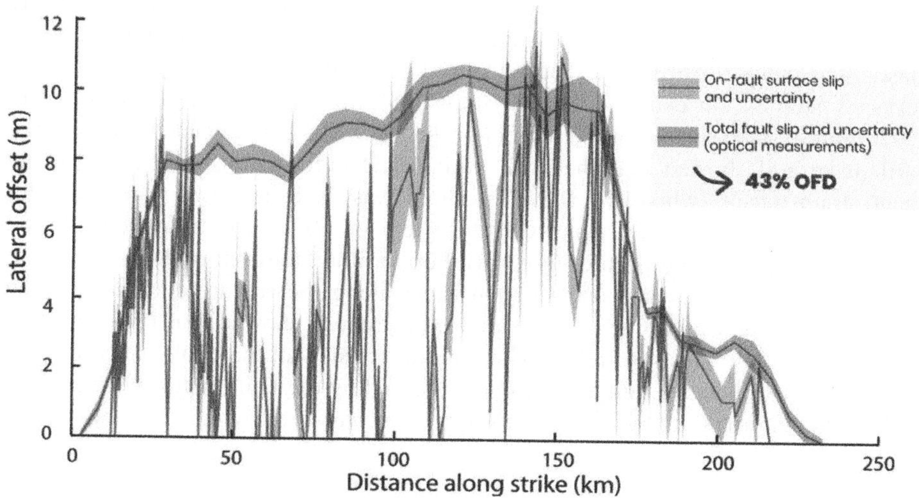

Figure 7.11. *Comparison of the along-strike offsets measured on the main fault strand (blue) and across the entire fault zone (red, optical offsets) for the 2013 Mw 7.7 Balochistan earthquake (modified from Zinke* et al. *(2014)). For a color version of this figure, see www.iste.co.uk/cavalie/images.zip*

Finally, the amount of OFD can be impacted by the local topography. Weaker sediments tend to dominate the surface geology in flat and low-lying areas (e.g. flood plains), while steep mountainous regions often feature strong bedrock materials; therefore, we may expect to see more OFD where fault ruptures cross flat, sediment-filled areas, and less OFD in mountainous settings. Nevertheless, Zinke *et al.* (2019) found the opposite (albeit relatively weak) relationship along the Kekerengu segment of the 2016 Mw 7.8 Kaikorua earthquake, with more OFD in the steep mountains compared to the low-slope coastal regions. One explanation proposed by Zinke *et al.* (2019) is topographic amplification of SH seismic waves, which could in turn enhance the damage within the fault zone, especially where it intersects the surface (e.g. Boore 1972; Ma 2008). Delorme *et al.* (2020) also noted the influence of topography on surface slip following the 2016 Norcia earthquake (normal), which featured locally enhanced (i.e. very shallow) slip on the primary fault plane along steep sections of the fault. Hollingsworth *et al.* (2019) also noted that low slope regions generally featured more OFD compared with steep slope regions for the Norcia event, with the steepest slopes featuring locally enhanced slip (shallow slip surplus). Gravitational effects have been postulated as a contributing factor for locally enhancing the slip where the fault breaks the surface (Delorme *et al.* 2020).

High-resolution optical displacement fields can also be used to infer surface strain tensors (e.g. Scott *et al.* 2018; Barnhart *et al.* 2020). The zone of inelastic failure can be readily observed by delineating the areas where the threshold between elastic and inelastic strain is exceeded (a ratio of 0.5% is commonly assumed for surface materials; for example, Scott *et al.* (2018). In a recent study, Barnhart *et al.* (2020) inferred the surface-strain tensors of the 2019 Mw 7.1 Ridgecrest, California, earthquake and showed that the zone of inelastic deformation inferred from the shear strain tensor (where the 0.5% threshold was exceeded) was narrow (width of $\sim$25 m) and did not correlate with the width of off-fault deformation inferred from fault-perpendicular profiles ($\sim$63 m). Furthermore, they showed that the width inferred from the strain tensors was consistent along the rupture compared to the OFD width, which was much more variable. They also showed that no correlation exists between the inelastic failure width and the near-fault materials. According to the authors, the inelastic failure width is influenced by the fault maturity (increasing width with maturity) and by the rupture speed (narrower width with increasing rupture speed).

Understanding the degree to which the deformation is localized within the fault zones has several important implications. First, if the deformation is distributed within hundreds of meters from the main fault trace, fault slip rates deduced from measurements of offset geological markers made within tens of meters of the main fault trace could be significantly underestimated (Dolan and Haravitch 2014). If real, this underestimation has several consequences. First, the estimation of seismic hazard using probabilistic approaches incorporating geological fault slip rates could lead to a systematic underestimation of the real hazard. Second, any interpretation of the differences observed between geological slip rate and geodetic loading rate would be biased. The degree to which the deformation is localized can also influence the frequency content and magnitude of strong ground motions during an earthquake (Dolan and Haravitch 2014). Better understanding the link between off-fault deformation and strong ground motion could therefore help to make more accurate predictions of strong ground motions in future earthquakes and thus better assess seismic hazard. Finally, partitioning of slip into damage zones has major implications for how we interpret information recorded in paleoseismic trenches, as offset markers only capture a component of the slip at depth, and may thus lead to an incorrect estimate of past earthquake magnitudes, which also form an integral part of seismic hazard assessment.

High-resolution optical maps of coseismic surface displacements can also be combined with dynamic rupture modeling incorporating off-fault damage to determine the most likely rupture scenario for an earthquake. Indeed, by comparing the pattern of off-fault damage observed in the optical displacement maps with the one simulated by the dynamic rupture modeling, Klinger *et al.* (2018) succeeded in determining the most likely rupture scenario at the Jordan–Kekerengu–Papatea triple junction during the 2016 Mw 7.8 Kaikoura earthquake.

7.3.4.2. *Improvement of fault slip model*

Over the last 40 years, fault slip distribution models are routinely inferred each time a significant earthquake occurs. As the estimation of fault slip from surface data is a fundamentally ill-posed problem, slip models for a given earthquake can be very different depending on the type and quantity of data, the inversion method and the assumed fault geometry and Earth model. In the last three decades, the development of space-based geodetic techniques, in particular InSAR, has given access to increasingly complete descriptions of earthquake surface displacements, thereby increasing the robustness of slip models. However, InSAR does not routinely provide near-fault measurements for large surface-rupturing earthquakes. By improving the near-fault coverage, optical correlation data can increase the reliability of fault slip models (e.g. Dominguez *et al.* 2003; Xu *et al.* 2016; Gombert *et al.* 2017; Marchandon *et al.* 2018b; Delorme *et al.* 2020; Lauer *et al.* 2020). For example, by re-estimating the slip model of the 1997 Mw 7.2 Zirkuh earthquake, Marchandon *et al.* (2018b) have shown that the addition of near-field optical data in the modeling changes the obtained fault geometry, the slip amplitude and the slip distribution significantly compared to the slip model based on InSAR data only.

Furthermore, near-fault optical data have been used to better understand the origin of the shallow slip deficit (SSD) commonly seen in fault slip models for large continental strike-slip earthquakes (Fialko *et al.* 2005). By reassessing the fault slip model of three large strike-slip earthquakes and incorporating near-field displacements from optical correlation data, Xu *et al.* (2016) have shown that a significant part of the previously inferred SSD for each event was an artifact due to a lack of near-fault coverage. The impact of the lack of near-fault coverage for SSD has also been shown by Lauer *et al.* (2020) for the Balochistan earthquake and by Marchandon *et al.* (2021) using synthetic tests. However, because optical data are typically much noisier compared to InSAR, noise-related model uncertainties should be systematically estimated when using optical data in fault slip inversions in order to avoid any misinterpretations of the inferred slip models, in particular concerning the distribution of slip with depth (Marchandon *et al.* 2021).

7.3.4.3. *Impact of the fault structural properties on the coseismic slip distribution*

It has long been noticed that fault complexities can impact rupture propagation during an earthquake (King and Nabelek 1985). This impact is reflected in the coseismic slip distribution of earthquakes and shows that fault complexities such as step-overs or bends correlate with (1) zones of low slip amplitude (e.g. Lasserre *et al.* 2005) and (2) the start and endpoint of ruptures (e.g. King and Nabelek 1985). By providing the geometry of the surface rupture and the distribution of surface fault slip, optical correlation results have helped to increase the number of examples that illustrate the impact of fault geometry on coseismic slip (e.g. Klinger *et al.* 2006; Milliner *et al.* 2015; Vallage *et al.* 2015). For example, analysis of the surface coseismic slip distributions of the Khuli-Boniabad and Zirkuh earthquakes obtained from optical correlation have shown that the ruptures of these earthquakes are characterized by several segments, with the inter-segments located in zones of large-scale structural complexity of the Dasht-e-Bayaz and Abiz faults (Marchandon *et al.* 2018a, 2018b).

Furthermore, high-resolution optical measurements allow a more in-depth analysis of the surface slip variability. In particular, the frequency content of surface coseismic slip distributions can be analyzed by computing the power spectral density of surface slip (e.g. Milliner *et al.* 2016; Bruhat *et al.* 2020). Power spectral density (PSD) is the measure of the power content of a signal versus frequency and can be used to characterize fractal properties of a signal. By computing the PSD of the surface slip distribution of various earthquakes, it has been shown that the surface slip distribution is fractal (Milliner *et al.* 2016; Bruhat *et al.* 2020) (i.e. a power law describes the relationship between the slip amplitude and the spatial wavelength). The fractal nature of surface slip echoes the fractal nature of fault surfaces. Indeed, although faults are typically treated as planes, faults are in reality irregular surfaces at all scales (e.g. Candela *et al.* 2012). Dynamic rupture simulations have shown that these geometric irregularities, i.e. roughness, induce local stress perturbations that affect the rupture prorogation (e.g. rapid accelerations and decelerations of the rupture front, variability in moment release and super-shear transitions) and induce slip heterogeneities (e.g. Dieterich and Smith 2009). By analyzing the coseismic slip variability of real earthquakes, it could thus be, in principle, possible to investigate the roughness properties of the faults with which they occurred. Bruhat *et al.* (2020) made a first attempt towards this goal. Using a catalog containing thousands of dynamic rupture simulations on a self-similar fractal fault (H=1) with varying roughness, they investigated the relationship between the fault roughness and the spectral characteristics of the resulting slip distribution. They found that the slip distribution deviates from self-similarity (Hurst coefficient decreases) with increasing fault roughness. They also analyzed the spectral characteristics of various real earthquakes and found that the real slip distributions were significantly more self-affine (lower Hurst coefficient) compared to the simulated slip distributions, thus making it hard to draw any conclusions about the fault roughness with which these real earthquakes occurred.

7.4. Conclusion

In recent decades, the development of InSAR and optical correlation, as well as the tremendous increase of SAR and optical satellite data available, has revolutionized the measurement of surface displacements induced by the earthquake deformation cycle. Far from being interchangeable, these two methods are highly complementary, as each technique addresses the limitations of the other (see Table 11.1 for a summary of the pros and cons of each method). For nearly 30 years, InSAR measurements of coseismic deformation have contributed to showing that fault ruptures are complex (e.g. multi-segmented) and that the structural complexities of faults can sometimes impact the propagation and arrest of ruptures. Furthermore, by constraining fault slip models, InSAR data have shown that the surface slip seems to be a poor indicator of the amount of slip at depth, implying that seismic hazard estimated from geological surface offsets could be systematically underestimated. More recently, InSAR coseismic observations have been complemented by near-fault optical correlation data, which have allowed a more direct observation of the surface rupture and surface fault slip. The inclusion of near-field data from optical correlation indicate that the discrepancy between the surface slip and slip at depth previously inferred using InSAR data alone could partly be a modeling artifact, further complicated by slip being accommodated by distributed deformation in the upper crustal volume surrounding the fault (among other processes), which is not taken into account in the slip inversion. Moreover optical data have also been used to investigate the fractal properties of faults, offering a new insights into fault mechanics and generation of strong ground motions.

In contrast to optical correlation, InSAR is very efficient in measuring small displacements. The capacity of InSAR to detect small displacements has been exploited to measure inter-seismic strain rates and deduce geodetic slip rates. The comparison of short-term geodetic and long-term geological slip rates for a number of major faults suggests that the inter-seismic strain rate varies little with time, implying that it could be used to constrain seismic hazard models. Finally, InSAR data have been widely used to measure transient aseismic deformation such as post-seismic deformation and inter-seismic fault creep. Thanks to InSAR studies, we now know that post-seismic deformation is commonly observed after moderate to large earthquakes. Modeling of the post-seismic response shows that multiple post-seismic mechanisms are often required to explain the observations (e.g. afterslip, viscoelastic relaxation and poro-elastic rebound), as well as placing of constraints on the frictional properties of faults and the rheology of the lithosphere. Finally, detailed studies of transient aseismic creep have highlighted the link between fault creep and fault zone lithology and the control of pore fluid pressures on the temporal variations of creep.

InSAR
Pros: – Able to measure very small deformation (e.g. produced by post-seismic or inter-seismic deformation) – Able to measure time series of ground movements over years to decades – High spatial resolution – High precision – Weather independent – Can acquire data at night – Global systematic acquisitions of free and open data – Sensitive to vertical and horizontal displacements Cons: – Cannot reliably measure near-fault deformation of surface rupturing earthquakes – Provides only one component of the displacement, although two components can be obtained by combining ascending and descending interferograms – Lower spatial resolution in along- or across-track directions (depending on the sensor) – Insensitive to deformation in the along-track direction – Depending on the wavelength, can be incoherent in vegetated areas – Can be difficult to process multi-year interferograms – Relatively expensive hardware means there have been relatively few satellites capable of InSAR – Can only go back as far as the 1990s
Optical correlation
Pros: – Able to measure the near-fault horizontal displacements of surface rupturing earthquakes – Provides two components of the horizontal displacements (plus the vertical component if stereo pre- and post-images are available) – Images with very long temporal baselines (decades) can be used in desert areas – Images from different sensors and acquired with very different geometries can be correlated – High spatial resolution (higher than InSAR if high-resolution optical images are used) – Optical sensors are more common-place than InSAR – Optical sensors are relatively cheap to deploy, especially in constellations and as secondary payloads, and on drones – Can go back to the early 20th century (1920s onward) Cons: – Much less precise than InSAR – Weather-dependent (need cloud-free imagery) – Very difficult to measure small displacements – High-resolution optical sensors are generally commercial and thus expensive – There is no background acquisition strategy for high-resolution data, which needs to be tasked following an event – Errors can result from illumination changes – Challenging to apply in vegetated environments – Stereo images are essential for resolving vertical deformation

Table 7.1. *Summary of pros and cons of InSAR and optical correlation*

In the future, the quality and quantity of SAR and optical images will continue to improve. Regarding the optical data, this increase associated with more effective processing methods will enable the measurement of small transient deformations. The analysis of future surface-rupturing earthquakes with optical data will also allow a more comprehensive understanding of the parameters controlling the distribution of deformation in the upper crust. Optical data can also provide constraints on models of dynamic rupture propagation incorporating off-fault deformation, which will further our understanding of how the upper crust behaves during an earthquake. Concerning InSAR, the processing of the huge Sentinel-1 data will shortly provide continent-scale maps of inter-seismic velocity. Future satellite missions, such as the L-band NISAR mission, will improve coverage and coherence, and there is the prospect of measuring 3D deformation using companion satellites of Sentinel-1. Automatic processing systems will mean that non-expert users can work with the data, and in the near future, we will begin to see inter-seismic deformation data from InSAR incorporated routinely into seismic hazard models. Collectively, the knowledge on earthquake and fault systems brought by InSAR and optical data (e.g. control of structural complexities in the rupture propagation, degree of localization of the deformation around faults in the upper crust and strain accumulation) is of great significance to both science and society and will ultimately lead to a seismically more resilient world through improved seismic hazard assessment.

7.5. Acknowledgments

The authors thank Yann Klinger for his careful reviews and comments that helped to improve this chapter.

7.6. References

Agram, P., Jolivet, R., Riel, B., Lin, Y., Simons, M., Hetland, E., Doin, M.-P., Lasserre, C. (2013). New radar interferometric time series analysis toolbox released. *Eos, Transactions of the American Geophysical Union*, 94(7), 69–70.

Andrews, D.J. and Ben-Zion, Y. (1997). Wrinkle-like slip pulse on a fault between different materials. *Journal of Geophysical Research: Solid Earth*, 102(B1), 553–571.

Attema, E., Desnos, Y.-L., Duchossois, G. (2000). Synthetic aperture radar in Europe: ERS, Envisat, and beyond. *Johns Hopkins APL Technical Digest*, 21(1), 155–161.

Avouac, J.-P. and Leprince, S. (2015). Geodetic imaging using optical systems. *Geodesy*, 3, 387–424

Avouac, J.-P., Ayoub, F., Leprince, S., Konca, O., Helmberger, D.V. (2006). The 2005, Mw 7.6 Kashmir earthquake: Sub-pixel correlation of ASTER images and seismic waveforms analysis. *Earth and Planetary Science Letters*, 249(3–4), 514–528.

Ayoub, F., Leprince, S., Avouac, J.-P. (2009). Co-registration and correlation of aerial photographs for ground deformation measurements. *ISPRS Journal of Photogrammetry and Remote Sensing*, 64(6), 551–560.

Ayoub, F., Leprince, S., Avouac, J.-P. (2017). User's Guide to COSI-CORR Co-registration of Optically Sensed Images and Correlation. Technical report, California Institute of Technology, CA.

Barnhart, W.D., Gold, R.D., Shea, H.N., Peterson, K.E., Briggs, R.W., Harbor, D.J. (2019a). Vertical coseismic offsets derived from high-resolution stereogrammetric DSM differencing: The 2013 Baluchistan, Pakistan earthquake. *Journal of Geophysical Research: Solid Earth*, 124(6), 6039–6055.

Barnhart, W.D., Hayes, G.P., Wald, D.J. (2019b). Global earthquake response with imaging geodesy: Recent examples from the USGS NEIC. *Remote Sensing*, 11(11), 1357.

Barnhart, W.D., Gold, R.D., Hollingsworth, J. (2020). Localized fault-zone dilatancy and surface inelasticity of the 2019 Ridgecrest earthquakes. *Nature Geoscience*, 13(10), 699–704.

Barrick, D.E. and Swift, C.T. (1980). The Seasat microwave instruments in historical perspective. *IEEE Journal of Oceanic Engineering*, 5(2), 74–79.

Bekaert, D.P., Karim, M., Linick, J.P., Hua, H., Sangha, S., Lucas, M., Malarout, N., Agram, P.S., Pan, L., Owen, S.E., Lai-Norling, J. (2019). Development of open-access standardized InSAR Displacement Products by the Advanced Rapid Imaging and Analysis (ARIA) project for natural hazards. *AGU Fall Meeting*, G23A–04.

Beyer, R.A., Alexandrov, O., McMichael, S. (2018). The Ames stereo pipeline: NASA's open source software for deriving and processing terrain data. *Earth and Space Science*, 5(9), 537–548.

Biggs, J., Burgmann, R., Freymueller, J.T., Lu, Z., Parsons, B., Ryder, I., Schmalzle, G., Wright, T. (2009). The postseismic response to the 2002 M 7.9 Denali Fault earthquake: Constraints from InSAR 2003–2005. *Geophysical Journal International*, 176(2), 353–367.

Bird, P. and Kreemer, C. (2015). Revised tectonic forecast of global shallow seismicity based on version 2.1 of the Global Strain Rate Map. *Bulletin of the Seismological Society of America*, 105(1), 152–166.

Boore, D.M. (1972). A note on the effect of simple topography on seismic SH waves. *Bulletin of the Seismological Society of America*, 62(1), 275–284.

Brodsky, E.E., Gilchrist, J.J., Sagy, A., Collettini, C. (2011). Faults smooth gradually as a function of slip. *Earth and Planetary Science Letters*, 302(1–2), 185–193.

Bruhat, L., Klinger, Y., Vallage, A., Dunham, E.M. (2020). Influence of fault roughness on surface displacement: From numerical simulations to coseismic slip distributions. *Geophysical Journal International*, 220(3), 1857–1877.

Buades, A., Coll, B., Morel, J.-M. (2005). A non-local algorithm for image denoising. *IEEE Computer Society Conference on Computer Vision and Pattern Recognition (CVPR'05)*, 60–65.

Bürgmann, R., Schmidt, D., Nadeau, R., d'Alessio, M., Fielding, E., Manaker, D., McEvilly, T., Murray, M. (2000). Earthquake potential along the northern Hayward fault, California. *Science*, 289(5482), 1178–1182.

Cakir, Z., Akoğlu, A.M., Belabbes, S., Ergintav, S., Meghraoui, M. (2005). Creeping along the Ismetpasa section of the North Anatolian fault (Western Turkey): Rate and extent from InSAR. *Earth and Planetary Science Letters*, 238(1–2), 225–234.

Cakir, Z., Ergintav, S., Akoğlu, A.M., Çakmak, R., Tatar, O., Meghraoui, M. (2014). InSAR velocity field across the North Anatolian Fault (eastern Turkey): Implications for the loading and release of interseismic strain accumulation. *Journal of Geophysical Research: Solid Earth*, 119(10), 7934–7943.

Candela, T., Renard, F., Klinger, Y., Mair, K., Schmittbuhl, J., Brodsky, E.E. (2012). Roughness of fault surfaces over nine decades of length scales. *Journal of Geophysical Research: Solid Earth*, 117(B8), 1–30.

Cavalié, O., Lasserre, C., Doin, M.-P., Peltzer, G., Sun, J., Xu, X., Shen, Z.-K. (2008). Measurement of interseismic strain across the Haiyuan fault (Gansu, China), by InSAR. *Earth and Planetary Science Letters*, 275(3–4), 246–257.

Cavalié, O., Pathier, E., Radiguet, M., Vergnolle, M., Cotte, N., Walpersdorf, A., Kostoglodov, V., Cotton, F. (2013). Slow slip event in the Mexican subduction zone: Evidence of shallower slip in the Guerrero seismic gap for the 2006 event revealed by the joint inversion of InSAR and GPS data. *Earth and Planetary Science Letters*, 367, 52–60.

Cetin, E., Cakir, Z., Meghraoui, M., Ergintav, S., Akoglu, A.M. (2014). Extent and distribution of aseismic slip on the Ismetpaşa segment of the North Anatolian Fault (Turkey) from persistent scatterer InSAR. *Geochemistry, Geophysics, Geosystems*, 15(7), 2883–2894.

Cowgill, E., Gold, R.D., Xuanhua, C., Xiao-Feng, W., Arrowsmith, J.R., Southon, J. (2009). Low Quaternary slip rate reconciles geodetic and geologic rates along the Altyn Tagh fault, northwestern Tibet. *Geology*, 37(7), 647–650.

Delorme, A., Grandin, R., Klinger, Y., Pierrot-Deseilligny, M., Feuillet, N., Jacques, E., Rupnik, E., Morishita, Y. (2020). Complex deformation at shallow depth during the 30 October 2016 Mw6. 5 Norcia earthquake: Interference between tectonic and gravity processes? *Tectonics*, 39(2), e2019TC005596.

Deng, J., Gurnis, M., Kanamori, H., Hauksson, E. (1998). Viscoelastic flow in the lower crust after the 1992 Landers, California, earthquake. *Science*, 282(5394), 1689–1692.

Dieterich, J.H. and Smith, D.E. (2009). Nonplanar faults: Mechanics of slip and off-fault damage. *Pure Applied Geophysics*, 166, 1799–1815.

Doin, M.-P., Guillaso, S., Jolivet, R., Lasserre, C., Lodge, F., Ducret, G., Grandin, R. (2011). Presentation of the small baseline NSBAS processing chain on a case example: The Etna deformation monitoring from 2003 to 2010 using Envisat data. *ESA SP-697 Fringe Symposium*, Frascati.

Dolan, J.F. and Haravitch, B.D. (2014). How well do surface slip measurements track slip at depth in large strike-slip earthquakes? The importance of fault structural maturity in controlling on-fault slip versus off-fault surface deformation. *Earth and Planetary Science Letters*, 388, 38–47.

Dominguez, S., Avouac, J.-P., Michel, R. (2003). Horizontal coseismic deformation of the 1999 Chi-Chi earthquake measured from SPOT satellite images: Implications for the seismic cycle along the western foothills of central Taiwan. *Journal of Geophysical Research*, 108(B2), 1–19.

Ekström, G., Stein, R.S., Eaton, J., Eberhart-Phillips, D. (1992). Seismicity and geometry of a 110-km-long blind thrust fault 1. The 1985 Kettleman Hills, California, earthquake. *Journal of Geophysical Research: Solid Earth*, 97(B4), 4843–4864.

Elliott, J. (2020). Earth observation for the assessment of earthquake hazard, risk and disaster management. *Surveys in Geophysics*, 41, 1323–1354.

Elliott, J., Parsons, B., Jackson, J., Shan, X., Sloan, R., Walker, R. (2011). Depth segmentation of the seismogenic continental crust: The 2008 and 2009 Qaidam earthquakes. *Geophysical Research Letters*, 38(6), 1–6.

Elliott, J., Nissen, E., England, P., Jackson, J.A., Lamb, S., Li, Z., Oehlers, M., Parsons, B. (2012). Slip in the 2010–2011 Canterbury earthquakes, New Zealand. *Journal of Geophysical Research: Solid Earth*, 117(B3), 1–36.

Elliott, J., Elliott, A., Hooper, A., Larsen, Y., Marinkovic, P., Wright, T. (2015). Earthquake monitoring gets boost from a new satellite. *Eos*, 96(3), 14–18.

Elliott, J., Jolivet, R., Gonzalez, P., Avouac, J., Hollingsworth, J., Searle, M., Stevens, V. (2016a). Himalayan megathrust geometry and relation to topography revealed by the Gorkha earthquake. *Nature Geoscience*, 9(2), 174–180.

Elliott, J., Walters, R., Wright, T. (2016b). The role of space-based observation in understanding and responding to active tectonics and earthquakes. *Nature Communications*, 7(1), 1–16.

Emre, Ö., Duman, T.Y., Özalp, S., Şaroğlu, F., Olgun, Ş., Elmaci, H., Can, T. (2018). Active fault database of Turkey. *Bulletin of Earthquake Engineering*, 16(8), 3229–3275.

Fattahi, H. and Amelung, F. (2016). InSAR observations of strain accumulation and fault creep along the Chaman Fault system, Pakistan and Afghanistan. *Geophysical Research Letters*, 43(16), 8399–8406.

Faulkner, D., Mitchell, T., Jensen, E., Cembrano, J. (2011). Scaling of fault damage zones with displacement and the implications for fault growth processes. *Journal of Geophysical Research: Solid Earth*, 116(B5), 1–11.

Fialko, Y. (2004). Evidence of fluid-filled upper crust from observations of postseismic deformation due to the 1992 Mw 7.3 Landers earthquake. *Journal of Geophysical Research: Solid Earth*, 109(B8), 1–21.

Fialko, Y., Sandwell, D., Simons, M., Rosen, P. (2005). Three-dimensional deformation caused by the Bam, Iran, earthquake and the origin of shallow slip deficit. *Nature*, 435, 295–299.

Fielding, E.J., Wright, T.J., Muller, J., Parsons, B.E., Walker, R. (2004). Aseismic deformation of a fold-and-thrust belt imaged by synthetic aperture radar interferometry near Shahdad, southeast Iran. *Geology*, 32(7), 577–580.

Floyd, M.A., Walters, R.J., Elliott, J.R., Funning, G.J., Svarc, J.L., Murray, J.R., Hooper, A.J., Larsen, Y., Marinkovic, P., Bürgmann, R., Johanson, I.A., Wright, T.J. (2016). Spatial variations in fault friction related to lithology from rupture and afterslip of the 2014 South Napa, California, earthquake. *Geophysical Research Letters*, 43(13), 6808–6816.

Foumelis, M., Papadopoulou, T., Bally, P., Pacini, F., Provost, F., Patruno, J. (2019). Monitoring geohazards using on-demand and systematic services on ESA's geohazards exploitation platform. *IEEE International Geoscience and Remote Sensing Symposium*, 5457–5460.

Freed, A.M. and Bürgmann, R. (2004). Evidence of power-law flow in the Mojave desert mantle. *Nature*, 430(6999), 548–551.

Fujiwara, S., Yarai, H., Kobayashi, T., Morishita, Y., Nakano, T., Miyahara, B., Nakai, H., Miura, Y., Ueshiba, H., Kakiage, Y., Une, H. (2016). Small-displacement linear surface ruptures of the 2016 Kumamoto earthquake sequence detected by ALOS-2 SAR interferometry. *Earth, Planets and Space*, 68(1), 160.

Fukahata, Y. and Wright, T.J. (2008). A non-linear geodetic data inversion using ABIC for slip distribution on a fault with an unknown dip angle. *Geophysical Journal International*, 173(2), 353–364.

Funning, G.J. and Garcia, A. (2019). A systematic study of earthquake detectability using Sentinel-1 interferometric wide-swath data. *Geophysical Journal International*, 216(1), 332–349.

Furlong, K.P. and Herman, M. (2017). Reconciling the deformational dichotomy of the 2016 Mw 7.8 Kaikoura New Zealand earthquake. *Geophysical Research Letters*, 44(13), 6788–6791.

Furuya, M. and Satyabala, S. (2008). Slow earthquake in Afghanistan detected by InSAR. *Geophysical Research Letters*, 35(6), 1–4.

Gabriel, A.K., Goldstein, R.M., Zebker, H.A. (1989). Mapping small elevation changes over large areas: Differential radar interferometry. *Journal of Geophysical Research: Solid Earth*, 94(B7), 9183–9191.

Gold, R.D., Reitman, N.G., Briggs, R.W., Barnhart, W.D., Hayes, G.P., Wilson, E. (2015). On- and off-fault deformation associated with the September 2013 Mw 7.7 Balochistan earthquake: Implications for geologic slip rate measurements. *Tectonophysics*, 660, 65–78.

Goldstein, R.M., Zebker, H.A., Werner, C.L. (1988). Satellite radar interferometry: Two-dimensional phase unwrapping. *Radio Science*, 23(4), 713–720.

Gombert, B., Duputel, Z., Jolivet, R., Doubre, C., Rivera, L., Simons, M. (2017). Revisiting the 1992 Landers earthquake: A Bayesian exploration of co-seismic slip and off-fault damage. *Geophysical Journal International*, 212(2), 839–852.

Grandin, R., Doin, M.-P., Bollinger, L., Pinel-Puysségur, B., Ducret, G., Jolivet, R., Sapkota, S.N. (2012). Long-term growth of the Himalaya inferred from interseismic InSAR measurement. *Geology*, 40(12), 1059–1062.

Grandin, R., Vallée, M., Satriano, C., Lacassin, R., Klinger, Y., Simoes, M., Bollinger, L. (2015). Rupture process of the Mw=7.9 2015 Gorkha earthquake (Nepal): Insights into Himalayan megathrust segmentation. *Geophysical Research Letters*, 42(20), 8373–8382.

Hamling, I.J., Hreinsdóttir, S., Clark, K., Elliott, J., Liang, C., Fielding, E., Litchfield, N., Villamor, P., Wallace, L., Wright, T.J., D'Anastasio, E., Bannister, S., Burbidge, D., Denys, P., Gentle, P., Howarth, J., Muller, C., Palmer, N., Pearson, C., Power, W., Barnes, P., Barrel, D.J.A., Van Dissen, R., Langridge, R., Little, T., Andrew, N., Pettinga, J., Rowland, J., Stirling, M. (2017). Complex multifault rupture during the 2016 Mw 7.8 Kaikōura earthquake, New Zealand. *Science*, 356(6334), 1–16.

Hetland, E. and Hager, B.H. (2006). The effects of rheological layering on post-seismic deformation. *Geophysical Journal International*, 166(1), 277–292.

Hollingsworth, J., Leprince, S., Ayoub, F., Avouac, J.P. (2012). Deformation during the 1975–1984 Krafla rifting crisis, NE Iceland, measured from historical optical imagery. *Journal of Geophysical Research: Solid Earth*, 117, 1–24.

Hollingsworth, J., Leprince, S., Ayoub, F., Avouac, J.-P. (2013). New constraints on dike injection and fault slip during the 1975–1984 Krafla rift crisis, NE Iceland. *Journal of Geophysical Research: Solid Earth*, 118, 1–21.

Hollingsworth, J., Lingling, Y., Avouac, J.-P. (2017). Dynamically triggered slip on a splay fault in the Mw 7.8, 2016 Kaikoura (New Zealand) earthquake. *Geophysical Research Letters*, 44, 3517–3525.

Hollingsworth, J., Maubant, L., Pathier, E., Socquet, A. (2019). Shallow slip deficit and off-fault deformation in the 2016 Norcia earthquake derived from 3D optical image correlation. *American Geophysical Union Fall Meeting*, December.

Hollingsworth, J., Daout, S., Doin, M.-P., Cantraine, M. (2020). Optical geodesy and the measurement of ground deformation by image correlation. In *Remote Sensing Applications to Characterization of Geohazards and Natural Resources*, Chaussard, E. (ed.). Springer, New York.

Huang, M.-H., Fielding, E.J., Dickinson, H., Sun, J., Gonzalez-Ortega, J.A., Freed, A.M., Bürgmann, R. (2017). Fault geometry inversion and slip distribution of the 2010 Mw 7.2 El Mayor-Cucapah earthquake from geodetic data. *Journal of Geophysical Research: Solid Earth*, 122(1), 607–621.

Hussain, E., Wright, T.J., Walters, R.J., Bekaert, D., Hooper, A., Houseman, G.A. (2016). Geodetic observations of postseismic creep in the decade after the 1999 Izmit earthquake, Turkey: Implications for a shallow slip deficit. *Journal of Geophysical Research: Solid Earth*, 121(4), 2980–3001.

Hussain, E., Wright, T.J., Walters, R.J., Bekaert, D.P., Lloyd, R., Hooper, A. (2018). Constant strain accumulation rate between major earthquakes on the North Anatolian Fault. *Nature Communications*, 9(1), 1–9.

Ingleby, T. and Wright, T. (2017). Omori-like decay of postseismic velocities following continental earthquakes. *Geophysical Research Letters*, 44(7), 3119–3130.

Ingleby, T., Wright, T., Hooper, A., Craig, T., Elliott, J. (2020). Constraints on the geometry and frictional properties of the main Himalayan thrust using coseismic, postseismic, and interseismic deformation in Nepal. *Journal of Geophysical Research: Solid Earth*, 125(2), e2019JB019201.

Johanson, I. and Bürgmann, R. (2005). Creep and quakes on the northern transition zone of the San Andreas fault from GPS and InSAR data. *Geophysical Research Letters*, 32(14), 1–5.

Jones, T. (1979). *Monty Python's Life of Brian*. Film, HandMade Films and Python (Monty) Pictures.

Jónsson, S., Segall, P., Pedersen, R., Björnsson, G. (2003). Post-earthquake ground movements correlated to pore-pressure transients. *Nature*, 424(6945), 179–183.

Kaneko, Y. and Fialko, Y. (2011). Shallow slip deficit due to large strike-slip earthquakes in dynamic rupture simulations with elasto-plastic off-fault response. *Geophysical Journal International*, 186(3), 1389–1403.

Karimzadeh, S., Cakir, Z., Osmanoğlu, B., Schmalzle, G., Miyajima, M., Amiraslanzadeh, R., Djamour, Y. (2013). Interseismic strain accumulation across the North Tabriz Fault (NW Iran) deduced from InSAR time series. *Journal of Geodynamics*, 66, 53–58.

Khoshmanesh, M. and Shirzaei, M. (2018). Episodic creep events on the San Andreas Fault caused by pore pressure variations. *Nature Geoscience*, 11(8), 610–614.

Khoshmanesh, M., Shirzaei, M., Nadeau, R. (2015). Time-dependent model of aseismic slip on the central San Andreas Fault from InSAR time series and repeating earthquakes. *Journal of Geophysical Research: Solid Earth*, 120(9), 6658–6679.

King, G. and Nabelek, J. (1985). Role of fault bends in the initiation and termination of earthquake rupture. *Science*, 228, 984–987.

Klinger, Y., Michel, R., King, G.C.P. (2006). Evidence for an earthquake barrier model from Mw 7.8 Kokoxili (Tibet) earthquake slip-distribution. *Earth and Planetary Science Letters*, 242(3–4), 354–364.

Kobayashi, T., Tobita, M., Nishimura, T., Suzuki, A., Noguchi, Y., Yamanaka, M. (2011). Crustal deformation map for the 2011 off the Pacific coast of Tohoku Earthquake, detected by InSAR analysis combined with GEONET data. *Earth, Planets and Space*, 63(7), 20.

Kreemer, C., Blewitt, G., Klein, E.C. (2014). A geodetic plate motion and global strain rate model. *Geochemistry, Geophysics, Geosystems*, 15(10), 3849–3889.

Kuo, Y.-T., Wang, Y., Hollingsworth, J., Huang, S.-Y., Chuang, R.Y., Lu, C.-H., Hsu, Y.-C., Tung, H., Yen, J.-Y., Chang, C.-P. (2019). Shallow fault rupture of the Milun Fault in the 2018 Mw 6.4 Hualien earthquake: A high-resolution approach from optical correlation of Pléiades satellite imagery. *Seismological Research Letters*, 90(1), 97–107.

Lacroix, P., Araujo, G., Hollingsworth, J., Taipe, E. (2019). Self-entrainment motion of a slow-moving landslide inferred from Landsat-8 time series. *Journal of Geophysical Research: Earth Surface*, 124(5), 1201–1216.

Lasserre, C., Peltzer, G., Crampé, F., Klinger, Y., Van Der Woerd, J., Tapponnier, P. (2005). Coseismic deformation of the 2001 Mw 7.8 Kokoxili earthquake in Tibet, measured by synthetic aperture radar interferometry. *Journal of Geophysical Research: Solid Earth*, 110(B12), 4–17.

Lauer, B., Grandin, R., Klinger, Y. (2020). Fault geometry and slip distribution of the 2013 Mw 7.7 Balochistan earthquake from inversions of SAR and optical data. *Journal of Geophysical Research: Solid Earth*, 125(7), e2019JB018380.

Lazecky, M., Spaans, K., González, P.J., Maghsoudi, Y., Morishita, Y., Albino, F., Elliott, J., Greenall, N., Hatton, E., Hooper, A., Juncu, D., McDougall, A., Walters, R.J., Watson, C.S., Weiss, J.R., Wright, T.J. (2020). LiCSAR: An automatic InSAR tool for measuring and monitoring tectonic and volcanic activity. *Remote Sensing*, 12(15), 2430.

Leprince, S., Ayoub, F., Klinger, Y., Avouac, J.-P. (2007a). Co-registration of optically sensed images and correlation (COSI-Corr): An operational methodology for ground deformation measurements. *IEEE International Geoscience and Remote Sensing Symposium*, 1943–1946.

Leprince, S., Barbot, S., Ayoub, F., Avouac, J.-P. (2007b). Automatic and precise orthorectification, coregistration, and subpixel correlation of satellite images, application to ground deformation measurements. *IEEE Transactions on Geoscience and Remote Sensing*, 45(6), 1529–1558.

Litchfield, N.J., Villamor, P., Dissen, R.J.V., Nicol, A., Barnes, P.M., Barrell, D.J.A., Pettinga, J.R., Langridge, R.M., Little, T.A., Mountjoy, J.J. *et al.* (2018). Surface rupture of multiple crustal faults in the 2016 Mw 7.8 Kaikōura, New Zealand, earthquake. *Bulletin of the Seismological Society of America*, 108(3B), 1496–1520.

Lohman, R.B. and Simons, M. (2005). Locations of selected small earthquakes in the Zagros mountains. *Geochemistry, Geophysics, Geosystems*, 6(3), 1–12.

Ma, S. (2008). A physical model for widespread near-surface and fault zone damage induced by earthquakes. *Geochemistry, Geophysics, Geosystems*, 9(11), 1–9.

Manighetti, I., Campillo, M., Bouley, S., Cotton, F. (2007). Earthquake scaling, fault segmentation, and structural maturity. *Earth and Planetary Science Letters*, 253, 429–438.

Marchandon, M., Vergnolle, M., Cavalié, O., Sudhaus, H., Hollingsworth, J. (2018a). Earthquake sequence in the NE Lut, Iran: Observations from multiple space geodetic techniques. *Geophysical Journal International*, 215, 1–37.

Marchandon, M., Vergnolle, M., Sudhaus, H., Cavalié, O. (2018b). Fault geometry and slip distribution at depth of the 1997 Mw 7.2 Zirkuh earthquake: Contribution of near-field displacement data. *Journal of Geophysical Research: Solid Earth*, 123, 1–21.

Marchandon, M., Hollingsworth, J., Radiguet, M. (2021). Origin of the shallow slip deficit on a strike slip fault: Influence of elastic structure, topography, data coverage, and noise. *Earth and Planetary Science Letters*, 554C, 116696.

Massonnet, D., Rossi, M., Carmona, C., Adragna, F., Peltzer, G., Feigl, K., Rabaute, T. (1993). The displacement field of the Landers earthquake mapped by radar interferometry. *Nature*, 364(6433), 138–142.

Massonnet, D., Thatcher, W., Vadon, H. (1996). Detection of postseismic fault-zone collapse following the Landers earthquake. *Nature*, 382(6592), 612–616.

Meade, B.J., Klinger, Y., Hetland, E.A. (2013). Inference of multiple earthquake-cycle relaxation timescales from irregular geodetic sampling of interseismic deformation. *Bulletin of the Seismological Society of America*, 103(5), 2824–2835.

Michel, R. and Avouac, J.-P. (2002). Deformation due to the 17 August 1999 Izmit, Turkey, earthquake measured from SPOT images. *Journal of Geophysical Research*, 107(0), 1–6.

Michel, R. and Avouac, J.-P. (2006). Coseismic surface deformation from air photos: The Kickapoo step over in the 1992 Landers rupture. *Journal of Geophysical Research*, 111, 1–13.

Milliner, C., Dolan, J., Hollingsworth, J., Leprince, S., Ayoub, F., Sammis, C.G. (2015). Quantifying near-field and off-fault deformation patterns of the 1992 Mw 7.3 Landers earthquake. *Geochemistry, Geophysics, Geosystems*, 16(5), 1577–1598.

Milliner, C., Dolan, J., Hollingsworth, J., Leprince, S., Ayoub, F. (2016a). Comparison of coseismic near-field and off-fault surface deformation patterns of the 1992 Mw 7.3 Landers and 1999 Mw 7.1 Hector Mine earthquakes: Implications for controls on the distribution of surface strain. *Geophysical Research Letters*, 43(19), 10–115.

Milliner, C., Sammis, C., Allam, A., Dolan, J., Hollingsworth, J., Leprince, S., Ayoub, F. (2016b). Resolving fine-scale heterogeneity of co-seismic slip and the relation to fault structure. *Scientific Reports*, 6, 27201.

Miranda, N., Rosich, B., Meadows, P.J., Haria, K., Small, D., Schubert, A., Lavalle, M., Collard, F., Johnsen, H., Monti-Guarnieri, A., D'Aria, D. (2013). The ENVISAT ASAR mission: A look back at 10 years of operation, *ESA-SP*, 772, 1–17.

Morishita, Y., Lazecky, M., Wright, T.J., Weiss, J.R., Elliott, J.R., Hooper, A. (2020). LiCSBAS: An open-source InSAR time series analysis package integrated with the LiCSAR automated Sentinel-1 InSAR processor. *Remote Sensing*, 12(3), 424.

Nelson, M.R. and Jones, C.H. (1987). Paleomagnetism and crustal rotations along a shear zone, Las Vegas range, southern Nevada. *Tectonics*, 6(1), 13–33.

Nissen, E., Maruyama, T., Arrowsmith, J.R., Elliott, J.R., Krishnan, A.K., Oskin, M.E., Saripalli, S. (2014). Coseismic fault zone deformation revealed with differential lidar: Examples from Japanese Mw 7 intraplate earthquakes. *Earth and Planetary Science Letters*, 405, 244–256.

Nissen, E., Elliott, J., Sloan, R., Craig, T., Funning, G., Hutko, A., Parsons, B., Wright, T. (2016). Limitations of rupture forecasting exposed by instantaneously triggered earthquake doublet. *Nature Geoscience*, 9(4), 330–336.

Nur, A. and Mavko, G. (1974). Postseismic viscoelastic rebound. *Science*, 183(4121), 204–206.

Okada, Y. (1985). Surface deformation due to shear and tensile faults in a half-space. *Bulletin of the Seismological Society of America*, 75(4), 1135–1154.

Oskin, M.E., Arrowsmith, J.R., Corona, A.H., Elliott, A.J., Fletcher, J.M., Fielding, E.J., Gold, P.O., Garcia, J.J.G., Hudnut, K.W., Liu-Zeng, J. *et al.* (2012). Near-field deformation from the El Mayor–Cucapah earthquake revealed by differential LIDAR. *Science*, 335(6069), 702–705.

Pagli, C., Pedersen, R., Sigmundsson, F., Feigl, K.L. (2003). Triggered fault slip on June 17, 2000 on the Reykjanes Peninsula, SW-Iceland captured by radar interferometry. *Geophysical Research Letters*, 30(6), 1–4.

Pagli, C., Wang, H., Wright, T.J., Calais, E., Lewi, E. (2014). Current plate boundary deformation of the Afar rift from a 3-D velocity field inversion of InSAR and GPS. *Journal of Geophysical Research: Solid Earth*, 119(11), 8562–8575.

Peltzer, G., Rosen, P., Rogez, F., Hudnut, K. (1996). Postseismic rebound in fault step-overs caused by pore fluid flow. *Science*, 273(5279), 1202–1204.

Peltzer, G., Rosen, P., Rogez, F., Hudnut, K. (1998). Poroelastic rebound along the Landers 1992 earthquake surface rupture. *Journal of Geophysical Research: Solid Earth*, 103(B12), 30131–30145.

Peltzer, G., Crampé, F., Hensley, S., Rosen, P. (2001). Transient strain accumulation and fault interaction in the Eastern California shear zone. *Geology*, 29(11), 975–978.

Plafker, G. (1965). Tectonic deformation associated with the 1964 Alaska Earthquake: The earthquake of 27 March 1964 resulted in observable crustal deformation of unprecedented areal extent. *Science*, 148(3678), 1675–1687.

Pousse Beltran, L., Pathier, E., Jouanne, F., Vassallo, R., Reinoza, C., Audemard, F., Doin, M.P., Volat, M. (2016). Spatial and temporal variations in creep rate along the El Pilar fault at the Caribbean–South American plate boundary (Venezuela), from InSAR. *Journal of Geophysical Research: Solid Earth*, 121(11), 8276–8296.

Qiu, Q., Hill, E.M., Barbot, S., Hubbard, J., Feng, W., Lindsey, E.O., Feng, L., Dai, K., Samsonov, S.V., Tapponnier, P. (2016). The mechanism of partial rupture of a locked megathrust: The role of fault morphology. *Geology*, 44(10), 875–878.

Reid, H.F. (1910). The California earthquake of April 18, 1906. *Report of the State Earthquake Investigation Commission*, 2, 16–18.

Rockwell, T.K., Lindvall, S., Dawson, T., Langridge, R., Lettis, W., Klinger, Y. (2002). Lateral offsets on surveyed cultural features resulting from the 1999 Izmit and Duzce earthquakes, Turkey. *Bulletin of the Seismological Society of America*, 92(1), 79–94.

Rosen, P.A., Kim, Y., Kumar, R., Misra, T., Bhan, R., Sagi, V.R. (2017). Global persistent SAR sampling with the NASA-ISRO SAR (NISAR) mission. *IEEE Radar Conference*, 8–12 May.

Rosenqvist, A., Shimada, M., Suzuki, S., Ohgushi, F., Tadono, T., Watanabe, M., Tsuzuku, K., Watanabe, T., Kamijo, S., Aoki, E. (2014). Operational performance of the ALOS global systematic acquisition strategy and observation plans for ALOS-2 PALSAR-2. *Remote Sensing of Environment*, 155, 3–12.

Rosu, A.-M., Pierrot-Deseilligny, M., Delorme, A., Binet, R., Klinger, Y. (2015). Measurement of ground displacement from optical satellite image correlation using the free open-source software MicMac. *ISPRS Journal of Photogrammetry and Remote Sensing*, 100, 48–59.

Rousset, B., Jolivet, R., Simons, M., Lasserre, C., Riel, B., Milillo, P., Çakir, Z., Renard, F. (2016). An aseismic slip transient on the North Anatolian fault. *Geophysical Research Letters*, 43(7), 3254–3262.

Ryder, I., Parsons, B., Wright, T.J., Funning, G.J. (2007). Post-seismic motion following the 1997 Manyi (Tibet) earthquake: InSAR observations and modelling. *Geophysical Journal International*, 169(3), 1009–1027.

Sacks, I.S., Linde, A.T., Suyehiro, S., Snoke, J.A. (1978). Slow earthquakes and stress redistribution. *Nature*, 275(5681), 599–602.

Salyards, S.L., Sieh, K.E., Kirschvink, J.L. (1992). Paleomagnetic measurement of nonbrittle coseismic deformation across the San Andreas fault at Pallett Creek. *Journal of Geophysical Research: Solid Earth*, 97(B9), 12457–12470.

Savage, J. (1983). A dislocation model of strain accumulation and release at a subduction zone. *Journal of Geophysical Research: Solid Earth*, 88(B6), 4984–4996.

Savage, J. and Burford, R. (1973). Geodetic determination of relative plate motion in central California. *Journal of Geophysical Research*, 78(5), 832–845.

Savage, J. and Prescott, W. (1978). Asthenosphere readjustment and the earthquake cycle. *Journal of Geophysical Research: Solid Earth*, 83(B7), 3369–3376.

Savage, J. and Svarc, J. (1997). Postseismic deformation associated with the 1992 M ω = 7.3 Landers earthquake, southern California. *Journal of Geophysical Research: Solid Earth*, 102(B4), 7565–7577.

Scott, C.P., Arrowsmith, J.R., Nissen, E., Lajoie, L., Maruyama, T., Chiba, T. (2018). The M7 2016 Kumamoto, Japan, earthquake: 3-D deformation along the fault and within the damage zone constrained from differential lidar topography. *Journal of Geophysical Research: Solid Earth*, 123(7), 6138–6155.

Shelef, E. and Oskin, M. (2010). Deformation processes adjacent to active faults: Examples from eastern California. *Journal of Geophysical Research: Solid Earth*, 115(B5), 1–24.

Socquet, A., Hollingsworth, J., Pathier, E., Bouchon, M. (2019). Evidence of supershear during the 2018 magnitude 7.5 Palu earthquake from space geodesy. *Nature Geosciences*, 12, 192–199.

Stein, R.S. and Barrientos, S.E. (1985). Planar high-angle faulting in the Basin and Range: Geodetic analysis of the 1983 Borah Peak, Idaho, earthquake. *Journal of Geophysical Research: Solid Earth*, 90(B13), 11355–11366.

Stein, S., Geller, R.J., Liu, M. (2012). Why earthquake hazard maps often fail and what to do about it. *Tectonophysics*, 562, 1–25.

Surazakov, A. and Aizen, V. (2010). Positional accuracy evaluation of declassified hexagon KH-9 mapping camera imagery. *Photogrammetric Engineering and Remote Sensing*, 76(5), 603–608.

Tapponnier, P., Zhiqin, X., Roger, F., Meyer, B., Arnaud, N., Wittlinger, G., Jingsui, Y. (2001). Oblique stepwise rise and growth of the Tibet Plateau. *Science*, 294(5547), 1671–1677.

Thomas, M.Y., Avouac, J.-P., Gratier, J.-P., Lee, J.-C. (2014). Lithological control on the deformation mechanism and the mode of fault slip on the Longitudinal Valley Fault, Taiwan. *Tectonophysics*, 632, 48–63.

Tong, X., Sandwell, D., Smith-Konter, B. (2013). High-resolution interseismic velocity data along the San Andreas Fault from GPS and InSAR. *Journal of Geophysical Research: Solid Earth*, 118(1), 369–389.

Torres, R., Snoeij, P., Geudtner, D., Bibby, D., Davidson, M., Attema, E., Potin, P., Rommen, B., Floury, N., Brown, M.M., Navas-Traver, I., Deghaye, P., Duesmann, B., Rosich, B., Miranda, N., Bruno, C., L'Abbate, M., Croci, R., Pietropaolo, A., Huchler, M., Rostan, F. (2012). GMES Sentinel-1 mission. *Remote Sensing of Environment*, 120, 9–24.

Utsu, T., Ogata, Y., Matsu'ura, R.S. (1995). The centenary of the Omori formula for a decay law of aftershock activity. *Journal of Physics of the Earth*, 43(1), 1–33.

Vallage, A., Klinger, Y., Grandin, R., Bhat, H.S. (2015). Inelastic surface deformation during the 2013 Mw 7.7 Balochistan, Pakistan, earthquake. *Geology*, 43(12), 1079–1082.

Van Puymbroeck, N., Michel, R., Binet, R., Avouac, J.-P. (2000). Measuring earthquakes from optical satellite images. *Applied Optics*, 39(20), 3486–3494.

Wang, K. and Bürgmann, R. (2020). Probing fault frictional properties during afterslip up and downdip of the 2017 Mw 7.3 Sarpol-e Zahab earthquake with space geodesy. *Journal of Geophysical Research: Solid Earth*, e2020JB020319.

Wang, H. and Wright, T. (2012). Satellite geodetic imaging reveals internal deformation of western Tibet. *Geophysical Research Letters*, 39(7).

Wang, H., Wright, T., Biggs, J. (2009). Interseismic slip rate of the northwestern Xianshuihe fault from InSAR data. *Geophysical Research Letters*, 36(3), 1–5.

Wang, H., Wright, T., Liu-Zeng, J., Peng, L. (2019). Strain rate distribution in south-central Tibet from two decades of InSAR and GPS. *Geophysical Research Letters*, 46(10), 5170–5179.

Weiss, J.R., Qiu, Q., Barbot, S., Wright, T., Foster, J.H., Saunders, A., Brooks, B.A., Bevis, M., Kendrick, E., Ericksen, T.L., Avery, J., Smalley, R., Cimbaro, S.R., Lenzano, L.E., Barón, J., Báez, J.C., Echalar, A. (2019). Illuminating subduction zone rheological properties in the wake of a giant earthquake. *Science Advances*, 5(12), 1–11.

Weiss, J.R., Walters, R.J., Morishita, Y., Wright, T.J., Lazecky, M., Wang, H., Hussain, E., Hooper, A.J., Elliott, J.R., Rollins, C., Yu, C., González, P.J., Spaans, K., Li, Z., Parsons, B. (2020). High-resolution surface velocities and strain for Anatolia from Sentinel-1 InSAR and GNSS data. *Geophysical Research Letters*, 47(17), 1–12.

Wells, D.L. and Coppersmith, K.J. (1994). New empirical relationships among magnitude, rupture length, rupture width, rupture area, and surface displacement. *Bulletin of the Seismological Society of America*, 84(4), 974–1002.

Weston, J., Ferreira, A., Funning, G. (2011). Global compilation of interferometric synthetic aperture radar earthquake source models 1: Comparisons with seismic catalogs. *Journal of Geophysical Research: Solid Earth*, 116(B8), 1–20.

Weston, J., Ferreira, A., Funning, G. (2012). Systematic comparisons of earthquake source models determined using InSAR and seismic data. *Tectonophysics*, 532, 61–81.

Wright, T. (2000). Crustal deformation in Turkey from synthetic aperture radar interferometry. PhD thesis, University of Oxford.

Wright, T., Parsons, B., Jackson, J., Haynes, M., Fielding, E., England, P., Clarke, P. (1999). Source parameters of the 1 October 1995 Dinar (Turkey) earthquake from SAR interferometry and seismic bodywave modelling. *Earth and Planetary Science Letters*, 172(1–2), 23–37.

Wright, T., Fielding, E., Parsons, B. (2001a). Triggered slip: Observations of the 17 August 1999 Izmit (Turkey) earthquake using radar interferometry. *Geophysical Research Letters*, 28(6), 1079–1082.

Wright, T., Parsons, B., Fielding, E. (2001b). Measurement of interseismic strain accumulation across the North Anatolian Fault by satellite radar interferometry. *Geophysical Research Letters*, 28(10), 2117–2120.

Wright, T., Parsons, B., England, P., Fielding, E. (2004). InSAR observations of low slip rates on the major faults of western Tibet. *Science*, 305(5681), 236–239.

Wright, T., Elliott, J., Wang, H., Ryder, I. (2013). Earthquake cycle deformation and the Moho: Implications for the rheology of continental lithosphere. *Tectonophysics*, 609, 504–523.

Xu, X., Tong, X., Sandwell, D.T., Milliner, C.W.D., Dolan, J.F., Hollingsworth, J., Leprince, S., Ayoub, F. (2016). Refining the shallow slip deficit. *Geophysical Journal International*, 204, 1867–1886.

Xu, X., Sandwell, D.T., Smith-Konter, B.R. (2020a). Coseismic displacements and surface fractures from Sentinel-1 InSAR: 2019 Ridgecrest earthquakes. *Seismological Research Letters*, 91(4), 1979–1985.

Xu, X., Sandwell, D.T., Ward, L.A., Milliner, C.W., Smith-Konter, B.R., Fang, P., Bock, Y. (2020b). Surface deformation associated with fractures near the 2019 Ridgecrest earthquake sequence. *Science*, 370(6516), 605–608.

Yamasaki, T., Wright, T.J., Houseman, G.A. (2014). Weak ductile shear zone beneath a major strike-slip fault: Inferences from earthquake cycle model constrained by geodetic observations of the western North Anatolian Fault Zone. *Journal of Geophysical Research: Solid Earth*, 119(4), 3678–3699.

Yielding, G., Jackson, J., King, G., Sinvhal, H., Vita-Finzi, C., Wood, R.M. (1981). Relations between surface deformation, fault geometry, seismicity, and rupture characteristics during the El Asnam (Algeria) earthquake of 10 October 1980. *Earth and Planetary Science Letters*, 56, 287–304.

Zebker, H.A. and Goldstein, R.M. (1986). Topographic mapping from interferometric synthetic aperture radar observations. *Journal of Geophysical Research: Solid Earth*, 91(B5), 4993–4999.

Zhou, Y., Walker, R.T., Hollingsworth, J., Talebian, M., Song, X., Parsons, B. (2016). Coseismic and postseismic displacements from the 1978 Mw 7.3 Tabas-e-Golshan earthquake in eastern Iran. *Earth and Planetary Science Letters*, 452, 185–196.

Zinke, R., Hollingsworth, J., Dolan, J.F. (2014). Surface slip and off-fault deformation patterns in the 2013MW 7.7 Balochistan, Pakistan earthquake: Implications for controls on the distribution of near-surface coseismic slip. *Geochemistry, Geophysics, Geosystems*, 15, 5034–5050.

Zinke, R., Hollingsworth, J., Dolan, J.F., Van Dissen, R. (2019). Three-dimensional surface deformation in the 2016 Mw 7.8 Kaikoura, New Zealand, earthquake from optical image correlation: Implications for strain localization and long-term evolution of the Pacific-Australian plate boundary. *Geochemistry, Geophysics, Geosystems*, 20(3), 1609–1628.

8

Volcanology: The Crucial Contribution of Surface Displacement Measurements from Space for Understanding and Monitoring Volcanoes

Virginie PINEL[1], Fabien ALBINO[1], Grace BATO[2] and Paul LUNDGREN[2]

[1] *ISTerre, Grenoble, France*

[2] *NASA Jet Propulsion Laboratory, California Institute of Technology, Pasadena, USA*

8.1. Introduction

Eight hundred million people currently live within a radius of 100 km around a volcano which has erupted in the last 10,000 years (Brown *et al.* 2015). This means that around 10% of the global population may be directly affected by a volcanic eruption. The most explosive eruptions, producing the highest volumes, are the most lethal; these are characterized by a volcanic explosivity index (VEI) (Newhall *et al.* 1982) greater than or equal to 4. Deaths generally result from pyroclastic flows: rock fragments suspended in hot gas flowing down the slopes of the volcano at speed. Secondary phenomena, such as lahars, which are mudflows triggered by volcanic activity, or tsunamis also have the potential to cause loss of

Surface Displacement Measurement from Remote Sensing Images,
coordinated by Olivier CAVALIÉ and Emmanuel TROUVÉ.
© ISTE Ltd 2022.

life on a significant scale (Voight 1990; Syamsidik *et al.* 2020). The consequences of volcanic activity in terms of human casualties have been significantly attenuated over the last few centuries with the development of modern volcanology (Auker *et al.* 2013). Volcanic eruptions cannot be prevented, but it is possible to evacuate the neighboring population in time (Surono *et al.* 2012), as precursors can often be observed before magma reaches the surface. The accumulation of magma in the upper part of the Earth's crust, along with its transport towards the surface, can create ground displacements, with stress modifications resulting in seismic events and gaseous emissions. The measurement and interpretation of these signals, based on in-depth knowledge of magmatic processes, offer a means of reducing volcanic risk. The problem is that only a tiny proportion (around 10%) of the 1500 known active volcanoes are correctly instrumented and monitored (Ferrucci *et al.* 2012). Over the last few decades, the development of remote sensing techniques has presented opportunities for improved knowledge and monitoring of volcanic systems (Sparks *et al.* 2012; Pyle *et al.* 2013). Radar satellite data can notably be used to measure surface displacement resulting from volcanic activity (Pinel *et al.* 2014), and optical data can be used to measure displacements of larger amplitudes (Hollingsworth *et al.* 2012). In this chapter, we describe different techniques for measuring surface displacement in volcanoes, essentially using radar data, but also using high-resolution optical data. The application of these techniques to the specific context of volcanology is discussed in detail, in terms of both signal processing, for example the management of atmospheric artifacts, and the spatio-temporal characteristics of the observed displacements. We show how remote sensing has been used to develop global statistical observations of volcanic activity, and how displacement measurements obtained from space can be combined with other remote sensing observations (gas emissions, thermal measurements, etc.) or ground instrumentation. Finally, we demonstrate the contribution made by satellite displacement measurements in managing recent volcanic crises.

8.2. Origin of surface displacement and topographic changes in a volcano

The displacements observed in a volcano can be of gravitational or tectonic origin, but are more often caused by volcanic activity (storage, transport and effusion of magma at the surface) (Pinel *et al.* 2014). Studying these displacements can thus provide us with crucial information concerning the state of the magma supply system and its future evolution. Magma, produced by partial fusion of the lithospheric mantle, does not generally rise to the surface directly. Instead, it is usually stored a few kilometers below the surface in the Earth's crust. When magma is stored at shallow level and the rate of accumulation is relatively fast, this results in an uplift, which can be observed from a few months to a few decades prior to an eruption. This displacement is thus a long-term precursor. The amplitude of the deformation is mostly dependent on the quantity of magma, the depth and the quantity of volatiles;

it may reach several tens of centimeters in size. The lateral extent of a deformation is essentially a function of the depth of storage, and also of the shape of the reservoir; it may reach scales of up to 10 kilometers.

At a certain point, when a large quantity of magma has been accumulated, the resulting overpressure or buoyancy may cause the reservoir to rupture. At this point, the magma will begin to ascend towards the surface, and an eruption may result. Magma transport occurs by magmatic intrusions (planar fractures are created and expanded through the injection of fluid) or through cylindrical conduits (the liquid flows through a pipe which is already open). This stage is generally fast, lasting from a few hours to a few days. The resulting surface displacement can be considered as a short-term precursor. The uplift which results from the arrival of magma can be accompanied by subsidence, over a wider area but of lower amplitude, due to the simultaneous emptying of the feeding storage area.

Volcanic products emitted at the surface, such as lava or pyroclastic flow, cause topographic changes over a surface of several tens of meters. Displacements are often observed in deposits and may continue to occur for several decades after their emplacement. This is a result of continued flow, thermal contraction and the reaction of the Earth's crust to the weight of the deposit. These signals are localized in, or in immediate proximity to, deposits. Their amplitude is generally a function of the thickness of the deposit and decreases exponentially over time. Studying these signals provides us with information concerning the physical (thermal and mechanical) properties of emitted lava and of the volcanic edifice itself.

Finally, volcanic activity can have a strong destabilizing effect on the volcano itself, resulting in major displacements, such as the collapse or destruction of the summit area or flank sliding. In cases where these displacements occur suddenly, they may reach one hundred meters and have visible effects over a kilometer. However, displacements of this type can also occur at slower rates and over longer periods.

Generally speaking, large amplitude topographic changes (from several tens to hundreds of meters) are a result of significant modifications in the morphology of the edifice (collapse of the crater zone, major destabilization of the flanks) or of the emergence of thick eruptive deposits (lava flows). The propagation of magmatic intrusions of several tens of kilometers beneath the surface may also result in surface displacements of several meters.

Furthermore, in many volcanoes, a hydrothermal system develops in the zone up to one hundred meters below the surface, resulting from interactions between rain or sea water and magmatic fluids. Perturbations of this system, due to changes in the heat source or in the arrival of magma, may also result in surface displacements; in this case, the spatial footprint is very close to that of the hydrothermal system, since the latter is very close to the surface.

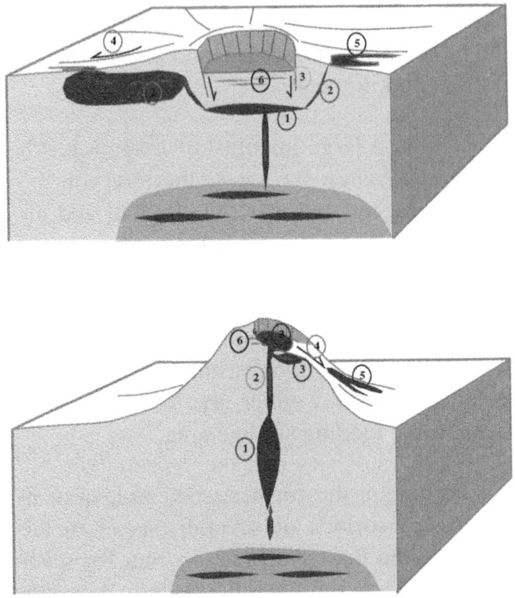

a)

b)

Figure 8.1. *Different sources of deformation and topographic change affecting volcanoes. a) Typical case of a basaltic (shield) volcano: 1. storage zone, 2. propagation of a magmatic intrusion, 3. subsidence of the caldera, 4. flank sliding, 5. creation of a lava flow and associated subsidence, 6. hydrothermal system. b) Typical case of an andesitic stratovolcano: 1. storage zone, 2. flow of magma through the conduit, 3. endogenous and exogenous growth of the dome, 4. flank sliding, 5. creation of a pyroclastic flow and associated subsidence, 6. hydrothermal system. For a color version of this figure, see www.iste.co.uk/cavalie/images.zip*

The morphology of a volcanic edifice and its eruptive deposits, along with the geometry of its magma supply system, are strongly affected by the type of magma being emitted. A distinction can be made between basaltic systems and andesitic stratovolcanoes. Basaltic systems mostly produce relatively homogeneous magma with a low silica content. These shield volcanoes are characterized by relatively gentle slopes, often with a crater or caldera zone at their summit (see Figure 8.1a). They typically occur in extension or rifting zones and at hotspots; their activity is essentially effusive in nature, and the magma which they produce is of low viscosity and relatively dense. Andesitic stratovolcanoes emit more differentiated magmas with a higher silica content; this magma has a lower density than basalt, but is much more viscous. The higher viscosity limits gas escape, making explosive activity more likely. Andesitic stratovolcanoes generally have steep flanks (see Figure 8.1b) and are mostly encountered in subduction zones.

8.2.1. *Typical deformations in a basaltic system*

Figure 8.1a shows the typical deformations observed in a basaltic volcano. The low viscosity of the magma favors the propagation of magmatic intrusions up to the surface, where they feed into eruptive fissures. Intrusions may extend laterally over several tens of kilometers, as in the case of the intrusive episode at Afar, Ethiopia, in 2005 (Grandin *et al.* 2009) or the 2014 eruption of Bardabunga, Iceland (Sigmundsson *et al.* 2015). Not all magmatic intrusions reach the surface, and even when this occurs, a significant quantity of magma remains trapped in the crust, leading to a surface displacement which can easily reach ten centimeters (Froger *et al.* 2004; Fukushima *et al.* 2010). Trapped magmatic intrusions contribute to the intrusive growth of the volcanic edifice. They may also trigger large-scale flank sliding events, as in the case of the Piton de la Fournaise volcano on Réunion Island (France) (Chaput *et al.* 2014; Froger *et al.* 2015). In cases where the storage zone is emptied to a significant extent by intrusions, there is also a possibility that the summit zone will collapse, with the formation of a caldera or its subsequent subsidence over several hundred meters (Froger *et al.* 2015; Sigmundsson *et al.* 2015; Anderson *et al.* 2019; Shreve *et al.* 2019). Deformations which are classically associated with recharging or emptying of shallow storage zones are sometimes linked to sliding along the faults marking the edge of the caldera (Liu *et al.* 2019). Basaltic lava flows are affected by subsidence caused by thermal contraction and mechanical loading effects (Bato *et al.* 2016; Wittmann *et al.* 2017). Hydrothermal systems can also cause localized surface deformations.

8.2.2. *Typical deformations in an andesitic stratovolcano*

Figure 8.1b shows the typical deformations observed in an andesitic stratovolcano. These volcanoes erupt less frequently than basaltic systems, with dormant periods lasting for decades or even centuries. For this reason, geophysical observations of volcanoes of this type are generally less complete. The greater viscosity of the magma and the maturity of the volcanic system mean that magma is generally transported over the final kilometers to the surface through the reactivation of more or less cylindrical permanent conduits. This form of magma transfer can be linked to transient displacements of low amplitude which evolve quickly over time (Voight *et al.* 1999). Viscous magma can settle just under the surface or in the crater, in thick, domed accumulations. The topographic changes resulting from the growth of these domes can reach hundreds of meters, and their temporal evolution provides direct information concerning the magma emission rate and thus the level of activity of the system (Watts *et al.* 2002). Domes can be destroyed very quickly during explosive phases, with the formation of excavations of tens or even hundreds of meters (Arambula-Mendoza *et al.* 2020). The lava flows that occur in these systems are generally thicker than those produced by basaltic volcanoes and take the form of fast, lethal pyroclastic flows. Studying these deposits provides further information regarding the physical properties of lava (Carrara *et al.* 2019).

8.3. Techniques used to measure displacements and topographic changes in volcanoes

Systematic monitoring of volcanoes, notably in the form of regular measurements of ground displacements, began in the late 19th century with the creation of the first volcanological observatories. Initially, very few volcanoes were monitored: the first observatory was built on the flanks of Vesuvius, in Italy, in 1841, followed by observatories on Etna, in Sicily, and Mount Pelée, in Martinique; in the latter case, monitoring essentially began in response to a lethal eruption in 1902. Other observatories included Mount Azama, in Japan, and Hawaii for the United States. The first measurements were generally leveling surveys, i.e. estimations of distances and angles across a network of designated benchmarks. Additional tilt measurements were often made, quantifying variations in the slope. The first time series featured only small numbers of points and were characterized by very weak temporal sampling, of the order of a few years (Dvorak and Dzurisin 1997). Extensometry was also used in some cases to characterize the opening and shear displacement of faults: the empirical information gathered in this way can provide interesting insights into volcanic systems (Peltier *et al.* 2006). The development of modern geodesy, with the appearance of GPS technology, made it possible to quantify displacements in 3D, with a degree of precision of within a millimeter for horizontal components and a centimeter for the vertical component; these techniques are crucial for modern volcanology. In terms of ground geodetic measurement, however, the main advance lies in the generalization of continuous measurements (with sampling periods as short as one second), leading to significant improvements in the temporal resolution of displacement measurements. Despite these advances, ground geodetic measurements are beset by intrinsic limitations due to the fact that measurements are punctual by nature and sensors must be installed and maintained on-site. In the late 1990s, surface deformation information was only available for around 50 volcanoes (Dvorak and Dzurisin 1997). This number was multiplied by at least four over the following 25 years through the use of satellite observations (Biggs *et al.* 2014). While warnings of an imminent eruption rely essentially on ground measurements, remote sensing offers significant advantages, notably in terms of spatial coverage, where surface displacement maps can be obtained for large areas. This means that deformations can be detected and quantified in regions where it is not possible to install instruments, either due to access problems relating to the political context, a lack of financial resources or a lack of personnel (Pritchard and Simons 2002; Wauthier *et al.* 2013). In addition, space archives open up the possibility of studying, *a posteriori*, volcanoes that have erupted and thus grant access to the observation of possible precursors. This type of highly informative study is never possible with ground-based instrumentation for all the volcanoes whose monitoring was not planned and carried out before the onset of the activity.

Technique	Precision	Phenomenon	Example
Pixel offset (optical, radar)	*m* (horizontal for optical, horizontal and vertical for radar)	Rifting, large dike Caldera collapse Flank sliding	Grandin *et al.* 2009 Casu *et al.* 2011 Yan 2011
Interferometry (radar)	*Single interferogram: cm* *Time series: mm/year* (horizontal and vertical)	Magma storage Magmatic intrusion Destabilization of the flank Lava flow subsidence Hydrothermal system	Champenois *et al.* 2014 Froger *et al.* 2004 Froger *et al.* 2015 Carrara *et al.* 2019 Vadon and Sigmundsson 1997
Differential DEM (optical, radar)	*m* (vertical)	Dome, laccolith growth Lava flow emplacement Caldera formation Partial destruction of the edifice	Kubanek *et al.* 2021 Bato *et al.* 2016 Carrara *et al.* 2019

Table 8.1. *Techniques for measuring surface displacement and topographic changes in a volcano with remote sensing. Note that, compared to optical data, radar data present the advantage of providing information even in cases of cloud cover*

With remote sensing, displacement measurements are obtained by comparing images acquired at different dates in order to estimate any displacements which may have taken place on the ground between these dates, over all or part of the selected imaging area. Displacement measurements always rely on the extraction of information from an image at pixel level. Measurements can be based on the difference between the positions of the pixels in the images, taking account of the geometry of the sensor, or on the phase difference, which relates to the difference in the path followed by the electromagnetic wave in the case of radar imagery. Table 8.1 provides a summary of different detection techniques used to measure surface displacement in volcanic edifices, specifying the accuracy of the measurements obtained and the types of phenomena which can be observed; for each case, an example from the literature is also cited. Brief descriptions of all measurement techniques using satellite data are provided in the following sections, alongside specific examples of applications in volcanic regions. Where the ground surface is modified, for example by the emplacement of volcanic deposits, we cannot speak of surface displacement in the strict sense, but rather of topographic changes. This variation can be quantified, where significant, by differences in digital elevation models (DEMs; see section 8.3.3). The term "topographic change" is also used to refer to large-scale vertical displacements without surface alteration, as in the case of caldera formation.

Figure 8.2 shows a comparison of the resolutions of different techniques, along with the spatial and temporal characteristics of surface displacements typically observed in volcanoes.

8.3.1. *Pixel offset tracking*

This technique consists of measuring pixel offsets (variations in the positions of pixels within the image) using a maximum similarity search. It can be applied to both optical and radar amplitude images (see Chapters 2 and 3, respectively). The accuracy achieved depends on the resolution of the sensor, which is in the order of one-tenth of the pixel size (Michel *et al.* 1999). However, there is a visible dependency relationship between accuracy and signal coherence for radar amplitude images (De Zan 2014). Ionospheric disturbances can also cause significant errors (Gray *et al.* 2000). Accuracy thus varies from a few decimeters, for HR images, to a few meters. Since this technique takes account of both axes of the image, a two-dimensional (2D) displacement measurement is obtained. The differences in the geometries of optical and radar images mean that two horizontal displacement components are obtained for optical images; in radar imagery, we obtain a horizontal displacement component in the direction of the satellite trajectory (azimuth) and a displacement in the direction of the line of sight (in range or distance). The displacement in distance thus combines a horizontal component and a vertical component, with respective weights depending on the angle of view of the satellite.

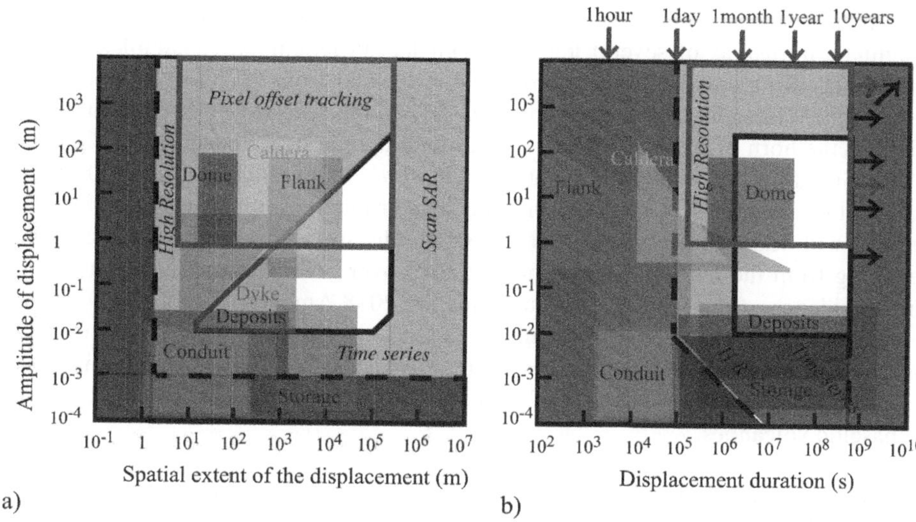

Figure 8.2. *Comparison of spatial and temporal characteristics of volcanic surface displacement and displacement which can be measured by remote sensing. a) Amplitude of the displacement as a function of spatial extent (deformation gradients), b) amplitude of the displacement as a function of duration (deformation rate). The zone shown in white with a full black outline shows the InSAR validity domain (the displacement amplitude considered is in the line of sight) with classical Sentinel-type data. The gray area bounded by dashed black lines shows the limits of SAR data, including pixel offset tracking, HR data and the use of time series. This illustration clearly shows the advances made possible by the use of high spatial and temporal resolution (HR) sensors and time-series data-processing methods and by the increasing amount of archival data available. The pixel-offset tracking technique is used as a complement to InSAR, extending the range of use of SAR data in the case of strong deformation gradients. The solid red lines indicate displacements which can be measured using optical imagery, in the form of photogrammetry for vertical displacements and pixel offset measurement for horizontal displacements. Storage: deformation caused by a magma reservoir, Dyke: deformation resulting from a magmatic intrusion, Conduit: deformation induced by the flow of magma in the cylindrical conduit at shallow depth, Deposits: deformation induced by magma deposits (the change in topography induced by lava flow emplacement is not taken into account here), Flank: deformation associated with flank sliding of the volcanic edifice, Caldera: vertical displacements induced by collapse of the summit zone, Dome: variation in topography induced by the emplacement of a viscous dome. For a color version of this figure, see www.iste.co.uk/cavalie/images.zip*

Only large amplitude displacements, in the order of several tens of decimeters to a meter, can be observed using this method (area shown within the red outline in Figure 8.2). In volcanology, amplitude correlation is essentially used to measure horizontal displacements induced by rifting episodes associated with the creation of magmatic intrusions of several tens of kilometers in length. Key examples include the Krafla rifting episode, which took place in Iceland from 1975 to 1984, observed using optical imagery (Hollingsworth *et al.* 2012); the Afar rifting episode that took place in the horn of Africa in 2005, studied with both optical imagery (Barisin *et al.* 2009; Grandin *et al.* 2009) and radar (Wright *et al.* 2006; Grandin *et al.* 2009); and the most recent episode associated with the 2014 eruption of Bardarbunga, in Iceland, studied with radar imagery (Ruch *et al.* 2016). The strong horizontal displacements resulting from the lateral propagation of a dike over several tens of kilometers and the subsidence of the summit caldera during the 2018 Ambrym eruption in Vanuatu, a relatively isolated and poorly instrumented volcanic region, were also measured by pixel-offset tracking in radar images (Shreve *et al.* 2019). This technique has also been used to document vertical displacements of the order of a meter in the calderas of basaltic volcanoes.

During the 2005 eruption of the Sierra Negra, in the Galapagos, a pre-eruptive uplift of the summit caldera along circumferential faults followed by co-eruptive subsidence (Casu *et al.* 2011) were described using radar images. The same technique was used to detect sliding down the eastern flank of the Piton de la Fournaise during the lateral eruption of January 2004, again using radar imagery (Yan 2011); prior to the installation of GNSS sensors in the Grandes Pentes area of the volcano in 2010, this type of flank sliding could only be detected using radar data.

8.3.2. *Measurement by interferometry (phase difference)*

Differential interferometry uses the phase of the radar signal. It provides a projection of the ground displacement in the direction of the line of sight (d) accurate to within a centimeter. The displacement measurement is obtained by calculating the difference between the phases of radar images acquired on two different dates, after coregistration. The basic principles of InSAR, or differential interferometry from synthetic aperture radar (SAR) images, can be found in the work by Hanssen (2001) or Massonnet and Souyris (2008). The difference between two phase images, commonly called an interferogram, contains information relating to ground displacement, and can be written as follows (see Chapter 4):

$$\psi = W\{\psi_{orb} + \psi_{topo} + \psi_{atm} + \psi_{noise} + \frac{4\pi d}{\lambda}\} \qquad [8.1]$$

where $W\{.\}$ is the wrapping operator of the measured phase modulo 2π. The ground displacement term is thus proportional to $4\pi/\lambda$, where λ is the wavelength of the radar emission used. The phase difference also contains an orbital term and a topography term that can be corrected using knowledge of orbits and a digital elevation model (DEM), respectively. This value is subject to noise, some of which is of atmospheric origin, induced by variations in the propagation speed of the radar wave in the atmosphere between two acquisitions (see Chapter 8, section 8.4.3), resulting in a delay in the reception of the wave.

The phase variation is known modulo 2π and represented as fringes. The phase must be unwrapped in order to obtain a displacement map. At each point in the image, the aim is to find the integer number of factor 2π to add to the phase difference measured between $-\pi$ and π (see Chapter 6). If done incorrectly, this step can result in errors in the displacement estimate. A preliminary filtering step is usually required.

The accuracy of the displacement measurement obtained by differential interferometry can be improved by applying noise reduction, using the redundancy of information available in a time series of images (see Chapter 5). In this way, ground displacements in the line of sight can be measured with an accuracy of the order of 1 mm/year.

Almost all surface deformations of volcanic origin have been observed using InSAR (for a full list of studies published before 2014, see Pinel *et al.* (2014)) with the exception of highly localized, transient phenomena, such as displacements triggered by magma flows through cylindrical conduits and very high-amplitude displacements (dome growth or caldera collapse, see Figure 8.2).

8.3.3. *Measurement of topographic changes from differences in digital elevation models*

Optical or radar satellite imagery can be used to obtain digital elevation models with a spatial resolution of a few meters, accurate to within around ten centimeters. Updating DEMs is crucial for volcanic risk management, since the topography of a volcano can evolve rapidly through partial destruction of the edifice or the emplacement of eruption products. A clear understanding of the topography of the volcano is key to modeling surface flows and their impacts, but is also essential for mechanical modeling, where the effects of topography must be taken into account (Cayol and Cornet 1998; Beauducel and Carbone 2015). The difference between two successive DEMs provides vertical surface displacements on a decimetric scale (see Figure 8.2). Cases where the change in elevation results from the addition of volcanic material on the surface are not counted as displacements in the strictest sense; however, the change in topography allows us to quantify deposit volumes and thus to determine the emission rate of the volcanic edifice.

8.3.3.1. *Building a digital elevation model*

A digital surface model (DSM) can be obtained by stereo-photogrammetry, using several optical images of the same area taken from different viewing angles, coregistered using image correlation techniques. In the absence of vegetation or human constructions, the DSM can be directly assimilated to a DEM. The resolution of the DEM depends directly on that of the source images, and its accuracy is a function of the relationship between the ground baseline and the altitude of the satellite (B/H ratio) and of the local slope. It is generally of the order of several tens of centimeters. Currently, the most widely used images of volcanic edifices are the Pléiades images, obtained in stereo or tri-stereo at a resolution of 0.7 m in panchromatic mode.

DEMs can also be obtained using two radar images, using the fact that the phase is affected by the topography via the term ψ_{topo} in the phase difference expression (see equation [8.1]). This term is a function of the altitude H according to the following expression:

$$\psi_{topo} = W\{\frac{4\pi}{\lambda R \sin(\theta_{inc})} B_{\perp} H\} = W\{\frac{2\pi}{h_{2\pi}} H\} \qquad [8.2]$$

where R is the distance between the surface and the satellite, θ_{inc} is the angle of incidence, $B_{\perp}$ is the perpendicular baseline and $h_{2\pi}$ is the altitude of ambiguity. The resolution of the DEM is a function of the original resolution of the radar images and the required multi-looking. Accuracy increases with the perpendicular baseline, and decreases as the altitude of ambiguity increases. A reference DEM is used to correct for topography and to ensure correct unwrapping. The residual fringes reveal variations in elevation with respect to this reference. Given that monostatic data are used, a time series needs to be processed in order to avoid atmospheric artifacts and to clearly distinguish between the topographic contribution and ground displacement. When available, bistatic data should be preferred, with two images acquired at the same time: this approach limits atmospheric errors and guarantees that no ground displacement will occur between the two acquisitions.

8.3.3.2. *Differential DEM*

DEMs obtained at successive time steps can be compared in order to estimate variations in altitude. In some cases, precise recalibration must be carried out first. The use of differential DEMs allows us to estimate lava emission volumes based on Pléiades optical data (Bagnardi *et al.* 2016), from TerraSAR-X and Tandem-X high-resolution radar data (Poland 2014; Albino *et al.* 2015; Kubanek *et al.* 2015a; Bato *et al.* 2016) or using a combination of Pléiades optical data and Tandem-X radar data (Carrara *et al.* 2019).

This technique can be used to quantify partial destructions of an edifice: for example, a volumetric loss of 4 Mm^3 was observed in the summit area of the Colima volcano in Mexico during an eruption in July 2015 (Carrara *et al.* 2019). It also allows estimation of the evolution of domes in a summit crater (e.g. Merapi, Indonesia, as shown in Figure 8.3 and described by Kubanek *et al.* (2015)); this information is crucial for assessing stability and determining the likelihood of future explosions.

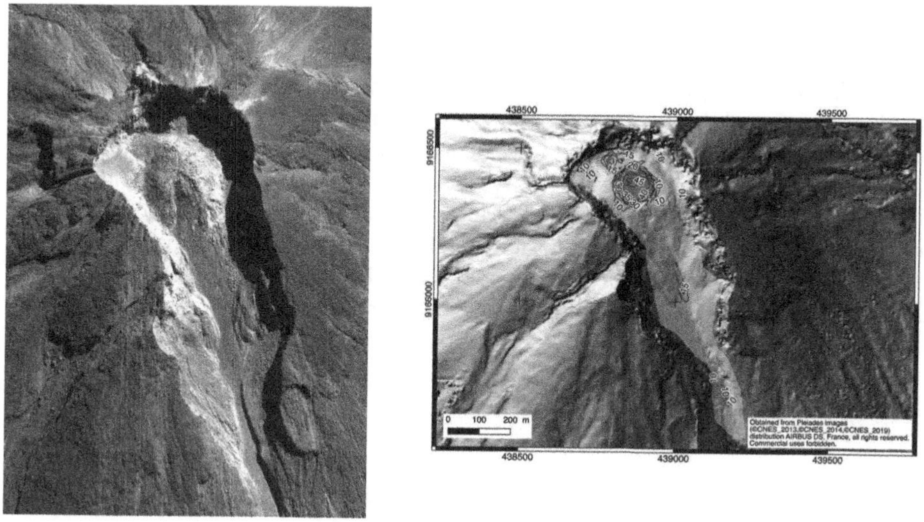

a) b)

Figure 8.3. *Use of Pléiades optical images (©CNES 2019, distribution AIRBUS, DS, images available through the DINAMIS/ISIS Science Support Programme) to quantify the growth of the dome in the summit crater of Merapi, Indonesia. a) Pléiades image from August 2019. b) Estimation of altitude variations caused by dome growth between 2014 and February 2019. The volume of the dome was estimated at approximately 0.5 million m*^{3}

8.4. Main limitations of measurements obtained by remote sensing

8.4.1. *Temporal resolution*

The frequency with which satellite data are acquired has increased in recent years, but the time step between two images with the same geometry covers at least a few days (every five days for Sentinel-2 optical images, and every six days, at best, for Sentinel-1 radar images). Some magmatic processes are too fast for their temporal evolution to be followed by remote sensing: notable examples include rapid destabilizations of the volcanic edifice and the propagation of magma to the surface, the dynamics of which can only be observed by continuous field measurements (see Figure 8.2). Satellite displacement measurements are still essential for quantifying permanent, static deformations, revealing information about the magma which is trapped underground in the form of magmatic intrusions. Recent improvements in acquisition frequencies mean that images acquired during magma propagation can now be exploited (e.g. in the case of the Piton de la Fournaise (Smittarello *et al.* 2019b)). However, the transient and localized deformations resulting from flows in the conduits cannot yet be assessed using remote approaches (Albino *et al.* 2011).

8.4.2. *Cloud cover (optical imagery)*

Unlike radar images, useful optical images can only be obtained when there is no cloud cover and during daylight hours. The proportion of acquired images which are actually useful is therefore strongly dependent on the weather conditions of the volcanic area in question; generally, this proportion is relatively low for volcanoes in the intertropical zone. In the case of the Merapi volcano, Indonesia, Sentinel-2 images are acquired every five days and systematically transmitted to the observatory. The proportion of images where the summit area is visible is relatively good (around 75%) during the dry season, from May to October, but extremely low (less than 10%) during the wet season, from November to April. In addition to these meteorological issues, the surface of volcanoes is often hidden by volcanic plumes during periods of eruptive activity.

8.4.3. *Atmospheric artifacts (radar imagery)*

The main limitation on the use of radar interferometry to measure displacement comes from atmospheric artifacts (Zebker *et al.* 1997). This problem does not exist when using bistatic data, which combine images acquired at the same time. Variations in the physical properties of the atmosphere over time, notably in terms of water vapor content, induce significant variations in the speed of propagation of electromagnetic waves and cause a variation in phase (term ϕ_{atm} in equation [8.1]), with an amplitude which may be well in excess of that of the geophysical signal. On the first occasion when interferometry techniques were applied to volcanology, a deflation signal was identified on Etna (Massonnet *et al.* 1995); this signal was later found to contain a significant portion of atmospheric effects (Beauducel *et al.* 2000). Atmospheric artifacts include a turbulent component that can be reduced by temporal filtering using time series (Hooper *et al.* 2012); however, they also contain a stratified component correlating with the topography. The spatial characteristics of the stratified artifact are very similar to the displacement field produced by a magma storage source located deep beneath the volcanic edifice.

Atmospheric fringes correlating with topography can be eliminated empirically by using the correlation between phase and altitude from deformation-free zones (Remy *et al.* 2003), albeit with the risk of partly eliminating the displacement signal itself. GNSS data can also be used to estimate and correct for atmospheric effects (Li *et al.* 2006). Another common approach consists of using temperature, pressure and water vapor content profiles taken from global meteorological models in order to calculate the tropospheric delay (Doin *et al.* 2009). These models have a temporal resolution of a few hours and a spatial resolution of a few tens of kilometers. The most widespread example is ERA5 re-analysis, supplied by the European Centre for Median-Range Weather Forecasting (ECMWF), available within five days. The Generic Atmospheric

Correction Online Service (GACOS) also publishes maps of estimated tropospheric phase delays, extrapolated from the ECMWF data in conjunction with GNSS data. This information is made available one to two days after acquisition (Yu *et al.* 2018).

8.5. Main contributions of spatial geodesy for monitoring and studying volcanoes

In the years since the first study in this area was carried out by Massonnet *et al.* (1995), spatial geodesy has come to play a major role in modern volcanology (Pinel *et al.* 2014; Biggs and Pritchard 2017). This approach can be used to detect magma storage in non-instrumented areas. The broad extent of spatial coverage has brought significant improvements in terms of our knowledge of the geometry of magmatic intrusions and has enabled the detection of storage zones far below the surface or at some distance from the main volcanic edifice.

8.5.1. *The development of global and statistical studies*

The use of remote sensing techniques means that more active volcanoes can be studied using geodetic techniques than ever before, increasing the feasibility of broad statistical studies at the planetary level (Biggs *et al.* 2014) or across a specific volcanic region (e.g. Pritchard and Simons 2002; Biggs *et al.* 2009; Chaussard and Amelung 2012; Ebmeier *et al.* 2013). This approach has enabled systematic investigations of the relationship between eruptive activity and ground deformations (Biggs *et al.* 2014), along with a characterization of the depth of magma storage zones as a function of the tectonic context (Chaussard and Amelung 2014; Pinel *et al.* 2014) and their position with respect to eruptive centers (Ebmeier *et al.* 2018). The findings of these studies provide new information to further our understanding of magma transport mechanisms, the respective roles of lateral and vertical transport and the localization of zones where transport slows or even ceases. The use of remote techniques has also promoted cooperation between different international teams: one example is an initiative launched by the Committee on Earth Observation Satellites (CEOS), with a pilot project focusing on Latin America (2013–2017; Pritchard *et al.* (2018)), extended in 2019 to cover southeast Asia and Africa, with the aim of promoting the use of remote sensing techniques for the study and monitoring of volcanoes.

8.5.2. *Improving knowledge of the magma supply system*

The surface displacement maps provided by the InSAR, inverted using analytical or numerical models, have provided key information on the subject of storage zones, their location (Ebmeier *et al.* 2018) and shape, and on the presence of multiple sources operating simultaneously beneath a single volcanic edifice (see Pinel *et al.* (2014), section 4.2.1). Trajectories from these zones up to the surface, which can be complex, can now be represented at a sufficiently high resolution (Bagnardi *et al.* 2013; Galetto *et al.* 2020) to increase our understanding of the way in which the stress field influences the trajectory of magmatic intrusions (Corbi *et al.* 2015).

8.5.3. *Growth and stability of volcanic edifices*

Remote sensing data is unique in that it can be used to quantify volumes of magma both beneath and above the surface (in the form of volcanic deposits). This provides valuable quantitative information concerning the volcano growth modes, making a distinction between endogenous growth, from intrusions beneath the surface, and exogenous growth resulting from the accumulation of deposits on the surface (e.g. studying five eruptions of the Piton de la Fournaise, Fukushima *et al.* (2010) showed that 17% of the volume of magma which began transiting towards the surface remained trapped). The mode of growth has an effect on edifice stability, and thus on volcanic risk.

8.6. Recent progress

8.6.1. *Integrating satellite imagery with field data*

In relation to dynamic phenomena such as the propagation of magmatic intrusions, it can be useful to combine the spatial resolution of InSAR data with GNSS data, which has high temporal resolution but lower spatial coverage. A method for joint inversion of InSAR and GNSS was recently proposed and applied to the Piton de la Fournaise (Smittarello *et al.* 2019a). In this case, co-eruptive interferograms were used to model the final trajectory of a magmatic intrusion, then the temporal evolution of the opening of the intrusion was modeled using GNSS time series. The validity of this method was confirmed using a radar image obtained during the propagation of the intrusion. It highlighted both deceleration and acceleration phases in magma propagation, probably linked to the sensitivity of the stress field surrounding a small volume of magma with limited driving force (Smittarello *et al.* 2019b).

8.6.2. *Automating processing chains for real-time detection*

As the number of SAR missions increased, a number of international initiatives were launched with the aim of automating InSAR processing chains, rapidly providing the scientific community with InSAR products (interferograms, coherency and amplitude) in response to natural hazards such as volcanic eruptions or earthquakes. Global efforts in favor of automation and near-real time production accelerated still further after 2014, when the ESA made Sentinel data freely available. Key examples of these initiatives include:

– the SARVIEWS online service, developed by the University of Alaska Fairbanks and Alaska Satellite Facility (ASF), produces interferograms for earthquakes and volcanic eruptions just a few hours after the acquisition of SAR data (Meyer *et al.* 2019);

– the Advanced Rapid Imaging and Analysis (ARIA) project, run by the NASA Jet Propulsion Laboratory (JPL), provides displacement maps and value-added products, such as damage maps, in response to earthquakes, hurricanes or floods (Agram 2016);

– the online Geohazard Exploitation Platform (GEP), developed by the European Space Agency (ESA), provides satellite data for use in evaluating different natural hazards such as earthquakes, volcanic eruptions or landslides. Unlike the other services listed above, the GEP allows users to process SAR data directly within the platform. The service also offers automatic production for certain priority zones, including a selection of 22 active volcanoes;

– the Center of Observation and Modeling of Earthquakes and Volcanoes (COMET), based in the United Kingdom, has developed an automatic processing chain for Sentinel-1 data, known as LiCSAR, for monitoring major tectonic zones (the Alps, Turkey, the Himalayas) along with all active volcanoes (Lazecky *et al.* 2020; Morishita *et al.* 2020);

– in France, the Observatoire InSAR de l'Océan Indien (OI2) works closely with the Observatoire Volcanologique du Piton de la Fournaise (OVPF). Using the Casoar online platform, the OI2 service provides the OVPF with displacement maps and lava flow tracking maps for each eruption of the Piton de la Fournaise volcano.

This systematic production of InSAR data has also contributed to the development of methods for automatically detecting and extracting displacement signals and for predicting the evolution of ground displacements. The long-term objective is to develop a valuable detection system for volcano observatories, with near-real time alerts each time a new displacement signal is detected.

Blind source separation methods, such as principal component analysis (PCA) or independent component analysis (ICA), can be used to extract spatial and temporal signal information from InSAR time series. In the context of volcanology, these methods have been applied to a number of case studies, such as Long Valley (USA), Cordon del Azufre and Calbuco (Chili) and the Galapagos archipelago (Lin *et al.* 2010; Remy *et al.* 2014; Ebmeier 2016; Gaddes *et al.* 2018, 2019). By extracting the different components of a signal, this method makes it possible to separate displacement signals and atmospheric signals without performing a pre-correction step. Other work, such as that of Albino *et al.* (2020), has demonstrated the effectiveness of simpler algorithms, such as CUSUM, which rely on anomaly detection in order to rapidly flag the start of inflation periods.

All of the methods cited above rely on the production of InSAR time series. In spite of recent efforts (Dalaison and Jolivet 2020), real-time and automatic time-series production continues to present a major challenge. This means that PCA and ICA methods are not entirely suitable for automatic detection of displacement signals on a global level.

Machine learning (ML) techniques can provide an answer to this limitation, taking advantage of the large number of raw interferograms routinely produced by InSAR processing chains. The value of ML techniques in a remote sensing context has already been proved for tasks such as image merging, scene classification, object detection and terrain classification (Ma *et al.* 2019). Work carried out at the University of Bristol has shown the value of ML for automatically detecting displacement signals in InSAR data (Anantrasirichai *et al.* 2018, 2019). The authors of this study used a dataset of 30,000 interferograms produced by LiCSAR for over 900 active volcanoes in order to develop and train a convolutional neural network (CNN), which successfully detected displacements in around 40 interferograms.

Finally, data assimilation strategies, commonly used for weather forecasting, have recently been applied to volcanology with the aim of increasing forecasting capacity (Bato *et al.* 2017). Combining observations with dynamic models of magma flow gives us information concerning magma pressure in storage zones, whilst providing insights into deep phenomena, such as variations in the deep magma inflow in volcanic systems (Bato *et al.* 2018).

8.6.3. *Integration of all satellite sources in multi-parameter studies*

Several recent studies have adopted a "multiparameter" approach, combining multiple satellite observations. Furtney *et al.* (2018) systematically compiled deformation signals, SO_2 emissions and thermal anomalies detected by satellite measurements before and during volcanic eruptions over the whole period covered by satellite observations, starting in the late 1970s. They found that those eruptions for which no signals are found in remote sensing data are essentially minor events, or date from the earlier part of the investigation period, prior to 2000, for which observations are much less frequent. This study demonstrates the potential offered by these observations in a volcanology context. A systematic comparison of thermal satellite data, gas emissions and InSAR data was also carried out for 47 volcanoes in Latin America over a period of 17 years (Reath *et al.* 2019, 2020). This work contributed to the development of a classification of volcanic systems and provided valuable information for databases. WOVOdat, an international database aimed at improving the detection and interpretation of pre-eruption precursors, initially contained compilations of measurements from volcanological laboratories, to which satellite data were then added. More focused studies also exist, such as Kilbride *et al.*'s (2016) comparison of the relationship between variations in magma volumes in the storage areas, estimated using geodetic techniques, and the volumes of magma emitted at the surface with gas emissions. The aim of this study was to better defining the effects of compressibility induced by the volatiles contained in the magma.

8.7. Volcanic crisis management: the contribution of displacement measurements obtained using spatial imagery

In the context of volcanic crises, as for any natural disaster, the authorities of the affected countries can invoke the International Charter for Space and Major Disasters. This triggers the rapid acquisition of satellite data covering the impact zone, which are then used to map the damage.

8.7.1. *Piton de la Fournaise, Réunion: benefits of spatial imagery for a highly active, closely monitored volcano*

The Piton de la Fournaise, a basaltic shield volcano associated with hot spot activity, is one of the most active volcanoes on Earth, averaging two to three eruptions per year since 1998 (Global Volcanism Program, https://volcano.si.edu/volcano.cfm?vn=233020). Most historic eruptions have taken place in the Enclos Fouqué-Grand Brûlé, a long horseshoe depression which extends over 13 km along an east–west axis, likely formed by the collapse of a summit caldera or by catastrophic flank sliding of the eastern part of the edifice. The Piton de la Fournaise is actively monitored by the Observatoire Volcanologique du Piton de la Fournaise (OPVF), which has been recording seismicity, monitoring gas emissions and collecting regular ground displacement measurements since the 1980s (Peltier *et al.* 2018). In 1998, ground displacement maps obtained by InSAR were added to this list. Initially produced on a punctual basis, these maps began to be created systematically (Froger *et al.* 2004) by the Observatoire InSAR de l'Océan Indien at the Observatoire de Physique du Globe in Clermont-Ferrand, France, following the eruption of August 2003. Using SAR data acquired by the different available satellites, the OI^2 was able to model the ground displacements generated by over 40 eruptions (Froger *et al.* 2020). The deformation of the volcano's central cone was clearly detected by a dense GPS network, but InSAR made significant contributions for areas further from the summit. InSAR data was used to model the geometry of intrusions and, more importantly, highlighted the presence of flank sliding in the Grandes Pentes zone to the east of the summit, an area in which field instrumentation was only installed in 2007 (Richter and Froger 2020). InSAR also offers a means of better understanding the intrusive and extrusive growth of volcanic edifices (Fukushima *et al.* 2010) and, as a result, their stability (Chen *et al.* 2017; Richter and Froger 2020). Radar data can be used in real time during eruptions for rapid and efficient mapping of the extent of lava flows, drawing on the loss of coherence which results from the emplacement of deposits.

This information, combined with satellite thermal data, can be used to validate lava flow models and to update predictions concerning the areas affected by an eruption in real time (Harris *et al.* 2019).

8.7.2. *Mount Agung, Indonesia: using InSAR data in real time in a crisis management situation*

Mount Agung is an andesitic stratovolcano in northeastern Bali, Indonesia (Global Volcanism Program, https://volcano.si.edu/volcano.cfm?vn=264020). Following a major explosive eruption in 1963 (Volcanic Explosivity Index (VEI) 5), the volcano remained inactive for over 50 years, before displaying signs of seismic activity in August 2017 (Figure 8.4a). The seismic crisis accelerated in September before reaching a plateau in October, with over 800 events per day (Figure 8.4b). Activity then dropped off until the start of the eruption on November 21, 2017. Believing eruption to be imminent, the Indonesian authorities, working in collaboration with the Centre for Volcanology and Geological Hazard Mitigation (CVGHM), began preventive evacuation of 140,000 inhabitants in late September and created an exclusion zone of around 10 kilometers around the volcano (Syahbana *et al.* 2019). Evidently, this decision prevented a major human catastrophe.

The magmatic origin of the seismic activity was discussed at length during the crisis management period. The collection and analysis of displacement data played a crucial role in developing an understanding of the processes leading to the reactivation of the volcano. While only five GNSS stations (installed in 2012) were available in September 2017, some of these instruments detected an inflation in the ground, interpreted as the pressurization of a deep-lying magma source (Syahbana *et al.* 2019). However, with so few stations, it was not possible to constrain the nature or depth of the source. In this context, InSAR data was key to characterizing the displacement field and identifying the source(s) of the observed signals.

The fact that Sentinel-1 SAR data (ESA) are freely available presents an advantage to the scientific community and volcanological observatories in crisis situations, as preliminary results can be produced and circulated rapidly. However, the free accessibility of InSAR products is not without its drawbacks. In the case of the Mount Agung eruption, certain Sentinel-1 interferograms were shared on social media sites, such as Twitter, in September 2017, along with interpretations suggesting that the volcano had expanded by over 10 cm in 12 days. In reality, the signal in question was a result of atmospheric effects, an issue that is frequently encountered in the case of stratovolcanoes (see section 8.4.3). The tropical island of Bali, where Mount Agung is located, is subject to strong variations in atmospheric conditions, notably in terms of water vapor content, due to proximity to the sea and the high topographical gradient between the coast (0 m) and the summit of the volcano (>3,000 m). The spread of false information, not recalibrated or contextualized by scientists, created problems in terms of crisis management; the Indonesian authorities were forced to communicate and reassure the population in a way which would not otherwise have been necessary.

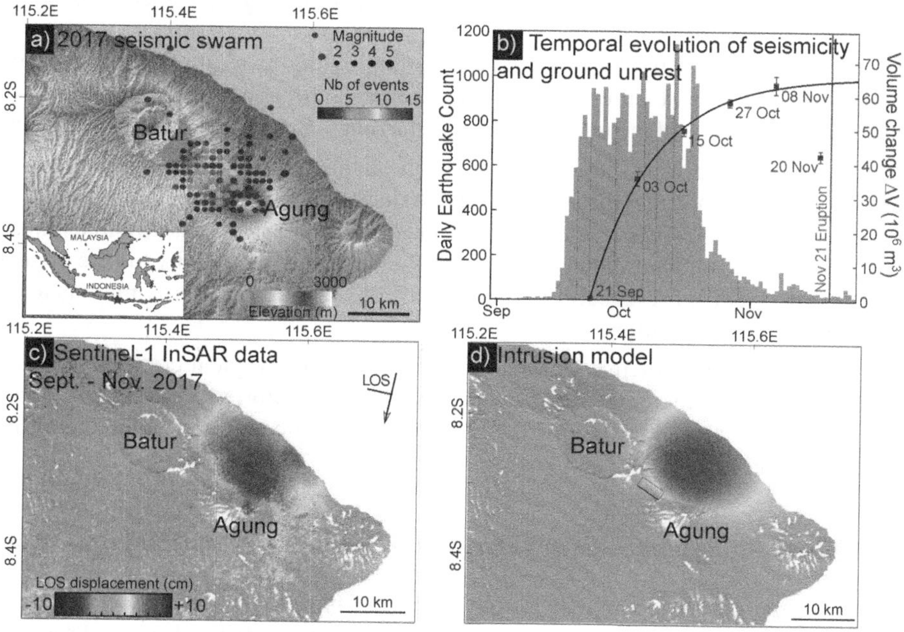

Figure 8.4. *Volcanic crisis (seismicity and surface deformation) observed at Mount Agung in the pre-eruptive period from September to November 2017. a) Location and magnitude of seismic activity; b) temporal evolution of seismicity and variations in the volume of the magmatic intrusion; c) cumulative displacements obtained using Sentinel-1 InSAR data; d) deep magmatic intrusion model adjusting the deformation signal. For a color version of this figure, see www.iste.co.uk/cavalie/images.zip*

In-depth analysis of InSAR data is crucial. First, atmospheric signals are systematically corrected using meteorological models (from the ECMWF), and then, time series are produced (Albino *et al.* 2020). Using this approach, an inflation signal with a radius of around 5 km was identified after the event between Mount Agung and Mount Batur, with displacements of around 10 cm over the period from September to October 2017 (Figure 8.4c). This deformation episode was concurrent with the seismic crisis (Figure 8.4b). Numerical modeling of the displacement signals showed that these precursors were caused by the propagation of a deep magmatic intrusion (7–12 km) between the Agung and Batur volcanoes (Albino *et al.* 2019) (Figure 8.4d).

The eruption essentially took the form of a dome growing in the summit crater, with a volume quantified at 24 ± 1 million m^3 by differential DEM using tristereo Pléiades images, acquired in February 2018 in response to a request from the CEOS Volcano supersite. The magma surface emission was combined with a deflation corresponding to the emptying of the deep source identified by the GNSS data.

The study of the 2017 Mount Agung eruption highlights a number of important points for discussion relating to crisis management, the contribution of InSAR to volcanic monitoring and the understanding of the magmatic systems of arc volcanoes:

– in a globalized world, the rapid diffusion of false information or “fake news” (such as mistaken interpretations of InSAR products) hinders correct management of volcanic crises by the competent authorities and organizations and has a visible impact on local populations. This shows the need for the scientific community to take greater responsibility and, more generally, raises questions concerning the communication and diffusion of information;

– in-depth analysis of InSAR data, while difficult in a tropical context, provides information to complement that provided by field monitoring networks (seismic stations, GNSS). This information is key to developing a better understanding of the magmatic processes leading up to eruption.

8.7.3. *Taal, Philippines: innovative use of InSAR information in real time*

Taal is the second most active volcano in the Philippines (Delos Reyes *et al.* 2018), located just south of Manila, the national capital (Global Volcanism Program, https://volcano.si.edu/volcano.cfm?vn=273070). At around 05:00 UTC on January 12, 2020, the Taal volcano underwent a phreatic eruption, spewing a column of ash mixed with water vapor up to an altitude of 15 km while producing a number of volcanic lightning blots, reported by the Philippine Institute of Volcanology and Seismology (PHIVOLCS). The eruption entered a magmatic phase at 18:49 UTC, with the emission of low-intensity lava fountains. Visibility of the main crater of the erupting volcanic island was extremely limited, and most of the on-site field instruments (GNSS stations, seismic stations and MultiGAS) operating at Taal were destroyed or disabled in the paroxysm of the crisis. The “Space and Major Catastrophes” charter was activated. Programmed optical acquisitions, notably Pléiades data, showed significant cloud cover. In this context, radar data from the Sentinel-1 satellite provided crucial information to the local observatory, directed by the PHIVOLCS, helping scientists to evaluate the state of the volcano with relatively high temporal resolution and wide spatial coverage. Figure 8.5a shows the very first interferogram shared by the Jet Propulsion Laboratory team at Caltech on January 16, 2020, mere hours after the second SAR image became available to download from the Copernicus platform. Each of the concentric fringes shown in the figure is equivalent to a displacement of around 2.8 cm on the ground in the line of sight (LOS) of the satellite. However, the large ground displacement gradients in the immediate vicinity of the eruptive crater meant that fringes were difficult to discern and impossible to unwrap: this phenomenon is known as aliasing. Figure 8.5b is a surface displacement map, produced by pixel offset analysis of the amplitude image (see Chapter 1, section 1.2.1). It shows how the displacement measurements obtained using pixel offset tracking can be used as a complement to InSAR results, notably in regions with large displacement gradients, such as the Taal volcanic island and the Pansipit valley (southwest of the caldera lake). A displacement of tens of meters in the direction of the line of sight of the satellite, moving away from the satellite, was detected for most regions of the volcanic island; in the Pansipit valley, around 1.5 m of positive and negative displacements in the direction of the line of sight of the satellite. Ground fissures, generally an indicator of subterranean magma activity, were also observed, mostly around the Pansipit valley area: these findings corroborated the results of the pixel offset analysis. This method also allows new deposits of eruptive products, shown as red pixels in the volcanic island in Figure 8.5b, to be detected and their areas to be measured. Note, however, that the accuracy of pixel offset results is sub-metric and, as such, is well below that of InSAR results.

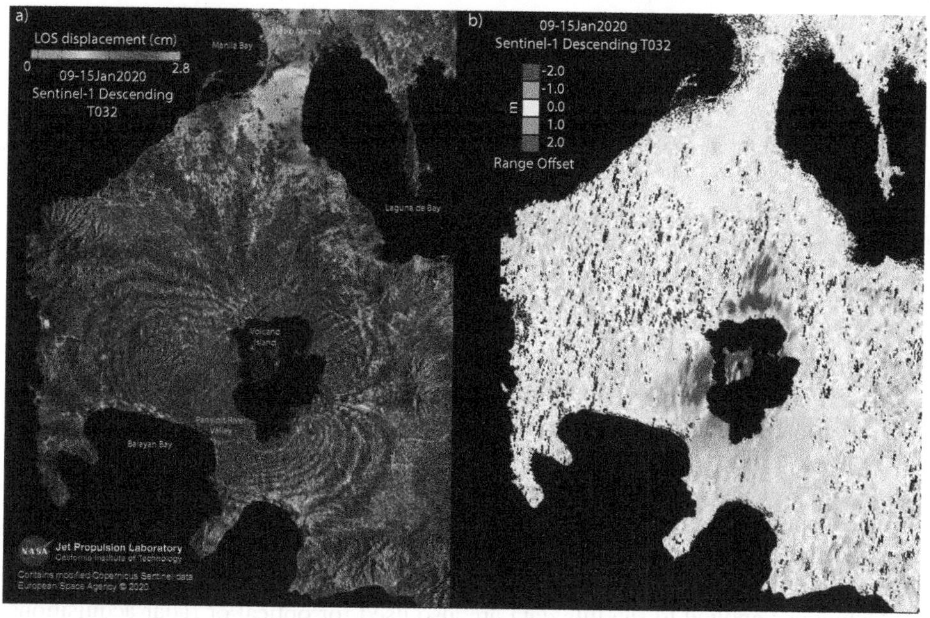

Figure 8.5. *Surface displacement measurements using a) interferometry and b) pixel offset tracking from the first Sentinel-1 image acquired after the start of the eruption. In addition to the displacement field centered on the volcanic island, there is a signal induced by a vertical intrusion extending south of the caldera, with boundaries corresponding to the lake surrounding the volcanic island. For a color version of this figure, see www.iste.co.uk/cavalie/images.zip*

The Sentinel-1 mission involves two satellites, meaning that interferograms can be produced every six days. However, this level of systematic acquisition in ascending and descending trajectories only applies to Europe. Thanks to the support of the Copernicus program (ESA), the JPL's request for a six-day acquisition frequency for Taal was accepted and implemented until February 22, 2020. This allowed the JPL to provide the PHIVOLCS with regular data and analysis as and when it became available.

The role that Sentinel-1 data played throughout the Taal eruption was unprecedented (Bato *et al.* 2021); this data was used to follow the progression of a magmatic intrusion of several tens of kilometers in length, just below the surface, in real time. This magmatic intrusion had not been detected by the field network as it had been in the case of the Bardarbunga eruption of August 2014 in Iceland. Although InSAR data provide a vast amount of information, it is important to note that the behavior of a volcano cannot be understood and predicted on this basis alone, especially during a crisis. Additional *in situ* measurements, ground observations and local and historic knowledge of the volcano are of critical importance. For this reason, in a crisis management context, data and analysis should be transmitted to the local scientific community and not sent directly to the authorities in charge of evacuation proceedings.

Sentinel-1 data allowed an ongoing crisis to be monitored from a distance and in near-real time. At the time of writing, Taal has not undergone any major eruptions since the instance discussed above (January 2020). The JPL continues to support the local volcanic observatory using time-series analysis, and the PHIVOLCS now include InSAR data in scientific discussions and in routine monitoring activity.

8.8. Conclusion

The role of remote sensing in volcanology continues to grow, notably thanks to the increasing number of observations of volcanic systems and to the development of multiparameter statistical studies on a global scale. Satellite ground displacement measurements have contributed to significant advances in imaging and in understanding magma supply systems, although the low temporal resolution of this data is currently a limiting factor for the observation of system dynamics. Note that the techniques applied to satellite data are also used for optical or radar acquisitions obtained at ground level, by drones or from the air (Derrien *et al.* 2015). They have the potential to be used in planetary volcanology and in analog laboratory experiments (Galland *et al.* 2016; Liu *et al.* 2019). In conclusion, radar data, which can be used irrespective of cloud cover, present a major advantage in comparison with optical data; furthermore, radar data offer the only means of measuring low amplitude displacements, which constitute the majority of precursors of volcanic activity. Satellite data are not yet being used to their full potential for real-time monitoring and crisis management; nevertheless, current initiatives should increase the impact of these techniques.

8.9. Acknowledgments

The authors thank Valérie Cayol and Jean-Luc Froger for reviewing this chapter and for their useful suggestions.

8.10. References

Agram, P. (2016). *ARIA: Delivering State-of-the-Art InSAR Products for End Users*. NASA Jet Propulsion Laboratory, Pasadena.

Albino, F., Pinel, V., Massol, H., Collombet, M. (2011). Conditions for detection of ground deformation induced by conduit flow and evolution. *Journal of Geophysical Research Solid Earth*, 116, B06201.

Albino, F., Smets, B., d'Oreye, N., Kervyn, F. (2015). High-resolution TanDEM-X DEM: An accurate method to estimate lava flow volumes at Nyamulagira Volcano (D. R. Congo). *Journal of Geophysical Research: Solid Earth*, 120(6), 4189–4207.

Albino, F., Biggs, J., Syahbana, D.K. (2019). Dyke intrusion between neighbouring arc volcanoes responsible for 2017 pre-eruptive seismic swarm at Agung. *Nature Communications*, 10(1), 748.

Albino, F., Biggs, J., Yu, C., Li, Z. (2020). Automated methods for detecting volcanic deformation using Sentinel-1 InSAR time series illustrated by the 2017–2018 unrest at Agung, Indonesia. *Journal of Geophysical Research: Solid Earth*, 125(2), e2019JB017908.

Anantrasirichai, N., Biggs, J., Albino, F., Hill, P., Bull, D. (2018). Application of machine learning to classification of volcanic deformation in routinely generated InSAR data. *Journal of Geophysical Research: Solid Earth*, 123(8), 6592–6606.

Anantrasirichai, N., Biggs, J., Albino, F., Bull, D. (2019). A deep learning approach to detecting volcano deformation from satellite imagery using synthetic datasets. *Remote Sensing of Environment*, 230, 111179.

Anderson, K.R., Johanson, I.A., Patrick, M.R., Gu, M., Segall, P., Poland, M.P., Montgomery-Brown, E.K., Miklius, A. (2019). Magma reservoir failure and the onset of caldera collapse at Kilauea Volcano in 2018. *Science*, 366(6470).

Arambula-Mendoza, R., Varley, N., Garcia-Flores, R., Vargas-Bracamontes, D., Navarro-Ochoa, C., Marquez-Ramirez, V., Capra, L., Gonzalez-Amezcua, M., Martinez-Fierros, A., Ramirez-Vazquez, C. (2020). Destruction of a lava dome observed with photogrammetry, acoustic and seismic sensors at Volcán de Colima, Mexico. *Journal of Volcanology and Geothermal Research*, 395, 106834.

Auker, M.R., Sparks, R.S.J., Siebert, L., Crosweller, H.S., Ewert, J. (2013). A statistical analysis of the global historical volcanic fatalities record. *Journal of Applied Volcanology*, 2(1), 2.

Bagnardi, M., Amelung, F., Poland, M.P. (2013). A new model for the growth of basaltic shields based on deformation of Fernandina volcano, Galápagos Islands. *Earth and Planetary Science Letters*, 377–378(0), 358–366.

Bagnardi, M., González, P.J., Hooper, A. (2016). High-resolution digital elevation model from tri-stereo Pleiades-1 satellite imagery for lava flow volume estimates at Fogo Volcano. *Geophysical Research Letters*, 43(12), 6267–6275.

Barisin, I., Leprince, S., Parsons, B., Wright, T. (2009). Surface displacements in the September 2005 Afar rifting event from satellite image matching: Asymmetric uplift and faulting. *Geophysical Research Letters*, 36(7).

Bato, M.G., Froger, J.L., Harris, A.J.L., Villeneuve, N. (2016). Monitoring an effusive eruption at Piton de la Fournaise using radar and thermal infrared remote sensing data: Insights into the October 2010 eruption and its lava flows. In *Detecting, Modelling and Responding to Effusive Eruptions*, Harris, A.J.L., De Groeve, T., Garel, F., Carn, S.A. (eds). Geological Society, London.

Bato, M.G., Pinel, V., Yan, Y. (2017). Assimilation of deformation data for eruption forecasting: Potentiality assessment based on synthetic cases. *Frontiers in Earth Science*, 5, 48.

Bato, M.G., Pinel, V., Yan, Y., Jouanne, F., Vandemeulebrouck, J. (2018). Possible deep connection between volcanic systems evidenced by sequential assimilation of geodetic data. *Scientific Reports*, 8(1), 11702.

Bato, M.G., Lundgren, P., Pinel, V., Solidum, R., Daag, A., Cahulogan, M. (2021). The 2020 eruption and large lateral dike emplacement at Taal volcano, Philippines: Insights from satellite radar data. *Geophysical Research Letters*, 48, e2021GL092803.

Beauducel, F. and Carbone, D. (2015). A strategy to explore the topography-driven distortions in the tilt field induced by a spherical pressure source: The case of Mt Etna. *Geophysical Journal International*, 1471–1481.

Beauducel, F., Briole, P., Froger, J.-L. (2000). Volcano wide fringe in ERS synthetic aperture radar interferograms of Etna (1992–1998): Deformation or tropospherique effect? *Journal of Geophysical Research: Solid Earth*, 105, 16391–16402.

Biggs, J. and Pritchard, M.E. (2017). Global volcano monitoring: What does it mean when volcanoes deform? *Elements*, 13(1), 17–22.

Biggs, J., Anthony, E.Y., Ebinger, C.J. (2009). Multiple inflation and deflation events at Kenyan volcanoes, East African Rift. *Geology*, 37, 979–982.

Biggs, J., Ebmeier, S.K., Aspinall, W.P., Lu, Z., Pritchard, M.E., Sparks, R.S.J., Mather, T.A. (2014). Global link between deformation and volcanic eruption quantified by satellite imagery. *Nature Communications*, 5, 3471.

Brown, S., Loughlin, S., Sparks, R., Vye-Brown, C., Barclay, J., Calder, E., Cottrell, E., Jolly, G., Komorowski, J.-C., Mandeville, C., Newhall, C., Palma, J., Potter, S., Valentine, G., Baptie, B., Biggs, J., Crossweller, H., Ilyinskaya, E., Kilburn, C., Mee, K., Pritchard, M. (2015). Global volcanic hazard and risk. In *Global Volcanic Hazards and Risk*, Loughlin, S.C., Sparks, S., Brown, S.K., Jenkins, S.F., Vye-Brown, C. (eds). Cambridge University Press, Cambridge.

Carrara, A., Pinel, V., Bascou, P., Chaljub, E., la Cruz-Reyna, S.D. (2019). Post-emplacement dynamics of andesitic lava flows at Volcán de Colima, Mexico, revealed by radar and optical remote sensing data. *Journal of Volcanology and Geothermal Research*, 381, 1–15.

Casu, F., Manconi, A., Pepe, A., Lanari, R. (2011). Deformation time-series generation in areas characterized by large displacement dynamics: The SAR amplitude pixel-offset SBAS technique. *IEEE Journal of Selected Topics in Applied Earth Observations and Remote Sensing*, 99, 1–12.

Cayol, V. and Cornet, F.H. (1998). Effects of topography on the interpretation of the deformation field of prominent volcanoes – Application to Etna. *Geophysical Research Letters*, 25, 1979–1982.

Champenois, J., Pinel, V., Baize, S., Audin, L., Jomard, H., Hooper, A., Alvarado, A., Yepes, H. (2014). Large-scale inflation of Tungurahua volcano (Ecuador) revealed by persistent scatterers SAR interferometry. *Geophysical Research Letters*, 41(16), 5821–5828.

Chaput, M., Pinel, V., Famin, V., Michon, L., Froger, J.-L. (2014). Cointrusive shear displacement by sill intrusion in a detachment: A numerical approach. *Geophysical Research Letters*, 41, 1937–1943.

Chaussard, E. and Amelung, F. (2012). Precursory inflation of shallow magma reservoirs at west Sunda volcanoes detected by InSAR. *Geophysical Research Letters*, 39.

Chaussard, E. and Amelung, F. (2014). Regional controls on magma ascent and storage in volcanic arcs. *Geochemistry, Geophysics, Geosystems*, 15(4), 1407–1418.

Chen, Y., Remy, D., Froger, J.-L., Peltier, A., Villeneuve, N., Darrozes, J., Perfettini, H., Bonvalot, S. (2017). Long-term ground displacement observations using InSAR and GNSS at Piton de la Fournaise volcano between 2009 and 2014. *Remote Sensing of Environment*, 194, 230–247.

Corbi, F., Rivalta, E., Pinel, V., Maccaferri, F., Bagnardi, M., Acocella, V. (2015). How caldera collapse shapes the shallow emplacement and transfer of magma in active volcanoes. *Earth and Planetary Science Letters*, 431, 287–293.

Dalaison, M. and Jolivet, R. (2020). A Kalman filter time series analysis method for InSAR. *Journal of Geophysical Research: Solid Earth*, 125(7), e2019JB019150.

De Zan, F. (2014). Accuracy of incoherent speckle tracking for circular Gaussian signals. *IEEE Geoscience and Remote Sensing Letters*, 11, 264–267.

Delos Reyes, P., Bornas, M., Dominey-Howes, D., Pidlaoan, A., Magill, C., Solidum, R. (2018). A synthesis and review of historical eruptions at Taal Volcano, Southern Luzon, Philippines. *Earth Science Reviews*, 177, 565–588.

Derrien, A., Villeneuve, N., Peltier, A., Beauducel, F. (2015). Retrieving 65 years of volcano summit deformation from multitemporal structure from motion: The case of Piton de la Fournaise (La Réunion Island). *Geophysical Research Letters*, 42(17), 6959–6966.

Doin, M.P., Lasserre, C., Peltzer, G., Cavalié, O., Doubre, C. (2009). Corrections of stratified tropospheric delays in SAR interferometry: Validation with global atmospheric models. *Journal of Applied Geophysics*, 69, 35–50.

Dvorak, J.J. and Dzurisin, D. (1997). Volcano geodesy: The search for magma reservoirs and the formation of eruptive vents. *Reviews of Geophysics*, 35, 343–384.

Ebmeier, S.K. (2016). Application of independent component analysis to multitemporal InSAR data with volcanic case studies. *Journal of Geophysical Research: Solid Earth*, 121(12), 8970–8986.

Ebmeier, S.K., Biggs, J., Mather, T.A., Amelung, F. (2013). Applicability of InSAR to tropical volcanoes: Insights from Central America. In *Remote Sensing of Volcanoes and Volcanic Processes: Integrating Observation and Modelling*, Pyle, D.M., Mather, T.A., Biggs, J. (eds). Geological Society, London.

Ebmeier, S.K., Andrews, B.J., Araya, M.C., Arnold, D.W.D., Biggs, J., Cooper, C., Cottrell, E., Furtney, M., Hickey, J., Jay, J., Lloyd, R., Parker, A.L., Pritchard, M.E., Robertson, E., Venzke, E., Williamson, J.L. (2018). Synthesis of global satellite observations of magmatic and volcanic deformation: Implications for volcano monitoring and the lateral extent of magmatic domains. *Journal of Applied Volcanology*, 7(2), 1–26.

Ferrucci, F., Prata, F., Amelung, F., Bawden, G., Biggs, J., Briole, P., Del Negro, C., Eineder, M., Jordan, C., Loughlin, S., Puglisi, G., Bianchi, M., Tait, S., Theys, N., Schneider, D., Norbury, D. (2012). Perspectives concerning satellite EO and geohazard risk management: Volcanic hazards. In *The International Forum on Satellite EO and Geohazards, The Santorini Conference*, Bally, P. (ed.). European Space Agency, Santorini.

Froger, J.-L., Fukushima, Y., Briole, P., Staudacher, T., Souriot, T., Villeneuve, N. (2004). The deformation field of the August 2003 eruption at Piton de la Fournaise, Reunion Island, mapped by ASAR interferometry. *Geophysical Research Letters*, 31, L14601.

Froger, J.-L., Famin, V., Cayol, V., Augier, A., Michon, L., Lénat, J.-F. (2015). Time-dependent displacements during and after the April 2007 eruption of Piton de la Fournaise, revealed by interferometric data. *Journal of Volcanology and Geothermal Research*, 296, 55–68.

Froger, J.-L., Pinel, V., Bato, M.-G., Tridon, M., Smittarello, D., Prival, J.-M., Hrysiewicz, A., Cayol, V., Guehenneux, Y. (2020). Apport des données sar à la compréhension et à la surveillance des volcans : exemple du piton de la fournaise. *Revue française de photogrammétrie et de télédétection*, (219–220), 83–90.

Fukushima, Y., Cayol, V., Durand, P., Massonnet, D. (2010). Evolution of magma conduits during the 1998–2000 eruptions of Piton de la Fournaise volcano, Réunion Island. *Journal of Geophysical Research: Solid Earth*, 115(B10).

Furtney, M.A., Pritchard, M.E., Biggs, J., Carn, S.A., Ebmeier, S.K., Jay, J.A., Kilbride, B.T., Reath, K.A. (2018). Synthesizing multi-sensor, multi-satellite, multi-decadal datasets for global volcano monitoring. *Journal of Volcanology and Geothermal Research*, 365, 38–56.

Gaddes, M.E., Hooper, A., Bagnardi, M., Inman, H., Albino, F. (2018). Blind signal separation methods for InSAR: The potential to automatically detect and monitor signals of volcanic deformation. *Journal of Geophysical Research: Solid Earth*, 123(11), 10226–10251.

Gaddes, M.E., Hooper, A., Bagnardi, M. (2019). Using machine learning to automatically detect volcanic unrest in a time series of interferograms. *Journal of Geophysical Research: Solid Earth*, 124(11), 12304–12322.

Galetto, F., Hooper, A., Bagnardi, M., Acocella, V. (2020). The 2008 eruptive unrest at Cerro Azul Volcano (Galápagos) revealed by InSAR data and a novel method for geodetic modelling. *Journal of Geophysical Research: Solid Earth*, 125(2), e2019JB018521.

Galland, O., Bertelsen, H.S., Guldstrand, F., Girod, L., Johannessen, R.F., Bjugger, F., Burchardt, S., Mair, K. (2016). Application of open-source photogrammetric software MicMac for monitoring surface deformation in laboratory models. *Journal of Geophysical Research: Solid Earth*, 121(4), 2852–2872.

Grandin, R., Socquet, A., Binet, R., Klinger, Y., Jacques, E., de Chabalier, J.-B., King, G.C.P., Lasserre, C., Tait, S., Tapponnier, P., Delorme, A., Pinzuti, P. (2009). September 2005 Manda Hararo-Dabbahu rifting event, Afar (Ethiopia): Constraints provided by geodetic data. *Journal of Geophysical Research: Solid Earth*, 114.

Gray, A.L., Mattar, K.E., Sofko, G. (2000). Influence of ionospheric electron density fluctuations on satellite radar interferometry. *Geophysical Research Letters*, 27(10), 1451–1454.

Hanssen, R.F. (2001). *Radar Interferometry – Data Interpretation and Error Analysis*, 1st edition. Kluwer Academic Publishers, Dordrecht.

Harris, A., Chevrel, M., Coppola, D., Ramsey, M., Hrysiewicz, A., Thivet, S., Villeneuve, N., Favalli, M., Peltier, A., Kowalski, P., Muro, A.D., Froger, J., Gurioli, L. (2019). Validation of an integrated satellite-data-driven response to an effusive crisis: The April–May 2018 eruption of Piton de la Fournaise. *Annals of Geophysics*, 61(0), 48.

Hollingsworth, J., Leprince, S., Ayoub, F., Avouac, J.-P. (2012). Deformation during the 1975–1984 Krafla rifting crisis, NE Iceland, measured from historical optical imagery. *Journal of Geophysical Research: Solid Earth*, 117(B11).

Hooper, A., Bekaert, D., Spaans, K., Arikan, M. (2012). Recent advances in SAR interferometry time series analysis for measuring crustal deformation. *Tectonophysics*, 514–517, 1–13.

Kilbride, B.M., Edmonds, M., Biggs, J. (2016). Observing eruptions of gas-rich compressible magmas from space. *Nature Communications*, 7, 13744 EP.

Kubanek, J., Richardson, J.A., Charbonnier, S.J., Connor, L.J. (2015a). Lava flow mapping and volume calculations for the 2012–2013 Tolbachik, Kamchatka, fissure eruption using bistatic TanDEM-X InSAR. *Bulletin of Volcanology*, 77, 106.

Kubanek, J., Westerhaus, M., Schenk, A., Aisyah, N., Sri Brotopuspito, K., Heck, B. (2015b). Volumetric change quantification of the 2010 Merapi eruption using TanDEM-X InSAR. *Remote Sensing of Environment*, 164, 16–25.

Kubanek, J., Poland, M., Biggs, J. (2021). Applications of bistatic radar to volcano topography – A review of 10 years of TanDEM-X. *IEEE Journal of Selected Topics in Applied Earth Observations and Remote Sensing*, 14, 3282–3302.

Lazecky, M., Hatton, E., Gonzalez, P.J., Hlavacova, I., Jirankova, E., Dvorak, F., Sustr, Z., Martinovic, J. (2020). Displacements monitoring over Czechia by IT4S1 system for automatised interferometric measurements using Sentinel-1 data. *Remote Sensing*, 12(18).

Li, Z., Fielding, E.J., Cross, P., Muller, J.-P. (2006). Interferometric synthetic aperture radar atmospheric correction: GPS topography-dependent turbulence model. *Journal of Geophysical Research: Solid Earth*, 111, B02404.

Lin, Y.-N., Kositsky, A.P., Avouac, J.-P. (2010). PCAIM joint inversion of InSAR and ground-based geodetic time series: Application to monitoring magmatic inflation beneath the Long Valley Caldera. *Geophysical Research Letters*, 37(23), L23301.

Liu, Y.-H., Ruch, J., Bathke, H., Jonsson, S. (2019). Influence of ring faulting in localizing surface deformation at subsiding calderas. *Earth and Planetary Science Letters*, 526, 115784.

Ma, L., Liu, Y., Zhang, X., Ye, Y., Yin, G., Johnson, B.A. (2019). Deep learning in remote sensing applications: A meta-analysis and review. *ISPRS Journal of Photogrammetry and Remote Sensing*, 152, 166–177.

Massonnet, D. and Souyris, J.C. (2008). *Imaging with Synthetic Aperture Radar*. CRC Press, Boca Raton.

Massonnet, D., Briole, P., Arnaud, A. (1995). Deflation of Mount Etna monitored by spaceborne radar interferometry. *Nature*, 375, 567–570.

Meyer, F.J., Whitley, M., Logan, T., McAlpin, D.B., Hogenson, K., Nicoll, J.B. (2019). The sarviews project: Automated processing of Sentinel-1 SAR data for geoscience and hazard response. *IEEE International Geoscience and Remote Sensing Symposium*, 5468–5471.

Michel, R., Avouac, J.-P., Taboury, J. (1999). Measuring ground displacements from SAR amplitude images: Application to the Landers earthquake. *Geophysical Research Letters*, 26, 875–878.

Morishita, Y., Lazecky, M., Wright, T.J., Weiss, J.R., Elliott, J.R., Hooper, A. (2020). LiCSBAS: An open-source InSAR time series analysis package integrated with the liCSAR automated Sentinel-1 InSAR processor. *Remote Sensing*, 12(3), 424.

Newhall, G.C. and Self, S. (1982). The volcanic explosivity index (VEI) an estimate of explosive magnitude for historical volcanism. *Journal of Geophysical Research: Oceans*, 87(C2), 1231–1238.

Page, E.S. (1954). Continuous inspection schemes. *Biometrika*, 41, 100–115.

Peltier, A., Staudacher, T., Catherine, P., Ricard, L.-P., Kowalski, P., Bachélery, P. (2006). Subtle precursors of volcanic eruptions at Piton de la Fournaise detected by extensometers. *Geophysical Research Letters*, 33, L06315.

Peltier, A., Villeneuve, N., Ferrazzini, V., Testud, S., Hassen, A.T., Boissier, P., Catherine, P. (2018). Changes in the long-term geophysical eruptive precursors at Piton de la Fournaise: Implications for the response management. *Frontiers in Earth Science*, 6(22), 104.

Pinel, V., Poland, M., Hooper, A. (2014). Volcanology: Lessons learned from synthetic aperture radar imagery. *Journal of Volcanology and Geothermal Research*, 289, 81–113.

Poland, M.P. (2014). Time-averaged discharge rate of subaerial lava at Kīlauea Volcano, Hawai'i, measured from TanDEM-X interferometry: Implications for magma supply and storage during 2011–2013. *Journal of Geophysical Research: Solid Earth*, 119(7), 5464–5481.

Pritchard, M.E. and Simons, M. (2002). A satellite geodetic survey of large-scale deformation of volcanic centres in the central Andes. *Nature*, 418, 167–171.

Pritchard, M.E., Biggs, J., Wauthier, C., Sansosti, E., Arnold, D.W.D., Delgado, F., Ebmeier, S.K., Henderson, S.T., Stephens, K., Cooper, C., Wnuk, K., Amelung, F., Aguilar, V., Mothes, P., Macedo, O., Lara, L.E., Poland, M.P., Zoffoli, S. (2018). Towards coordinated regional multi-satellite InSAR volcano observations: Results from the Latin America pilot project. *Journal of Applied Volcanology*, 7(1), 5.

Pyle, D.M., Mather, T.A., Biggs, J. (2013). Remote sensing of volcanoes and volcanic processes: Integrating observation and modelling – Introduction. In *Remote Sensing of Volcanoes and Volcanic Processes: Integrating Observation and Modelling*, Pyle, D.M., Mather, T.A., Biggs, J. (eds). Geological Society, London.

Reath, K., Pritchard, M., Poland, M., Delgado, F., Carn, S., Coppola, D., Andrews, B., Ebmeier, S.K., Rumpf, E., Henderson, S., Baker, S., Lundgren, P., Wright, R., Biggs, J., Lopez, T., Wauthier, C., Moruzzi, S., Alcott, A., Wessels, R., Griswold, J., Ogburn, S., Loughlin, S., Meyer, F., Vaughan, G., Bagnardi, M. (2019). Thermal, deformation, and degassing remote sensing time series (CE 2000-2017) at the 47 most active volcanoes in Latin America: Implications for volcanic systems. *Journal of Geophysical Research: Solid Earth*, 124(1).

Reath, K., Pritchard, M., Biggs, J., Andrews, B., Ebmeier, S.K., Bagnardi, M., Girona, T., Lundgren, P., Lopez, T., Poland, M. (2020). Using conceptual models to relate multiparameter satellite data to subsurface volcanic processes in Latin America. *Geochemistry, Geophysics, Geosystems*, 21(1), e2019GC008494.

Remy, D., Bonvalot, S., Briole, P., Murakami, M. (2003). Accurate measurements of tropospheric effects in volcanic areas from SAR interferometry data: Application to Sakurajima volcano (Japan). *Earth and Planetary Science Letters*, 213, 299–310.

Remy, D., Froger, J.-L., Perfettini, H., Bonvalot, S., Gabalda, G., Albino, F., Cayol, V., Legrand, D., Saint Blanquat, M.D. (2014). Persistent uplift of the Lazufre volcanic complex (Central Andes): New insights from PCAIM inversion of InSAR time series and GPS data. *Geochemistry, Geophysics, Geosystems*, 15(9), 3591–3611.

Richter, N. and Froger, J.-L. (2020). The role of interferometric synthetic aperture radar in detecting, mapping, monitoring, and modelling the volcanic activity of Piton de la Fournaise, La Réunion: A review. *Remote Sensing*, 12(6).

Ruch, J., Wang, T., Xu, W., Hensch, M., Jónsson, S. (2016). Oblique rift opening revealed by reoccurring magma injection in Central Iceland. *Nature Communications*, 7(1), 12352.

Shreve, T., Grandin, R., Boichu, M., Garaebiti, E., Moussallam, Y., Ballu, V., Delgado, F., Leclerc, F., Vallée, M., Henriot, N., Cevuard, S., Tari, D., Lebellegard, P., Pelletier, B. (2019). From prodigious volcanic degassing to caldera subsidence and quiescence at Ambrym (Vanuatu): The influence of regional tectonics. *Scientific Reports*, 9(1), 18868.

Sigmundsson, F., Hooper, A., Hreinsdottir, S., Vogfjord, K.S., Ofeigsson, B.G., Heimisson, E.R., Dumont, S., Parks, M., Spaans, K., Gudmundsson, G.B., Drouin, V., Arnadottir, T., Jonsdottir, K., Gudmundsson, M.T., Hognadottir, T., Fridriksdottir, H.M., Hensch, M., Einarsson, P., Magnusson, E., Samsonov, S., Brandsdottir, B., White, R.S., Agustsdottir, T., Greenfield, T., Green, R.G., Hjartardottir, A.R., Pedersen, R., Bennett, R.A., Geirsson, H., La Femina, P.C., Bjornsson, H., Palsson, F., Sturkell, E., Bean, C.J., Mollhoff, M., Braiden, A.K., Eibl, E.P.S. (2015). Segmented lateral dyke growth in a rifting event at Bárðarbunga volcanic system, Iceland. *Nature*, 517(7533), 191–195.

Smittarello, D., Cayol, V., Pinel, V., Froger, J.-L., Peltier, A., Dumont, Q. (2019a). Combining InSAR and GNSS to track magma transport at Basaltic volcanoes. *Remote Sensing*, 11(19), 2236.

Smittarello, D., Cayol, V., Pinel, V., Peltier, A., Froger, J.-L., Ferrazzini, V. (2019b). Magma propagation at Piton de la Fournaise from joint inversion of InSAR and GNSS. *Journal of Geophysical Research: Solid Earth*, 124(2), 1361–1387.

Sparks, R.S.J., Biggs, J., Neuberg, J.W. (2012). Monitoring volcanoes. *Science*, 335, 1310–1311.

Surono, M., Jousset, P., Pallister, J., Boichu, M., Buongiorno, M.F., Budisantoso, A., Costa, F., Andreastuti, S., Prata, F., Schneider, D., Clarisse, L., Humaida, H., Sumarti, S., Bignami, C., Griswold, J., Carn, S., Oppenheimer, C., Lavigne, F. (2012). The 2010 explosive eruption of Java's Merapi volcano – A "100-year" event. *Journal of Volcanology and Geothermal Research*, 241–242, 121–135.

Syahbana, D.K., Kasbani, K., Suantika, G., Prambada, O., Andreas, A.S., Saing, U.B., Kunrat, S.L., Andreastuti, S., Martanto, M., Kriswati, E., Suparman, Y., Humaida, H., Ogburn, S., Kelly, P.J., Wellik, J., Wright, H.M.N., Pesicek, J.D., Wessels, R., Kern, C., Lisowski, M., Diefenbach, A., Poland, M., Beauducel, F., Pallister, J., Vaughan, R.G., Lowenstern, J.B. (2019). The 2017–19 activity at Mount Agung in Bali (Indonesia): Intense unrest, monitoring, crisis response, evacuation, and eruption. *Scientific Reports*, 9(1), 8848.

Syamsidik, Benazir, Luthfi, M., Suppasri, A., Comfort, L.K. (2020). The 22 December 2018 Mount Anak Krakatau volcanogenic tsunami on Sunda Strait coasts, Indonesia: Tsunami and damage characteristics. *Natural Hazards and Earth System Sciences*, 20(2), 549–565.

Vadon, H. and Sigmundsson, F. (1997). Crustal deformation from 1992 to 1995 at the Mid-Atlantic ridge, Southwest Iceland, mapped by satellite radar interferometry. *Science*, 275(5297), 194–197.

Voight, B. (1990). The 1985 Nevado del Ruiz volcano catastrophe: Anatomy and retrospection. *Journal of Volcanology and Geothermal Research*, 42(1), 151–188.

Voight, B., Sparks, R.S.J., Miller, A.D., Stewart, R.C., Hoblitt, R.P., Clarke, A., Ewart, J., Aspinall, W.P., Baptie, B., Calder, E.S., Cole, P., Druitt, H., Hartford, C., Herd, R.A., Jackson, P., Lejeune, A.M., Lockhart, A.B., Loughlin, S.C., Luckett, R., Lynch, L., Norton, G.E., Robertson, R., Watson, I.M., Watts, R., Young, S.R. (1999). Magma flow instability and cyclic activity at Soufriere Hills Volcano, Montserrat, British West Indies. *Science*, 283, 1138–1142.

Watts, R.B., Herd, R.A., Sparks, R.S.J., Young, S.R. (2002). Growth patterns and emplacement of the andesitic lava dome at Soufriére Hills volcano, Montserrat. In *The Eruption of the Soufriére Hills Volcano, Montserrat, from 1995 to 1999*, Druitt, T.H. and Kolelaar, B.P. (eds). Geological Society, London.

Wauthier, C., Cayol, V., Poland, M., Kervyn, F., D'Oreye, N., Hooper, A., Samsonov, S., Tiampo, K., Smets, B. (2013). Nyamuragira's magma plumbing system inferred from 15 years of InSAR. In *Remote Sensing of Volcanoes and Volcanic Processes: Integrating Observation and Modelling*, Pyle, D.M., Mather, T.A., Biggs, J. (eds). Geological Society, London.

Wittmann, W., Sigmundsson, F., Dumont, S., Lavallée, Y. (2017). Post-emplacement cooling and contraction of lava flows: InSAR observations and a thermal model for lava fields at Hekla volcano, Iceland. *Journal of Geophysical Research: Solid Earth*, 122(2), 946–965.

Wright, T., Ebinger, C., Biggs, J., Ayele, A., Yirgu, G., Keir, D., Stork, A. (2006). Magma-maintained rift segmentation at continental rupture in the 2005 Afar dyking episode. *Nature*, 442.

Yan, Y. (2011). Fusion de mesures de déplacement issues d'imagerie SAR : application aux modélisations séismo-volcaniques. PhD thesis, University of Grenoble, Grenoble.

Yu, C., Li, Z., Penna, N.T., Crippa, P. (2018). Generic atmospheric correction model for interferometric synthetic aperture radar observations. *Journal of Geophysical Research: Solid Earth*, 123(10), 9202–9222.

Zebker, H.A., Rosen, P.A., Hensley, S. (1997). Atmospheric effects in interferometric synthetic aperture radar surface deformation and topographic maps. *Journal of Geophysical Research*, 102, 7547–7563.

9

Anthropogenic Activity: Monitoring Surface-Motion Consequences of Human Activities with Spaceborne InSAR

Bénédicte FRUNEAU[1], Dinh HO TONG MINH[2]
and Daniel RAUCOULES[3]

[1] *LaSTIG, Gustave Eiffel University, Champs-sur-Marne, France*
[2] *UMR-TETIS, INRAE, France*
[3] *BRGM, Orléans, France*

9.1. Introduction

Human activities ranging from water pumping for drinking water or irrigating purposes, to oil and gas extraction from buried reservoirs or underground structures often result in depletion or uplift of the ground. Subsurface anthropogenic activity can also amplify natural subsidence and can even trigger significant earthquakes (Grandin *et al.* 2017). Human-related deformation has been widely observed and covers a broad range of scales, both spatial and temporal: it includes diffuse and low ground subsidence (i.e. mainly vertical downward displacement) covering large areas, as well as fast and punctual subsidence, depending on the involved phenomena. Monitoring the spatial extent and temporal evolution of surface deformations is critical

Surface Displacement Measurement from Remote Sensing Images,
coordinated by Olivier CAVALIÉ and Emmanuel TROUVÉ.
© ISTE Ltd 2022.

for the mitigation of related hazards, since they may affect populations, buildings and infrastructures. This is especially a major issue in urban areas, and more so in densely populated ones such as megacities, but also in coastal areas where subsidence can increase vulnerability to flooding and impact sea level rise (Chaussard *et al.* 2013). Several cities in the world are facing significant land subsidence due to groundwater over-extraction. In addition to threatening infrastructures, deformation can have important consequences on clean water access and on economic sustainability. The monitoring of human-induced surface deformation is sometimes mandatory and imposed by regulations. Land surface deformation is usually measured with ground-based methods, such as leveling, tiltmeters, inclinometers, total stations and extensometers, often complemented with GNSS measurements. These techniques, which are well recognized by authorities, are characterized by high precision but they are pointwise, i.e. they can only measure deformation at a limited number of discrete points and a limited spatial extent. In addition, they require ground-based instrumentation and thus only give access to measurement after installation. In contrast, remotely sensed observations do not depend on *in situ* deployment and are essential for mapping large regions. In particular, the use of satellite-based synthetic aperture radar (SAR) imagery processed by interferometry (InSAR) allows the measurement of large-scale land motion with unprecedented spatial resolution from space. Moreover, the current data archive – covering more than 25 years – allows the analysis of non-instrumented past motions. InSAR, as a space-based geodetic tool, thus provides interesting imaging capability for the mapping and monitoring of land motion in response to anthropogenic activities. In the 1990s, InSAR was first used to complete ground-based monitoring, providing access to a synoptic view of deformation. Over the years, it has evolved, improving its capability to map, monitor and analyze deformations, and provided new insights into deformation and physical-related processes, in particular through modeling. Thanks to the development of advanced time-series interferometric methods (see Chapter 5), especially since the early 2000s with persistent scatterer interferometry, InSAR has become much more precise. It is nowadays close to being systematically used in addition to *in situ* instrumentation as an operational tool for low-cost, high-density and large-coverage surface deformation monitoring.

9.2. Characteristics of subsidence/uplift phenomena associated with human activity

9.2.1. *Underground activities*

Underground construction works, such as tunneling projects and construction of new underground infrastructures, can lead to possible deformations at the surface and potential damage to buildings and infrastructures. More generally, ground deformation related to human activities can be the direct result of excavation (creating underground voids, i.e. cavities) or a consequence of the compaction of underground materials due to fluid extraction. In the case of underground construction works such as tunneling projects, water pumping is often required to enable the construction process in dry conditions. This water pumping reduces the piezometric head and induces ground subsidence, thus leading to possible damage in surrounding structures and buildings.

Groundwater extraction for industrial, agricultural or domestic purposes can also trigger land subsidence (Motagh *et al.* 2008). Oil and gas production, another activity aimed at underground fluid extraction, result in similar (in terms of size and motion rate) surface motion phenomena as water pumping. In both cases, the upper crust above the reservoir or aquifer is compacted due to the reduction in pressure caused by the removal of fluid, resulting in surface damage (e.g. to buildings or infrastructures required for the oil/gas extraction). Thus, there is a strong need to understand the dynamics and extent of subsidence, including for the requirements of exploitation itself (e.g. for the safety of the extraction wells). It is worth noting that, in contrast to other phenomena presented in this book, the evolution of deformation (and thus the consequences of the motion) can be modified by human action. Moreover, gas storage is specific to the production activity from the point of view of the surface motion's response. Due to consumption needs, the resulting motion has seasonal periodicity: gas is mostly injected (resulting in uplift) in summer for storage and extracted (resulting in subsidence) in winter for consumption. In addition to damage to infrastructures and buildings, land subsidence may also have dramatic consequences, as it may increase the risk of flooding in coastal areas (Tung and Hu 2012). Ground subsidence depends on the compressibility of the aquifer and can be partly recovered if the underground water level rises again.

Urban subsidence due to underground activities can constitute a major hazard. In many cities, subsidence-related risk has become increasingly common, particularly Mexico, Bangkok, Bandung and Hanoi (Aobpaet *et al.* 2013). Subsidence resulting from extensive groundwater extraction is a major concern for these cities facing rapidly developing urbanization and an increasing water demand. Planning responses to ground subsidence requires knowledge about its spatial extent and its evolution over time. For example, Castellazzi *et al.* (2016) have shown that land subsidence in Querétaro City (Central Mexico) of up to 8 cm/year has been almost entirely mitigated by major changes in the water management practices of the city.

9.2.2. *Cavities*

Subsidence resulting from artificial cavities has plagued the extractive industries (mines and quarries) for a long time. Particularly, underground mining (including dissolution mining for salt and potash) and quarrying activities produce caverns (empty or filled with water or brine) from where material is extracted. Noteworthy is the fact that before the Industrial Age, material for construction – when these materials were locally available – were extracted directly below the urban centers. This has led to the presence of quarrying cavities underground the historical centers of numerous European cities, inducing a risk of cavity collapse. In some cases, land subsidence is the consequence of the collapse of underground cavities weakened by the dissolution of limestone or other materials (e.g. pillars in underground mines) by groundwater. These phenomena weaken the support structure for rocks surrounding pumping wells or created voids and eventually cause them to collapse. These activities hinder the ability to extract underground resources and damage the ground infrastructure, affecting the water management of large land areas. Surface deformations associated with extraction are generally monitored by operating companies for safety reasons (to protect infrastructures and surroundings and anticipate potential issues) and also to adapt the production means in order to minimize the surface consequences.

9.2.3. *Suitability of the interferometric SAR techniques*

The conventional spaceborne InSAR takes advantage of two SAR acquisitions to obtain the interferometric phase, which contains information on any potential surface displacement occurring between the acquisitions. This interferometric phase is composed of different components related to deformation, residual topography and atmosphere and is hampered by decorrelation noise. The goal is to estimate the deformation phase from the interferometric phase. The biggest challenge is certainly the fact that the interferometric phase can only be measured as a wrapped phase (see Chapter 6). InSAR has several issues relative to spatial and temporal decorrelations, as well as the atmosphere, which cannot be efficiently eliminated, resulting in interferograms not representing the ground deformation perfectly. This deficiency can be overcome by a specific analysis considering phase changes in a series of SAR images acquired at different times over the same region. The rationale of the InSAR time series is to combine multiple flight tracks to minimize the different issues: InSAR time-series analysis is a powerful SAR processing technique that makes it possible to track millimeter-scale surface deformation over long periods, some years typically (see Chapter 5). Two different approaches are generally used. The first approach is carried out from the wrapped phase, such as permanent/persistent scatterer interferometry (PSI) techniques, which use so-called point-like targets for deformation analysis. In the second approach, the unwrapped phase is required, as in small baseline (SB) techniques that exploit distributed targets (DSs) for deformation measurements.

In urban areas, PSI techniques are very powerful and, in general, outperform SB approaches. If the density of these point targets is high enough, it is possible to measure surface deformations with millimeter-scale precision. A drawback is that the computational effort required for processing point-like targets is high. However, for non-urban applications (e.g. volcanoes, landslides and permafrost), SB techniques, which use distributed targets, are more efficient. More details on these techniques can be found in Chapter 5 and in the study by Ho Tong Minh *et al.* (2020).

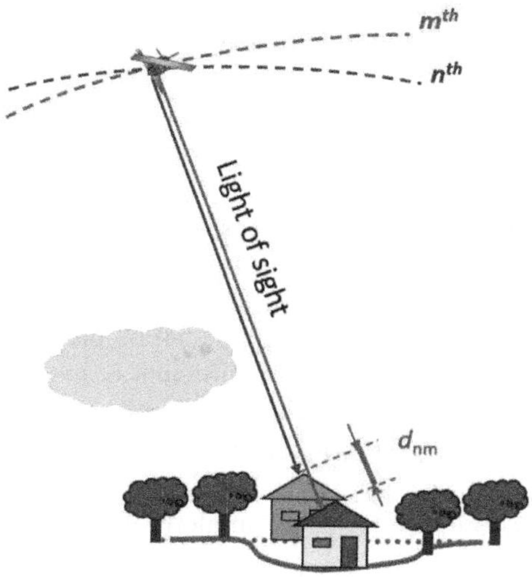

Figure 9.1. *An example of target motion captured using the InSAR technique. For a color version of this figure, see www.iste.co.uk/cavalie/images.zip*

In Figure 9.1, an example of target motion between the n-th and m-th acquisition times is shown. Once the displacement phase ϕ_{dis}^{nm} is estimated, its phase value is transformed into a metric as follows:

$$d_{nm} = \frac{\lambda}{4\pi}\phi_{dis}^{nm} = vt_{nm} \qquad [9.1]$$

where d_{nm} is the light-of-sight (LOS) distance change of the target[1], v is the constant velocity and t_{nm} is the temporal baseline between the n-th and m-th images.

The development of advanced multi-temporal interferometric techniques (PSI, PSDS and SBAS; see Chapter 5) and improved processing algorithms, combined with increasing accessibility of SAR images (including free and open access to the data acquired by the European Sentinel-1) and sensors with shorter revisit times and/or higher resolution, has greatly improved the monitoring of subsidence related to anthropogenic activities.

1. Note that InSAR measurements give access only to the LOS component of the displacement, whereas land subsidence or uplift, mainly vertical displacement, may include a horizontal component.

9.2.4. Optimum SAR data properties for monitoring anthropogenic activities

Due to the expected characteristics (geometry/size and motion rate) of the ground surface deformation resulting from human activities, some parameters of the data to be used can be emphasized.

9.2.4.1. *Satellite repeat cycle*

Typical anthropogenic surface deformation velocity can range from mm/year (e.g. residual post-mining motions) to m/year (e.g. collapses induced by underground activities). Slow motions are better assessed by relying on long time-spans to ensure that enough deformation occurs within the revisit time. PSI techniques, working on points little affected by temporal decorrelation, can be particularly efficient for long time-spans in areas where PS densities are sufficient (e.g. urban areas). In contrast, constraining faster motions requires shorter repeat cycles in order to avoid unwrapping errors. The shortest revisit time currently available for spaceborne SAR is less than a week (e.g. Sentinel-1, COSMO SkyMed). Despite the quick access to data and fast processing, InSAR still cannot be used as an alert system for real-time detection of abnormal movement.

9.2.4.2. *Spatial resolution*

The spatial extent of subsidence/uplift related to human activities ranges from a few tens of square meters for superficial collapses due to underground work to hundreds of square kilometers for motion related to water table changes (induced by pumping). Accordingly, depending on the context and objectives, the appropriate resolution can range from the meter scale to ~100 m. With high-resolution data, such as TerraSAR-X and COSMO SkyMed data, it is now even possible to measure through PSI the gradient of displacement within structures such as bridges or buildings (Lazecky *et al.* 2017; Weissgerber *et al.* 2017).

9.2.4.3. *Archive continuity*

Some phenomena resulting from human activity (such as post-mining deformation) can last for decades. Data availability over a long period is therefore required. Archived data (e.g. ERS-1/2 SAR, Envisat ASAR and ALOS-1 PalSAR data) are therefore of great interest for the calculation of the evolution of deformation over time. Moreover, since SAR data have been available since 1991, they can be exploited for retro-analysis of a phenomenon, even prior to the installation of ground instruments. Strategic planning of future missions should follow the Copernicus Sentinel-1 initiative, which guarantees continuity of free and open SAR data for the next decade through the development of a SAR satellite constellation.

9.2.4.4. *Radar wavelength*

The capabilities of radar sensors differ depending on the land-cover type (vegetated vs. urban). Sensors with longer wavelengths (e.g. L-band) are preferable in vegetated areas as they allow for greater penetration through the vegetation. Extractive (e.g. gas/oil production, mining) activities are often located in non- or partially urbanized areas. L-band SAR sensors (as provided by ALOS-1/2 and the future NISAR mission) are the most valuable. Similarly, L-band SAR sensors enable capture of rapid deformation as – due to the longer wavelength – L-band InSAR accepts larger motions between acquisitions than C-band (see e.g. Plattner *et al.* (2010) or Raucoules *et al.* (2007)).

9.3. Application examples: urban underground activities

9.3.1. *Urban underground construction works*

In urban areas, underground constructions require continuous monitoring to anticipate unexpected events and mitigate related risks.

In the late 1990s, InSAR was successfully used to map surface deformation related to construction works. The first examples of the application of SAR interferometry in urban areas were based on conventional InSAR, and it has shown its efficiency in the urban context, where coherence loss is not critical. There was, for example, the case of the Saint-Lazare area in Paris (France), which underwent important water pumping for the construction of the underground Haussmann–Saint-Lazare station for the Eole subway line in 1993. InSAR allowed the mapping of a large subsiding area near the station related to the water pumping (Fruneau and Sarti 2000). Temporal variations of the vertical displacements monitored in this area enabled the seperation of the different phases of the construction works – namely, subsidence, stability and uplift – linked to the pumping activity (Le Mouélic *et al.* 2002; Fruneau *et al.* 2005) (Figure 9.2). It was also shown that InSAR offered a better estimate of the spatial extension of the deforming area, which had been underestimated with leveling.

In Naples (Italy), where intensive construction of an underground railway was conducted between 1992 and 1995, deformations observed by InSAR were correlated with the excavation of the galleries (Tesauro *et al.* 2000). The use of advanced InSAR techniques, in particular PSI, has been generalized for underground construction monitoring. There are many examples of its application in tunneling projects. In the city of Bologna, in the framework of the high-speed Milan–Naples railway, the large construction project included a double-track tunnel crossing an urban area at shallow depths of about 10 m, with a high density of residential buildings and commercial activities. In this context, InSAR study made it possible to measure displacements in the order of cm/year (Figure 9.3) and also to isolate displacement induced by the excavation work from the natural subsidence induced by natural sediment compaction and groundwater exploitation.

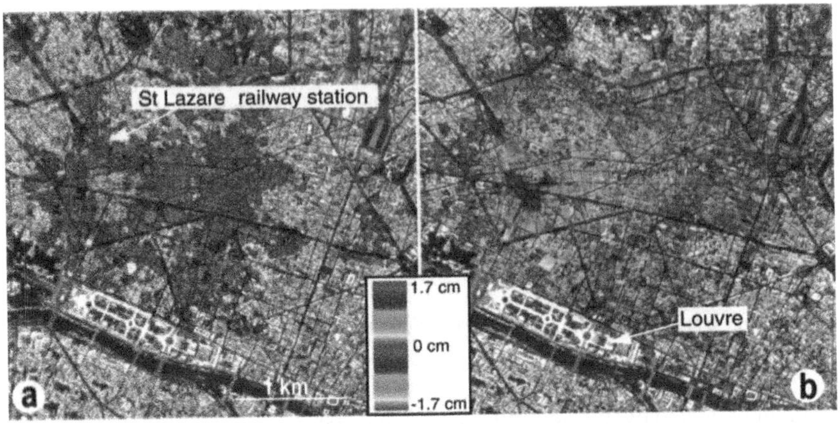

Figure 9.2. *Deformation measured by InSAR with ERS images in the Saint-Lazare area (Paris) during the construction of the Haussmann–Saint-Lazare station (RER Eole). (a) Map of cumulate subsidence due to the start of water pumping from May 1994 to April 1995. (b) Map of cumulate uplift due to the end of water pumping and the slow recharge of the nappe (from June to October 1998) (Fruneau* et al. *(2005)). For a color version of this figure, see www.iste.co.uk/cavalie/images.zip*

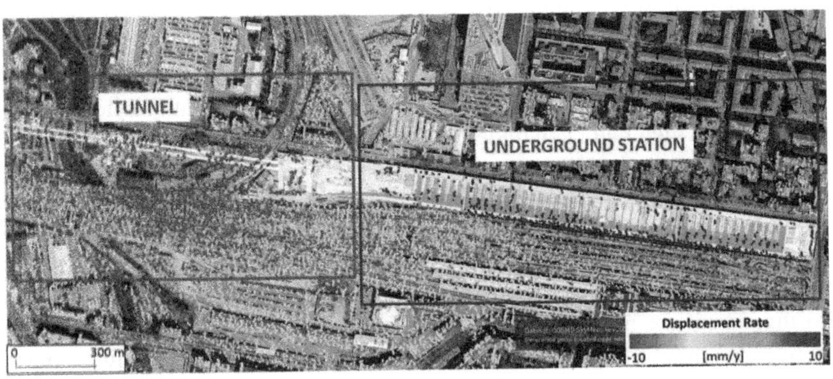

Figure 9.3. *Bologna high-speed railway site. Average displacement rate measured during the post-construction period (2009–2011; courtesy of TRE ALTAMIRA). Some subsidence (on the left) is still evident over the tunnel advance section that was only completed in early 2010, while uplift on the east side is related to the rebound effect on the completed underground station (MacDonald* et al. *(2015)). For a color version of this figure, see www.iste.co.uk/cavalie/images.zip*

Many cities around the world have exploited PSI for their tunneling projects: Madrid (Garcia *et al.* 2019); Lyon, for the construction of TEO urban highway (Thierry *et al.* 2005); Rennes, for the construction of subway line B (Nahli *et al.* 2018); the Vancouver metropolitan area, for the Canada Line LRT (Falorni and Iannacone 2014); Shanghai (Perissin *et al.* 2012); Beijing (Ge *et al.* 2016); and Paris, with the Grand Paris Express (Koudogbo *et al.* 2018). These measurements show the relationship between tunneling and surface settlements (Barla *et al.* 2016).

InSAR can be applied during all phases of tunneling projects: before, to identify pre-existing ground movements; during the work; and after the completion of the work for residual deformation monitoring. It allows coverage wider than the surroundings, in which most *in situ* instruments are deployed.

A recent study underlines the importance of using InSAR prior to tunneling. In east London, patterns of deformation during the construction of the Lee Tunnel were observed using InSAR with Radarsat-2 images in the 2010–2015 period, in particular subsidence at and around Newton hospital. This could be interpreted as a consequence of an unexpected geological feature discovered during tunneling. Thanks to an InSAR analysis of the period before tunneling (with 1999–2010 ERS and Envisat data), unusual and long-term deformation could be noted. This shows the importance of InSAR analysis prior to tunneling work (Scoular *et al.* 2020).

9.3.2. *Urban groundwater extraction*

InSAR is a valuable tool for monitoring surface displacements in response to aquifer deformation, both subsidence and uplift, accompanying groundwater discharge and recharge. InSAR also plays a growing role in evaluating water-management practices and in implementing appropriate water-management strategies that may include subsidence mitigation.

This has been shown from the first examples of InSAR applications, which involved mapping substantial and large-scale subsidence resulting from overdrafting of water resources in Antelope Valley, California (Galloway *et al.* 1998); Mexico City (Strozzi and Wegmuller 1999); and Las Vegas (Amelung *et al.* 1999). It has been proven that InSAR can measure the spatial extension and the magnitude of surface deformation associated with aquifer-system compaction. In Las Vegas Valley, where groundwater was pumped at a rate of two to three times more than the natural recharge and represented 25% of the water supply, detailed InSAR-derived ground deformation maps have shown that the extent of the subsidence was controlled by geological structures (Quaternary faults) and sediment composition (Figure 9.4). The temporal variability of surface displacements also highlighted seasonal subsidence and uplift associated with seasonal water decline and recovery.

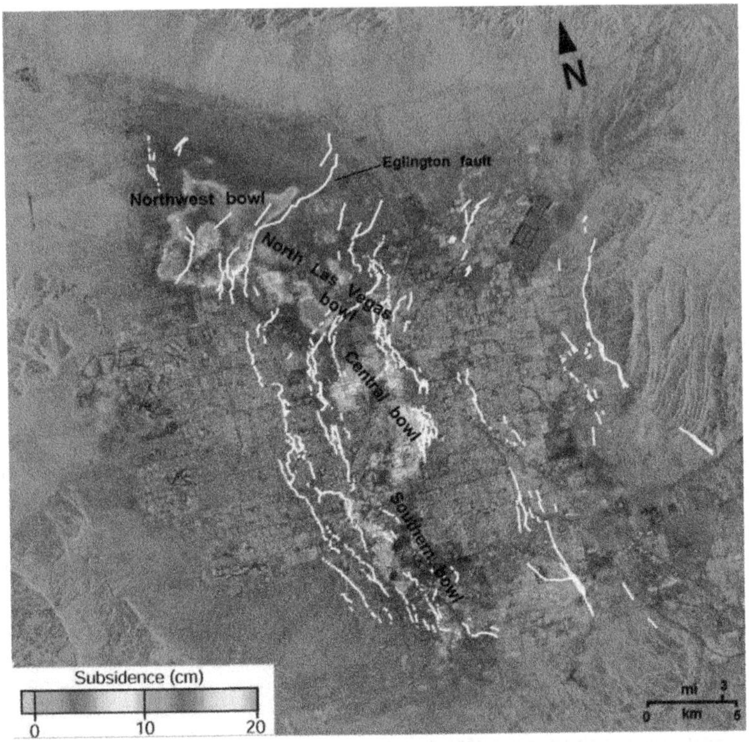

Figure 9.4. *A five-year interferogram (April 1992 to December 1997) shows the full extent of the aquifer system and 190 mm of localized subsidence in a distinct bowl in the northwest part of the valley. Faults are subsidence barriers (Amelung* et al. *(1999)). For a color version of this figure, see www.iste.co.uk/cavalie/images.zip*

In southeastern Iran, decades of extensive groundwater extraction in the Rafsanjan plain, which is used mainly for agricultural activities, have caused substantial land subsidence, monitored by InSAR, in an area of 1000 km^2, with subsidence larger than 5 cm/year and exceeding 20–30 cm/year during the 2000s, degrading large parts of the agricultural lands (Motagh *et al.* 2017). The average volume storage loss of the aquifer system has also been estimated to be 300 million m^3/year over the last decade. InSAR has greatly facilitated the monitoring of land subsidence in many metropolitan areas where vulnerability has continuously increased over the years, such as Las Vegas (Amelung *et al.* 1999; Hoffmann *et al.* 2001), Mexico City (Osmanoğlu *et al.* 2011; Yan *et al.* 2012; Chaussard *et al.* 2014), Venice (Bock *et al.* 2012) and Ho Chi Minh City (Ho Tong Minh *et al.* 2015).

In Central Mexico, as well as the well-known case of Mexico City, Castellazzi *et al.* (2016) addressed land subsidence problems in five other cities located within the Trans-Mexican Volcanic Belt, where compaction of an over-exploited aquifer causes significant structural damage to urban infrastructures with costly consequences. Water sewage structures must be constantly upgraded due to loss of gradient, and transitional areas between lacustrine beds and slope deposits are prone to severe differential subsidence, damaging housing and urban infrastructures.

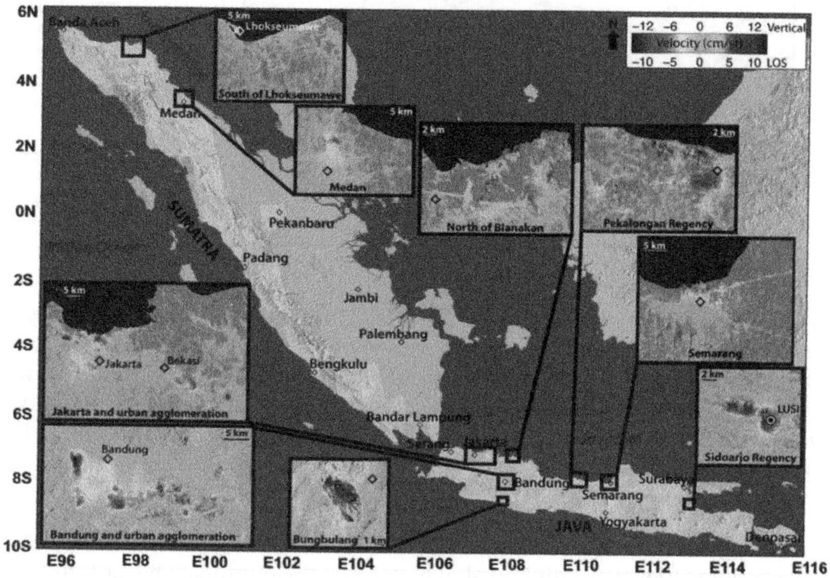

Figure 9.5. *Averaged 2006–2009 LOS velocity map of Sumatra, Java and Bali from ALOS InSAR time-series analysis. Insets: magnification of the subsiding areas. Major cities are shown by black diamonds and labeled for reference. The color scale shows red colors for negative LOS velocities (subsidence) and blue colors for positive LOS velocities (uplift). Vertical subsidence rates converted from the LOS rates are labeled on the color scale (Chaussard* et al. *2013). For a color version of this figure, see www.iste.co.uk/cavalie/images.zip*

This phenomenon is especially critical in coastal areas. In Western Indonesia, InSAR time-series analysis has revealed subsidence in nine locations, including six major cities, including Jakarta, the capital and largest city of Indonesia (Chaussard *et al.* 2013) (Figures 9.5 and 9.6). Subsidence rates up to 22 cm/year have been measured. Rapid subsidence observed in cities is due to industrial groundwater extraction, while groundwater extraction for agricultural use and gas field exploitation induce subsidence in other places. Land subsidence is a major concern in those areas, as it increases the risk of major flooding: they could be below relative sea level within decades (or less). Manilla in the Philippines is another example of a city presenting the same kind of significant deformation (>10 cm/year) (Raucoules *et al.* 2013).

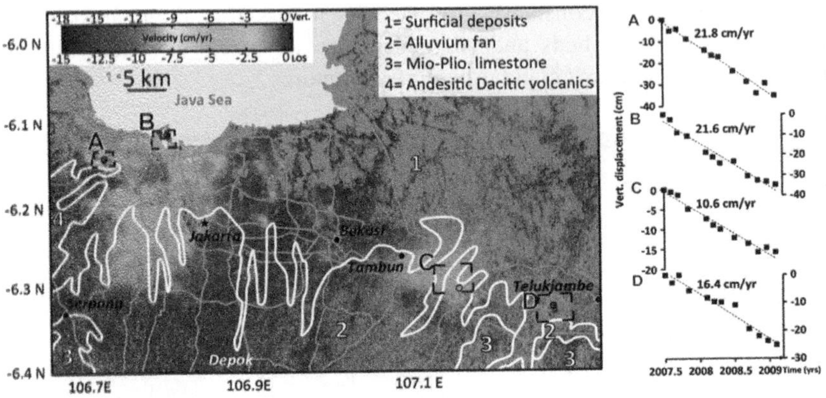

Figure 9.6. *Subsidence observed near Jakarta. a) Average 2007–2009 velocity map. The contours of the geological units are shown by white lines and referred to using white numbers. Black circles within the boxes show the locations of the pixels whose time series are shown in b). b) Vertical displacement time series and corresponding linear rates of subsidence (Chaussard* et al. *2013). For a color version of this figure, see www.iste.co.uk/cavalie/images.zip*

These studies underline the importance of continuous InSAR monitoring of subsidence related to water withdrawal, which can help to mitigate hazards associated with this phenomenon. It can help water management slow down the subsidence processes and reduce the risk of inundation. This is crucial, as permanent inundation of densely populated coastal areas is a major threat for the coming decades.

Numerous studies have shown that the precise monitoring of subsidence offered by InSAR provides key elements that enhance the hydrogeologic understanding on different points: it enables the identification of groundwater flow barriers, characterization of seasonal surface deformation and estimation of aquifer-system storage and it provides additional constraints for numerical groundwater flow models (Galloway and Hoffmann 2007). InSAR can help characterize aquifer system properties and also evaluate management practices (Hoffmann *et al.* 2001; Chaussard *et al.* 2014, 2017; Bell *et al.* 2008; Motagh *et al.* 2008). In particular, InSAR allows better characterization of the elastic and inelastic properties of aquifer systems (Hoffmann *et al.* 2001; Tomás *et al.* 2011; Tung *et al.* 2016). Chaussard and Farr (2019) have proposed a new method, based on an independent component analysis, for isolating elastic from inelastic deformations in aquifer systems. Separating recoverable from permanent deformation of aquifer systems is of prime interest for resource management since elastic properties are used to define sustainable pumping rates. In this context, InSAR can be seen as a valuable tool assisting in groundwater resources management.

In the following, two applications of InSAR are detailed: one study conducted on Mexico City using Sentinel-1 images with high temporal sampling and the other on Hanoi with high-resolution TerraSAR-X and COSMO SkyMed data.

9.3.2.1. *Mexico case study*

Mexico City is the capital of Mexico and the most populated city in North America, with more than 21 million people in 2020. The Mexico City metropolitan area consumes over 65 m^3/s of water and more than 70% of it comes from the aquifer beneath the city (Castellazzi *et al.* 2016). In a typical year, consumption exceeds recharge, lowering the water table by up to 1.5 m/year, reducing pore fluid pressure in the aquifer and overlying aquitard and leading to compaction of lacustrine shales and surface subsidence. The subsidence induced by deep well pumping has been affecting Mexico City since the end of the 19th century. Subsidence rates in many of Earth's urban areas are relatively low (< 10 mm/year) and local (< 100 km^2). However, Mexico City areas have undergone accelerated ground subsidence with much higher rates over a larger region since the 1950s, with an average subsidence rate of 90 mm/year for the 20 year period from 1965 to 1985 in the downtown area (Castellazzi *et al.* 2016).

Many studies measuring Mexico City subsidence by radar interferometry have focused on time-series analysis, which is capable of precisely mapping Mexico City subsidence in time and space (Osmanoğlu *et al.* 2011; Yan *et al.* 2012; Chaussard *et al.* 2014). These studies show that the Mexico subsidence zone has a Pac-Man-shape and that the largest subsidence rate can easily exceed 30 cm/year in many regions.

To obtain the recent displacement history, we processed 245 Sentinel-1 images (April 2015–April 2020) using the PSI technique. Figure 9.7 depicts the mean velocity maps in Mexico City from 2015 to 2020. The total PS pixels are more than four million, with corresponding densities of 2,500 point/km^2. The velocity profile and displacement history are shown in Figure 9.8.

The subsidence zone fits well with the lacustrine unit composed of soft and compressible silts and clays with relatively low permeability. This indicates that subsidence is due to pressure loss in the shallow aquitard (clay-rich lake sediments) associated with groundwater overdraft (withdrawal in excess of recharge). This has implications for water management in the Mexico City metropolitan area because compaction of clay-rich aquitards is often associated with permanent loss of porosity and reservoir capacity.

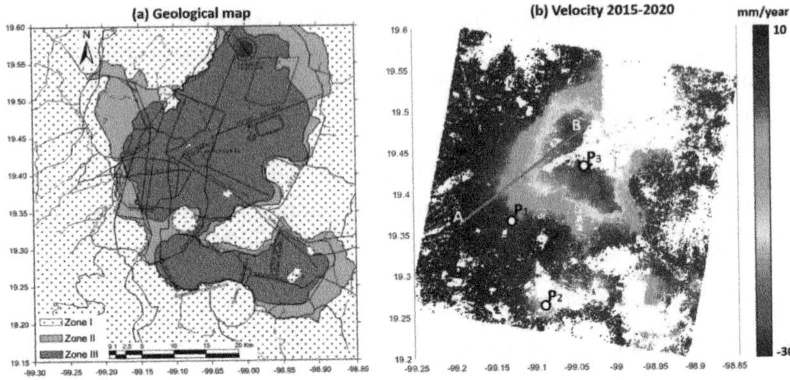

Figure 9.7. *Mexico city mapping. (a) Geological spatial information. Mexico Basin stratigraphy can be classified in three zones: (1) hard rock, (2) transitional and (3) lacustrine. (b) Averaged velocity (mm/year) from 2015 to 2020. A Pac-Man-shaped subsidence zone can be observed with subsidence over 30 cm/year. For a color version of this figure, see www.iste.co.uk/cavalie/images.zip*

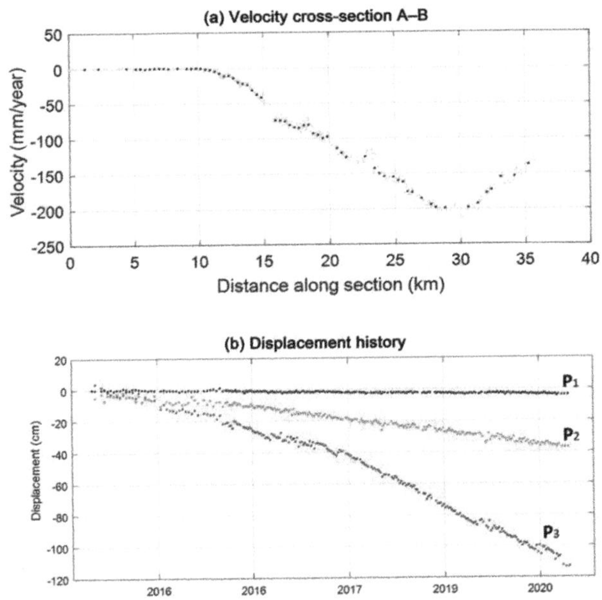

Figure 9.8. *The velocity and displacement history in Mexico City at the locations in Figure 9.7. (a) Velocity cross-section along points A and B. (b) Displacement history for three points. Subsidence is not primarily seasonal, which agrees well with findings presented by Osmanoğlu* et al. *(2011), Yan* et al. *(2012) and Chaussard* et al. *(2014). For a color version of this figure, see www.iste.co.uk/cavalie/images.zip*

9.3.2.2. *Hanoi case study*

Hanoi is the capital of Vietnam and is located in the Red River delta. It is a megacity with more than eight million people (in 2020) and great potential for the development of industry, exports, tourism and services. Hanoi is a good example of a quickly urbanized city in Vietnam, where water for domestic and industrial use comes from wells located within and around the city. In this city, the rapid increase in groundwater use started in the late 2010s (the groundwater abstraction was approximately 350,000 m^3/day in 1999 and 630,000 m^3/day in 2012, whereas the sustainable level of abstraction is estimated to be at most 300,000 m^3/day (Ho Tong Minh *et al.* 2019)). Heavy pumping of groundwater has produced a serious subsidence problem, which in turn has affected surface structures in Hanoi.

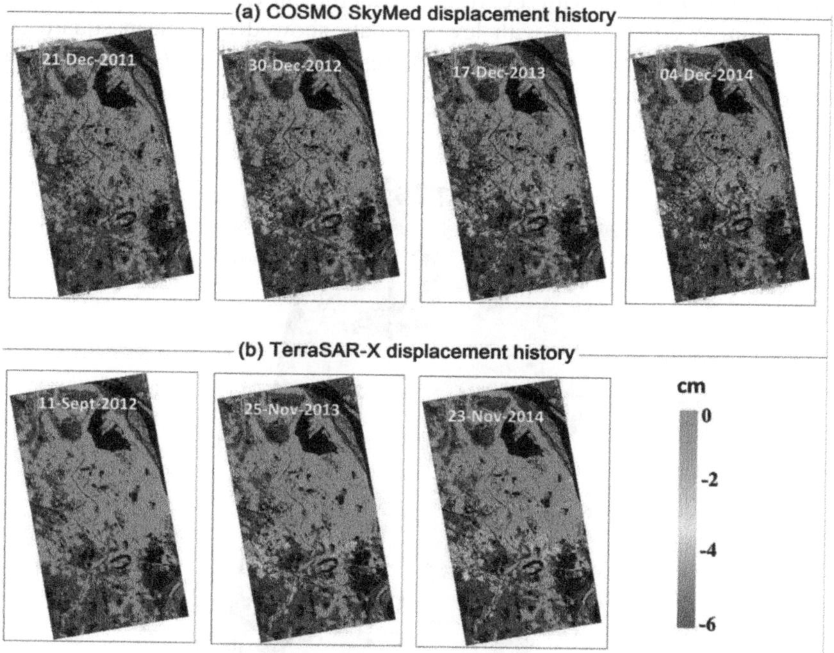

Figure 9.9. *The vertical displacement history (cm) in Hanoi city. The background is the average backscatter TerraSAR-X data. (a) Displacement derived from COSMO SkyMed data. (b) Displacement derived from TerraSAR-X data (Ho Tong Minh* et al. *2019). For a color version of this figure, see www.iste.co.uk/cavalie/images.zip*

In the Hanoi area, the InSAR approach has demonstrated its ability for mapping ground deformation using L-band (Dang *et al.* 2014) and X-band (Tran *et al.* 2015; Le *et al.* 2016; Ho Tong Minh *et al.* 2019). Studies so far have estimated ground subsidence using both classical and multi-temporal InSAR methods. Ho Tong Minh *et al.* (2019) showed that the InSAR technique can be effective in detecting and estimating subsidence phenomena, even in X-band and under conditions typical of tropical regions.

Two SAR stacks were exploited for the period 2011–2014: a TerraSAR-X dataset composed of 19 ascending acquisitions from April 10, 2012, to November 23, 2014, and a second one with 27 ascending COSMO SkyMed images acquired between May 27, 2011, and December 4, 2014. All images were processed using the PSDS method. More than 810,000 PSDS measuring points in the TerraSAR-X (LOS angle 34.4 degrees) and 420,000 PSDS measuring points in the COSMO SkyMed (LOS angle 25.2 degrees) datasets were identified within an area of about 200 km^2. In Figure 9.9, the displacement results obtained from TerraSAR-X and COSMO SkyMed datasets appear consistent, with a correlation coefficient (R^2) of 0.91 for the period of observation. A comparison with leveling data showed root mean square errors (RMSEs) of 3.4 mm/year for TerraSAR-X results and 2.9 mm/year for COSMO SkyMed ones.

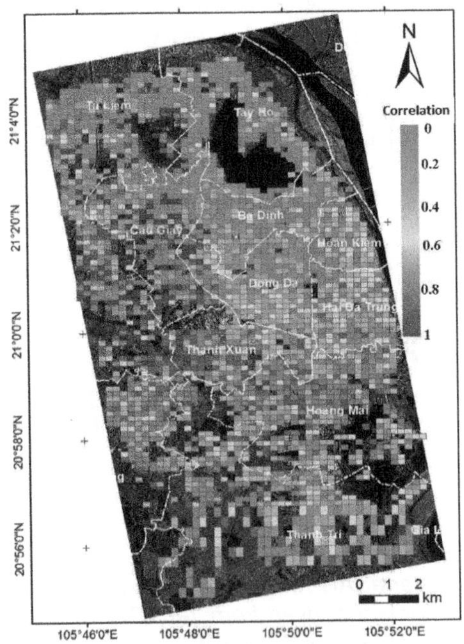

Figure 9.10. *The correlation between ground subsidence value as estimated by InSAR and groundwater fluctuation in the Pleistocene aquifer (qp) (Ho Tong Minh* et al. *2019). For a color version of this figure, see www.iste.co.uk/cavalie/images.zip*

To correlate the subsidence with the groundwater level, all well fields in Hanoi were used to compute the Pleistocene aquifer (qp) water level with a groundwater flow model. Figure 9.10 illustrates the correlation between the ground subsidence value as estimated by InSAR and groundwater fluctuation in the Pleistocene aquifer (qp). High correlation coefficients of greater than 0.6 were found in four areas: Ha Dong, Phap Van, Linh Dam and Hoang Mai. All of them are in the southern part of Hanoi and high subsidence values were found in the cone of depression (the area influenced by exploited groundwater). In contrast, in the Hoan Kiem district in central Hanoi, including sites such as the Ngo Si Lien and Yen Phu well fields, strong subsidence was not confined in the cone of depression and the correlation coefficient was quite low. This again indicates that the cone of depression from groundwater exploitation is not the single significant cause of ground subsidence in central Hanoi, where subsoil characteristics can be another factor. We observed that the risk of ground subsidence in Hanoi was found in the south of city areas where thick soft-soil layers are present.

9.4. Applications in mineral resources extractive activities

9.4.1. *Brine, oil and gas extraction/injection*

9.4.1.1. *Brine extraction*

Among the techniques applied for extracting mineral salt, solution mining is the most commonly used for industrial salt production (Klingman 1975). It is based on the circulation of water between boreholes: freshwater is injected to dissolve salt mineral and brine is pumped from another borehole several meters away. As with other types of industrial pumping of underground fluid resources, brine extraction results in ground motion. The resulting subsidence depends on the local properties of the site (such as the salt layer depth and cap-rock characteristics) and on the production (pumping rates and positions of the boreholes). In addition, this mining method produces caverns (located at the injection and pumping points) filled with brine that can evolve in the long term, including after the closure of the mine (Berest *et al.* 2005). Monitoring is therefore required to forecast potential surface consequences.

An example of the application of interferometric techniques is the site of the salt mine in Vauvert, France, which has been used for assessments of the relevance of InSAR techniques (Raucoules *et al.* 2003). Subsidence of about 2 cm/year with a 7 km diameter was observed (Figure 9.11). In this example, the exploited salt layer was located at depths of 1,800–3,000 m.

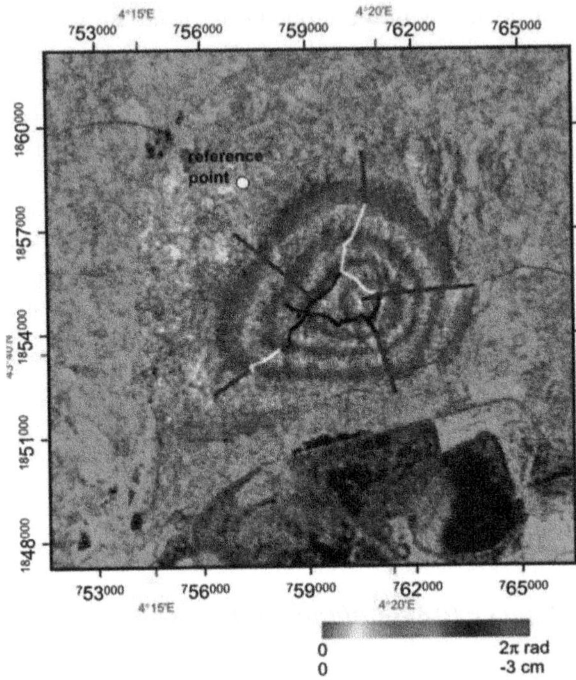

Figure 9.11. *Stack of interferograms showing the total deformation observed between 1993 and 1999 at the salt mine in Vauvert, France. Ground-based leveling networks are shown with black lines for the monitoring network and white lines for the stability network. Blue lines show the proposed increase in coverage of the leveling network based on InSAR analysis (Raucoules* et al. *2003; Furst* et al. *2021). For a color version of this figure, see www.iste.co.uk/cavalie/images.zip*

9.4.1.2. *GAS injection/extraction*

As with all exploitation of underground fluids, natural gas production and storage can result in small (mm to several cm) vertical motions. Monitoring the associated ground deformation allows the detection of any unexpected motions (e.g. different from the motion modeled on the basis on the production values and underground characteristics) due to integrity issues in the reservoir (leakages being the more concerning issues). InSAR techniques have been used for such monitoring of gas production and storage sites. If different techniques can produce useful information for natural gas production (resulting generally only in subsidence), monitoring of the seasonal evolution of gas storage with InSAR requires use of time series with sufficient SAR acquisitions per year. From this perspective, missions with short repeat cycles (such as Sentinel-1, with a six-day repeat cycle) offer an interesting tool.

The example of the CO_2 injection demonstrator at Krechba/In Salah (Algeria) is particularly significant. A series of CO_2 injection tests were performed at this natural gas production site between 2004 and 2011 (http://www.insalahco2.com). In addition to ground-based observations, spaceborne InSAR (PSI and conventional interferometry) made it possible to map the resulting deformation (Vasco *et al.* 2008; Onuma and Ohkawa 2009). Over the period monitored, uplift of about 5–7 mm/year over an area of 4–5 km was observed. The asymmetry (Figure 9.12) of the surface deformation geometry has been interpreted as related to faults.

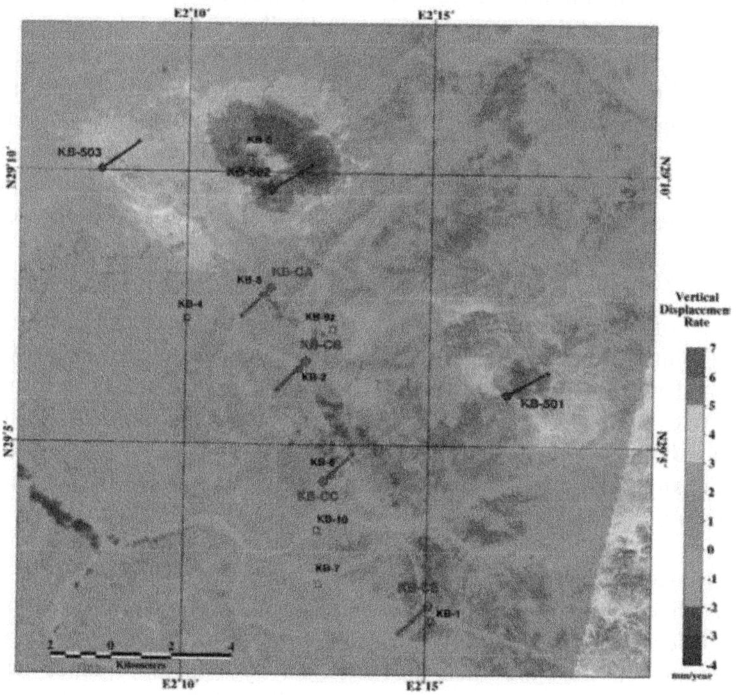

Figure 9.12. *Vertical displacement (adapted from Onuma and Ohkawa (2009)) in the In Salah gas production site between July 31, 2004, and May 31, 2008, resulting from CO_2 injection. Positions of the injection wells are indicated. For a color version of this figure, see www.iste.co.uk/cavalie/images.zip*

Another example of gas storage ground-surface-motion characterization is given by Burnol *et al.* (2019). In this case, annual vertical displacements of a few millimeters were detected at the storage site of Izaute-Lussagnet (Southwestern France). Acquisitions from the Sentinel-1 satellite were processed by persistent scatterer interferometry in order to observe and quantify the ground displacement during a three-year period above the gas storage site. In this study, the cyclic (annual) motion was characterized by the amplitude and phase of the annual component of the obtained time series. The amplitude quantifies (in millimeters) the annual displacement, the phase (radians that can be converted in days) when – over the year – the displacement is maximal. Such information can be derived from the time series by fitting it by a sine function or with Fourier or wavelet analysis. The resulting amplitudes of the vertical periodic displacements range between 4 and 8 mm (values representative of a gas storage in European), which are close to the noise level but sufficient to show the seasonal variations due to storage activity during summer and winter periods. The ground deformation was simulated with a simple mechanical model (taking into account production data and reservoir characteristics). An interesting aspect of the study was the use of soil moisture data to investigate the possibility of separating motion related to the storage from shallow phenomena (clay shrinking/swelling) based on the phase. Indeed, both phenomena have annual cyclicity with similar amplitudes but are shifted in time in related to the production (for the gas storage) or the surface soil moisture (for the clay shrinking). This asynchronicity can be detected in the phase information, making it possible to distinguish persistent scatterers affected by superficial phenomena from those related to the storage activity.

9.4.1.3. *Oil extraction*

Extraction of oil, like other forms of extraction of fluids from buried reservoirs, may cause ground subsidence. Fielding *et al.* (1998) were the first to use InSAR to map the subsidence over two oilfields (Lost Hills and Belridge) in central California. Rapid subsidence rates of up to 40 mm in 35 days (i.e. an annual rate of about 400 mm/year) were measured thanks to ERS data. Initial elastic strain modeling, giving information on volume change in the rock units at depth, confirmed the relevance of InSAR for such applications.

In recent decades, oil production activities have started to use novel recovery methods (enhanced oil recovery (EOR)) based on hydraulic fracturation, horizontal drilling and deep injection of fluid or steam to drive the oil out of the reservoir. Oil production activities are then characterized by both subsidence and uplift. In particular, monitoring surface uplift can deliver information about injection distribution and impact. It can indicate a potential steam/fluid excess and thus an increasing risk. Ground measurement can be seen as a tool to maximize injection efficiency, while limiting the risks to the extraction area. In northern Alberta (Canada), the oil sands contain one of the world's largest deposits of crude oil in the form of viscous bitumen embedded within uncemented sand. For reserves deeper than about 65 m (more than 80% of the recoverable bitumen in Alberta), a steam-assisted gravity drainage technique is used to extract bitumen, with horizontal wells that are drilled into the reservoir side by side, maximizing contact between the injected steam and the reservoir. An InSAR deformation survey has shown surface uplift at rates of 1–4 cm/year, associated with active steam injection at all sampled sites (Pearse *et al.* 2014). Rates of uplift are well correlated with rates of steam injection despite a net loss of total reservoir fluids with time.

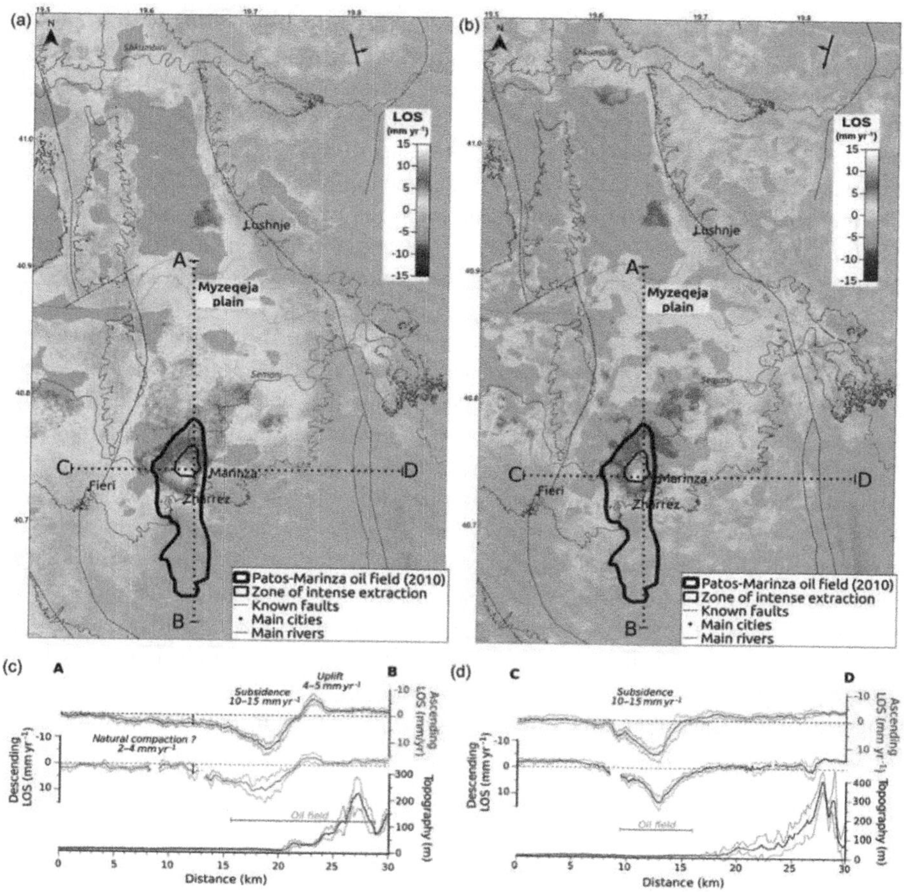

Figure 9.13. *LOS average-velocity maps for the ascending (a) and descending (b) tracks over the Patos-Marinza oil field. Positive values correspond to motion away from the satellite, i.e. mostly subsidence. Dotted lines indicate the locations of the profiles in (c) and (d). North–south (c) and west–east (d) velocity profiles across the oil field's most intense deformation zone together with topography profiles (c and d) (Métois* et al. *(2020)). For a color version of this figure, see www.iste.co.uk/cavalie/images.zip*

Another hazard induced by oil production is related to the significant and rapid changes in the stress field at depth caused by enhanced oil recovery: extensive extraction and injection of fluid may also induce earthquakes. It has been observed that EOR techniques may trigger non-tectonic seismic activity. Grandin *et al.* (2017) has shown that the 2016 Oklahoma Mw 5.7 Pawnee earthquake that broke a previously unknown basement fault was probably induced by wastewater disposal under the Oklahoma oil and gas field. In the year preceding the mainshock, it was reported that at least two wells were injecting saltwater at rates exceeding 500 m^3/day within 7 km of the epicenter, and there were also increases in the injection rates at two wells located near the epicenter. At the Karamay oilfield (China), Yang *et al.* (2019) monitored deformation related to oil production by exploiting Sentinel-1 images from March 2017 to August 2018. On this site, where a previous study conducted before 2010 with Envisat and ALOS images showed a maximum accumulated uplift of 200 mm between 2006 and 2011, the maximum uplift rate was > 160 mm/year, while subsidence was about 32 mm/year. However, no evidence could be given that low-magnitude earthquakes within and around the oilfield were related to oil production activities.

Patos-Marinza oil field (Albania), one of the largest onshore oil fields in Europe with more than 7 million oil barrels extracted every year, is another example where an InSAR time-series analysis was conducted during the 2014–2018 period. Based on Sentinel-1 data, slow regional subsidence of the entire basin of about 2.5 mm/year was measured. It was interpreted as a natural compaction phenomenon of this alluvial-to-deltaic plain, most probably amplified by human activities (Métois *et al.* 2020). However, another deformation pattern, roughly circular and superimposed on the overall subsidence of the plain, is centered over the oil field, and exhibits large subsidence rates of up to 15 mm/year, resulting from EOR techniques (Figure 9.13). One concern was to investigate if the apparent recent increase of local seismicity could be a consequence of the use of EOR techniques in this area. However, local seismological arrays were too sparse to answer this question and suggest that stress changes associated with the oil extraction were sufficient to trigger earthquakes.

9.4.2. *Mining activities*

Mining activities have a significant impact on the surrounding areas (environment, infrastructures, etc.), requiring monitoring from the beginning of the extraction work to decades after the end of the extraction (the period of mitigation of post-mining consequences). Ground-based monitoring techniques are generally required by local regulations. Spaceborne SAR interferometry can contribute to such monitoring by complementing and partially replacing ground-based measurements. During the active phase of the work, monitoring of ground surface deformation allows the mining company to anticipate potential surface damages and adapt the on-going operations in order to reduce their impact. In addition, information on ground motion can be valuable for assessing damages and identify a possible link with underground work from the perspective of financial compensations for the damaged building owners. However, as pointed out by Raucoules and Foumelis (2020), the deformation (about 0.8% slope in the cited example, i.e. about 10 cm of cumulated differential subsidence at the scale of a house) required for justifying legal financial compensations is generally higher than displacements observable by spaceborne InSAR – due to the limitations of InSAR techniques with high deformations values.

For this reason, InSAR provides useful information but needs to be complemented by other – ground-based – measurements/observations to be used in a risk management perspective.

After the closure of the mine, the liability on the site can be transferred to the authorities, which have to manage the long-term (up to centuries) consequences and rehabilitate the mining area. Residual ground motion is one of the consequences of past mining works. They are generally related to water table variations (after the end of mining works, pumping operations are stopped, resulting in an underground water level increase) that change the characteristics of the underground materials. The resulting motion may be subsidence, but uplift can also occur due to expansion of materials related to filling by water from the abandoned galleries (Figure 9.14 shows an example of post-mining uplift).

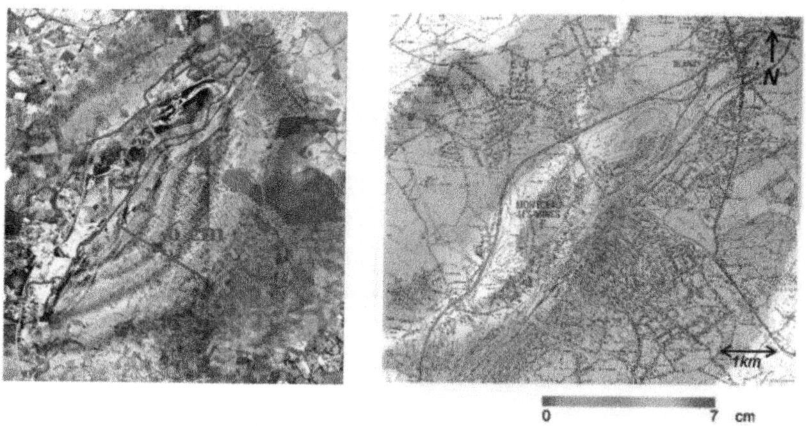

Figure 9.14. *Example of post-mining uplift in Montceau-les-Mines, France, between 1996 and 1998 (adapted from Carnec* et al. *(2004)). Left: sum of two interferograms of ERS-1–2 data. Right: unwrapped interferogram stack converted into cumulative vertical motion (about 2.5 cm/year). The deformation stopped after 1998. Severe damage to buildings occurred on this area. For a color version of this figure, see www.iste.co.uk/cavalie/images.zip*

In both cases, the resulting ground motion is related to the extraction techniques used during the mine's activity.

In the case of exploitation with chambers and pillars, the ore layer is partially excavated, creating pillars that support the roof. Except in cases of accidental collapse, the resulting surface displacement is limited to the active phase. However, during the post-mining phase, due to the inundation of the mine, pillars can decay and filling materials' characteristics can be modified, resulting in motions of significant amplitude (centimeters to meters). Plattner *et al.* (2010) presented an example of subsidence due to pillar collapse in 2007 in an active coal mine in Utah, which resulted in fatalities. ALOS-1 InSAR data enabled the mapping of 20 cm of ground deformation in an area of 1 km^2.

When the mine is worked using longwall mining techniques, the panel is fully exploited. The roof of the panel is collapsed in a controlled way with the progression of the work. The resulting ground surface motion is therefore extensive during the active phase (about 75–85% of the total motion occurs during the several months after the excavation). The residual phase due to the compaction of the fillings lasts several years. In contrast to chamber and pillar techniques, most of the impact on the surface occurs within a few years. Perski (1999), Carnec and Delacourt (2000), Guéguen *et al.* (2009) and Engelbrecht *et al.* (2011) were among the firsts to use the InSAR method to monitor this type of ground motion (Figure 9.15).

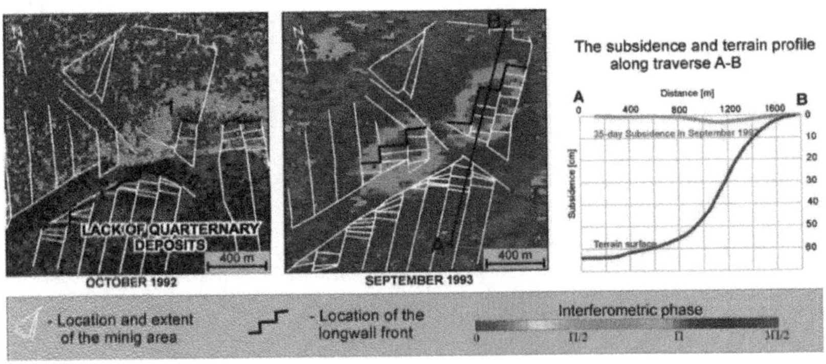

Figure 9.15. *Example of InSAR monitoring applied to longwall mining activities in the Katowice area, Poland (adapted from Perski (1999)). For a color version of this figure, see www.iste.co.uk/cavalie/images.zip*

9.5. Conclusion

In this chapter, we presented the application of spaceborne SAR interferometric techniques to ground surface (subsidence or uplift) deformation resulting from human activity. Such motions involve various scales in space and time – sizes ranging from a few meters to kilometers and motion rates from mm/year to sudden collapses – related to a diversity of phenomena. Fluid extraction (water pumping for human usage, agriculture or industry and oil/gas production) and cavity excavation (underground construction, mining and quarries) are the main anthropogenic sources of subsidence. Intensive groundwater withdrawal is becoming one of the major concerns for societies in all countries of the world, with growing water demand and increasing pressure on groundwater resources. Precise monitoring techniques of these ground surface motions are needed. This is essential to evaluate their consequences in terms of risk. It is also needed to characterize the involved processes and to help in providing exploitation/production management tools for resources, in particular for sustainable policies. From this perspective, spaceborne SAR interferometry can be a major source of information and is in use in an operational way in the concerned industrial activities. The increasing resolution and temporal sampling in the recent space missions (e.g. TerraSAR-X with high spatial resolution and Sentinel-1 with a high repeat cycle) and improvements expected in the future will make it possible to access more local and faster motions, offering a wider range of applications related to anthropogenic ground surface motion.

9.6. Acknowledgments

We thank Pierre Briole for his comments and edits to improve the quality of this chapter.

9.7. References

Amelung, F., Galloway, D.L., Bell, J.W., Zebker, H.A., Laczniak, R.J. (1999). Sensing the ups and downs of Las Vegas: InSAR reveals structural control of land subsidence and aquifer-system deformation. *Geology*, 27(6), 483–486.

Aobpaet, A., Caro Cuenca, M., Hooper, A., Trisirisatayawong, I. (2013). InSAR time-series analysis of land subsidence in Bangkok, Thailand. *International Journal of Remote Sensing*, 34, 2969–2982.

Barla, G., Tamburini, A., Del Conte, S., Giannico, C. (2016). InSAR monitoring of tunnel-induced ground movements. *Geomechanics and Tunnelling*, 9(1), 15–22.

Bell, J.W., Amelung, F., Ferretti, A., Bianchi, M., Novali, F. (2008). Permanent scatterer InSAR reveals seasonal and long-term aquifer-system response to groundwater pumping and artificial recharge. *Water Resources Research*, 44(2).

Berest, P., Brouard, B., Karimi-Jafari, M. (2005). Deep salt caverns abandonment. *Post-Mining Symposium*, Nancy, France.

Bock, Y., Wdowinski, S., Ferretti, A., Novali, F., Fumagalli, A. (2012). Recent subsidence of the Venice Lagoon from continuous GPS and interferometric synthetic aperture radar. *Geochemistry, Geophysics, Geosystems*, 13(3).

Burnol, A., Aochi, H., Raucoules, D., Veloso, F.M.L., Koudogbo, F.N., Fumagalli, A., Chiquet, P., Maisons, C. (2019). Wavelet-based analysis of ground deformation coupling satellite acquisitions (Sentinel-1, SMOS) and data from shallow and deep wells in Southwestern France. *Scientific Reports*, 9, 8812.

Carnec, C. and Delacourt, C. (2000). Three years of mining subsidence monitored by SAR interferometry, near Gardanne, France. *Journal of Applied Geophysics*, 43(1), 43–54.

Carnec, C., Le Mouelic, S., Raucoules, D., Colesanti, C., King, C. (2004). Avancées des technologies spatiales radar appliquées à la détection des déformations de surface d'origine minière. *Revue française de géotechnique*, 106–107, 15–22.

Castellazzi, P., Arroyo-Domínguez, N., Martel, R., Calderhead, A.I., Normand, J.C., Gárfias, J., Rivera, A. (2016). Land subsidence in major cities of Central Mexico: Interpreting InSAR-derived land subsidence mapping with hydrogeological data. *International Journal of Applied Earth Observation and Geoinformation*, 47, 102–111.

Chaussard, E. and Farr, T.G. (2019). A new method for isolating elastic from inelastic deformation in aquifer systems: Application to the San Joaquin Valley, CA. *Geophysical Research Letters*, 46(19), 10800–10809.

Chaussard, E., Amelung, F., Abidin, H., Hong, S.-H. (2013). Sinking cities in Indonesia: ALOS PALSAR detects rapid subsidence due to groundwater and gas extraction. *Remote Sensing of Environment*, 128, 150–161.

Chaussard, E., Bürgmann, R., Shirzaei, M., Fielding, E.J., Baker, B. (2014a). Predictability of hydraulic head changes and characterization of aquifer-system and fault properties from InSAR-derived ground deformation. *Journal of Geophysical Research: Solid Earth*, 119(8), 6572–6590.

Chaussard, E., Wdowinski, S., Cabral-Cano, E., Amelung, F. (2014b). Land subsidence in Central Mexico detected by ALOS InSAR time-series. *Remote Sensing of Environment*, 140, 94–106.

Chaussard, E., Milillo, P., Bargmann, R., Perissin, D., Fielding, E.J., Baker, B. (2017). Remote sensing of ground deformation for monitoring groundwater management practices: Application to the Santa Clara Valley during the 2012–2015 California drought. *Journal of Geophysical Research: Solid Earth*, 122(10), 8566–8582.

Dang, V.K., Doubre, C., Weber, C., Gourmelen, N., Masson, F. (2014). Recent land subsidence caused by the rapid urban development in the Hanoi region (Vietnam) using ALOS InSAR data. *Natural Hazards and Earth System Sciences*, 14(3), 657–674.

Engelbrecht, J., Inggs, M., Makusha, G. (2011). Detection and monitoring of surface subsidence associated with mining activities in the Witbank coalfields, South Africa, using differential radar interferometry. *South African Journal of Geology*, 114(1), 77–94.

Falorni, G. and Iannacone, J. (2014). Signature of tunneling-induced ground deformation in urban areas revealed by time series analysis. *TAC 2014: Tunneling in a Resource Driven World*, Vancouver, Canada.

Fielding, E.J., Blom, R.G., Goldstein, R.M. (1998). Rapid subsidence over oil fields measured by SAR interferometry. *Geophysical Research Letters*, 25(17), 3215–3218.

Fruneau, B. and Sarti, F. (2000). Detection of ground subsidence in the city of Paris using radar interferometry: Isolation of deformation from atmospheric artifacts using correlation. *Geophysical Research Letters*, 27(24), 3981–3984.

Fruneau, B., Deffontaines, B., Rudant, J.-P., Le Parmentier, A.-M. (2005). Monitoring vertical deformation due to water pumping in the city of Paris (France) with differential interferometry. *Comptes Rendus Geoscience*, 337(13), 1173–1183.

Furst, S.L., Doucet, S., Vernant, P., Champollion, C., Carme, J.-L. (2021). Monitoring surface deformation of deep salt mining in Vauvert (France), combining InSAR and leveling data for multi-source inversion. *Solid Earth*, 12(1), 15–34.

Galloway, D.L. and Hoffmann, J. (2007). The application of satellite differential SAR interferometry-derived ground displacements in hydrogeology. *Hydrogeology Journal*, 15(1), 133–154.

Galloway, D.L., Hudnut, K.W., Ingebritsen, S.E., Phillips, S.P., Peltzer, G., Rogez, F., Rosen, P.A. (1998). Detection of aquifer system compaction and land subsidence using interferometric synthetic aperture radar, Antelope Valley, Mojave Desert, California. *Water Resources Research*, 34(10), 2573–2585.

Garcia, A., Marchamalo, M., Martinez, R., Gonzalez-Rodrigo, B., Gonzalez, C. (2019). Integrating geotechnical and SAR data for the monitoring of underground works in the Madrid urban area: Application of the persistent scatterer interferometry technique. *International Journal of Applied Earth Observation and Geoinformation*, 74, 27–36.

Ge, D., Zhang, L., Li, M., Liu, B., Wang, Y. (2016). Beijing subway tunnelings and high-speed railway subsidence monitoring with PSInSAR and TerraSAR-X data. *2016 IEEE International Geoscience and Remote Sensing Symposium (IGARSS)*, 6883–6886.

Grandin, R., Vallée, M., Lacassin, R. (2017). Rupture process of the Mw 5.8 Pawnee, Oklahoma, earthquake from Sentinel-1 InSAR and seismological data. *Seismological Research Letters*, 88(4), 994–1004.

Guéguen, Y., Deffontaines, B., Fruneau, B., Al Heib, M., de Michele, M., Raucoules, D., Guise, Y., Planchenault, J. (2009). Monitoring residual mining subsidence of Nord/Pas-de-Calais coal basin from differential and persistent scatterer interferometry (Northern France). *Journal of Applied Geophysics*, 69(1), 24–34.

Ho Tong Minh, D., Van Trung, L., Toan, T.L. (2015). Mapping ground subsidence phenomena in Ho Chi Minh city through the radar interferometry technique using ALOS PALSAR data. *Remote Sensing*, 7(7), 8543–8562.

Ho Tong Minh, D., Tran, Q.C., Pham, Q.N., Dang, T.T., Nguyen, D.A., El-Moussawi, I., Le Toan, T. (2019). Measuring ground subsidence in Ha Noi through the radar interferometry technique using TerraSAR-X and Cosmos SkyMed data. *IEEE Journal of Selected Topics in Applied Earth Observations and Remote Sensing*, 12(10), 3874–3884.

Ho Tong Minh, D., Hanssen, R., Rocca, F. (2020). Radar interferometry: 20 years of development in time series techniques and future perspectives. *Remote Sensing*, 12(9).

Hoffmann, J., Zebker, H.A., Galloway, D.L., Amelung, F. (2001). Seasonal subsidence and rebound in Las Vegas Valley, Nevada, observed by synthetic aperture radar interferometry. *Water Resources Research*, 37(6), 1551–1566.

Klingman, C. (1975). *Mineral Facts and Problems*. U.S. Department of the Interior, Bureau of Mines, Washington, DC, USA.

Koudogbo, F., Urdiroz, A., Robles, J., Chapron, G., Lebon, G., Fluteaux, V., Priol, G. (2018). Radar interferometry as an innovative solution for monitoring the construction of the Grand Paris Express metro network – First results. *World Tunnel Conference*, 2–25 April, Dubai, United Arab Emirates.

Lazecky, M., Hlavacova, I., Bakon, M., Sousa, J.J., Perissin, D., Patricio, G. (2017). Bridge displacements monitoring using space-borne X-band SAR interferometry. *IEEE Journal of Selected Topics in Applied Earth Observations and Remote Sensing*, 10(1), 205–210.

Le, T., Chang, C., Nguyen, X., Yhokha, A. (2016). TerraSAR-X data for high-precision land subsidence monitoring: A case study in the historical Centre of Hanoi, Vietnam. *Remote Sensing*, 8(4), 338.

Le Mouélic, S., Raucoules, D., Carnec, C., King, C., Adragna, F. (2002). A ground uplift in the city of Paris (France) detected by satellite radar interferometry. *Geophysical Research Letters*, 29(17), 1–34.

MacDonald, B., Iannacone, J., Falorni, G., Gianniaco, C. (2015). InSAR-derived time series analysis of tunnel construction-induced ground deformation in urban landscapes. In *Proceedings of the Ninth Symposium on Field Measurements in Geomechanics*, Dight, P. (ed.). Australian Centre for Geomechanics, Crawley, Australia.

Métois, M., Benjelloun, M., Lasserre, C., Grandin, R., Barrier, L., Dushi, E., Koçi, R. (2020). Subsidence associated with oil extraction, measured from time series analysis of Sentinel-1 data: Case study of the Patos-Marinza oil field, Albania. *Solid Earth*, 11(2), 363–378.

Motagh, M., Walter, T.R., Sharifi, M.A., Fielding, E., Schenk, A., Anderssohn, J., Zschau, J. (2008). Land subsidence in Iran caused by widespread water-reservoir overexploitation. *Geophysical Research Letters*, 35, L16403.

Motagh, M., Shamshiri, R., Haghshenas Haghighi, M., Wetzel, H.-U., Akbari, B., Nahavandchi, H., Roessner, S., Arabi, S. (2017). Quantifying groundwater exploitation induced subsidence in the Rafsanjan Plain, southeastern Iran, using InSAR time-series and *in situ* measurements. *Engineering Geology*, 218, 134–151.

Nahli, A., Simonetto, E., Merrien-Soukatchoff, V., Durand, F., Rangeard, D. (2018). Sentinel-1 for monitoring tunnel excavations in Rennes, France. *Procedia Computer Science*, 138, 393–400.

Onuma, T. and Ohkawa, S. (2009). Detection of surface deformation related with CO2 injection by DInSAR at In Salah, Algeria. *Energy Procedia*, 1(1), 2177–2184.

Osmanoğlu, B., Dixon, T.H., Wdowinski, S., Cabral-Cano, E., Jiang, Y. (2011). Mexico city subsidence observed with persistent scatterer InSAR. *International Journal of Applied Earth Observation and Geoinformation*, 13(1), 1–12.

Pearse, J., Singhroy, V., Samsonov, S., Li, J. (2014). Anomalous surface heave induced by enhanced oil recovery in northern Alberta: InSAR observations and numerical modeling. *Journal of Geophysical Research: Solid Earth*, 119(8), 6630–6649.

Perissin, D., Wang, Z., Lin, H. (2012). Shanghai subway tunnels and highways monitoring through Cosmo-SkyMed persistent scatterers. *ISPRS Journal of Photogrammetry and Remote Sensing*, 73, 58–67.

Perski, Z. (1999). ERS InSAR data for geological interpretation of mining subsidence in Upper Silesian Coal Basin in Poland. *FRINGE'99: Advancing ERS SAR Interferometry from Applications towards Operations*, 10–12 November, Liège, Belgium.

Plattner, C., Wdowinski, S., Dixon, T.H., Biggs, J. (2010). Surface subsidence induced by the Crandall Canyon Mine (Utah) collapse: InSAR observations and elasto-plastic modelling. *Geophysical Journal International*, 183(3), 1089–1096.

Raucoules, D. and Foumelis, M. (2020). Monitoring mining/post-mining ground surface motion. *Remote Sensing Applications to Geohazards and Natural Resources*. Springer, New York, USA.

Raucoules, D., Maisons, C., Carnec, C., Le Mouelic, S., King, C., Hosford, S. (2003). Monitoring of slow ground deformation by ERS radar interferometry on the Vauvert salt mine (France): Comparison with ground-based measurement. *Remote Sensing of Environment*, 88(4), 468–478.

Raucoules, D., Colesanti, C., Carnec, C. (2007). Use of SAR interferometry for detecting and assessing ground subsidence. *Comptes Rendus Geoscience*, 339(5), 289–302.

Raucoules, D., Le Cozannet, G., Woppelmann, G., de Michele, M., Gravelle, M., Daag, A., Marcos, M. (2013). High nonlinear urban ground motion in Manila (Philippines) from 1993 to 2010 observed by DInSAR: Implications for sea-level measurement. *Remote Sensing of Environment*, 139, 386–397.

Scoular, J., Ghail, R., Mason, P., Lawrence, J., Bellhouse, M., Holley, R., Morgan, T. (2020). Retrospective InSAR analysis of East London during the construction of the Lee Tunnel. *Remote Sensing*, 12(5).

Strozzi, T. and Wegmuller, U. (1999). Land subsidence in Mexico City mapped by ERS differential SAR interferometry. *IEEE International Geoscience and Remote Sensing Symposium*, 1940–1942.

Tesauro, M., Berardino, P., Lanari, R., Sansosti, E., Fornaro, G., Franceschetti, G. (2000). Urban subsidence inside the city of Napoli (Italy) observed by satellite radar interferometry. *Geophysical Research Letters*, 27(13), 1961–1964.

Thierry, P., Deverly, F., Reppelin, M., Simonetto, E., Lembezat, C., Arandu, A., Raucoules, D., Closset, L., King, C. (2005). Survey of linear subsidence in an urban area using a 3D geological model and satellite differential InSAR. *International Symposium Geoline*, 23–25 May, Lyon, France.

Tomás, R., Herrera, G., Cooksley, G., Mulas, J. (2011). Persistent scatterer interferometry subsidence data exploitation using spatial tools: The Vega Media of the Segura River Basin case study. *Journal of Hydrology*, 400, 411–428.

Tran, Q.C., Ho Tong Minh, D., Le, V.T., Le Toan, T. (2015). Ground subsidence monitoring in Vietnam by multi-temporal InSAR technique. *IEEE International Geoscience and Remote Sensing Symposium*, 3540–3543.

Tung, H. and Hu, J.-C. (2012). Assessments of serious anthropogenic land subsidence in Yunlin county of central Taiwan from 1996 to 1999 by persistent scatterers InSAR. *Tectonophysics*, 578, 126–135.

Tung, H., Horng-Yue, C., Hu, J.-C., Ching, K.-E., Chen, H., Yang, K.-H. (2016). Transient deformation induced by groundwater change in Taipei metropolitan area revealed by high resolution X-band SAR interferometry. *Tectonophysics*, 692, 265–277.

Vasco, D.W., Ferretti, A., Novali, F. (2008). Reservoir monitoring and characterization using satellite geodetic data: Interferometric synthetic aperture radar observations from the Krechba field, Algeria. *Geophysics*, 73(6), WA113.

Weissgerber, F., Colin-Koeniguer, E., Nicolas, J.-M., Trouvé, N. (2017). 3D monitoring of buildings using TerraSAR-X InSAR, DInSAR and PolSAR capacities. *Remote Sensing*, 9(10), 1010.

Yan, Y., Doin, M., Lopez-Quiroz, P., Tupin, F., Fruneau, B., Pinel, V., Trouve, E. (2012). Mexico City subsidence measured by InSAR time series: Joint analysis using PS and SBAS approaches. *IEEE Journal of Selected Topics in Applied Earth Observations and Remote Sensing*, 5(4), 1312–1326.

Yang, C., Zhang, D., Zhao, C., Han, B., Sun, R., Du, J., Chen, L. (2019). Ground deformation revealed by Sentinel-1 MSBAS-InSAR time-series over Karamay oilfield, China. *Remote Sensing*, 11(17), 2027.

10

Measuring Kinematics of Slow-Moving Landslides from Satellite Images

Pascal LACROIX, Benedetta DINI and Aya CHEAIB

ISTerre, Grenoble, France

10.1. Introduction

Landslides result in at least several thousands of victims every year worldwide (Froude and Petley 2018). This global number hides a variety of mechanisms (fall, flow, slide, topple, spread), with different physical processes and associated kinematics (Hungr *et al.* 2014). One of the main characteristics of landslides is the variability of their kinematics over time. The same landslide may undergo short or long periods of quiescence before suddenly accelerating to reach catastrophic failure (e.g. Intrieri *et al.* 2018; Lacroix *et al.* 2020). Therefore, being able to detect landslides and monitor their displacement over time is a major objective of landslide science, with the operational goal of predicting their failure. Remote sensing is a key tool for this purpose.

The specificities of landslides compared to other active geomorphological objects, such as tectonic faults, glaciers or volcanoes, are their rather small size (a 1 km^2 landslide is already a large object); their wide range of velocities (from mm/yr to m/sec), including the temporal variability of their kinematics; and their presence in

Surface Displacement Measurement from Remote Sensing Images,
coordinated by Olivier CAVALIÉ and Emmanuel TROUVÉ.
© ISTE Ltd 2022.

steep and high-altitude slopes, where shadows or snow can occur. These different properties make the detection and monitoring of these objects from remote sensing data a challenging task. Moreover, due to the variety of sizes and velocities, no unique method or type of data is best suited to study them. For instance, a large branch of remote sensing studies on landslides focuses on the algorithms for the detection of disrupted landslides (i.e. when the rock mass or soil is fragmented due to a rapid propagation (m/min to m/sec) of the mass downslope), which aim to identify the locations where failures have occurred using change detection methods (Guzzetti *et al.* 2012) (Figure 10.1). The frequency of satellite revisit, however, might allow the monitoring of displacements during phases prior to failure.

As long as the surface state does not vary excessively over time (i.e. the landslide mass moves as a whole and thus stays coherent over time, which occurs mostly with low velocities), it is possible to detect and monitor ground motions from remote sensing images (Figure 10.1). Despite being too slow to produce casualties, slow-moving landslides have attracted the attention of the remote sensing community for a number of reasons. First, slow-moving landslides can damage infrastructures or agricultural fields. Moreover, these slow ground motions can be precursory of more rapid slides or falls with potential catastrophic consequences. Finally, they provide an opportunity to monitor different physical parameters of the landslide mass (kinematics, damage, pore water pressure (e.g. Schulz *et al.* 2009)) over time, allowing scientists to better understand the physical properties of landslides in general (Lacroix *et al.* 2020).

Satellite remote sensing has proven to be very useful in detecting and monitoring slow-moving landslides over wide and sometimes remote areas. This detection/monitoring is based on two main techniques: image correlation applied to SAR amplitude images or optical images and InSAR. These two techniques are complementary in terms of the velocity/size/orientation of the landslides detected (Figure 10.1). For instance, InSAR is mostly sensitive to vertical and east–west component of the ground motion, whereas optical image correlation is mostly sensitive to horizontal ground motion. Moreover, InSAR is better suited to retrieving slow velocities (mm/yr to cm/yr), depending on the frequency of revisit and the radar wavelength, whereas image correlation has an inherent uncertainty of approximately a tenth of a pixel, which makes it applicable only to landslides (or phases) of larger velocities (at least 15 cm of ground motion is typically required between two acquisitions in very high-resolution images (Stumpf *et al.* 2014)). Several other limitations are inherent to this type of data. Clouds are of specific concern in the use of optical images, given also that precipitation is one of the main drivers of the landslide kinematics. For SAR images, geometrical distortions and topographical shadows can render their use in mountainous areas relatively complex.

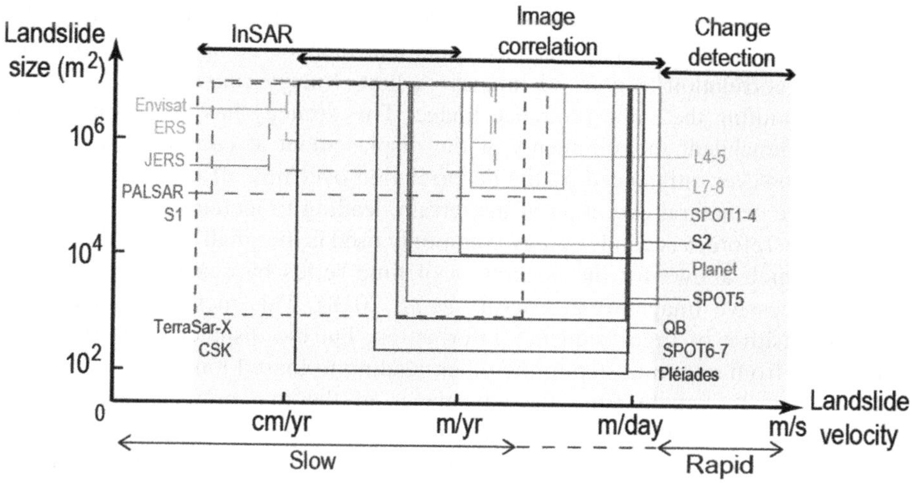

Figure 10.1. *Application domains of different techniques (InSAR, image correlation, change detection) and sets of satellite data in landslide studies as a function of the landslide size and velocity. Change detection methods are not applicable for slow-moving landslides. The satellite data for this technique are therefore not detailed here. Continuous and dashed lines represent optical and SAR sensors, respectively. For a color version of this figure, see www.iste.co.uk/cavalie/images.zip*

The launch of several generations of satellites in the last 20 years, with both very high resolution and high revisit rates, and the development of image correlation and InSAR techniques have allowed us to overcome some of the challenges posed by these objects, which are both small and kinematically variable over time. Different methods of image processing have been developed or adapted for the specific problems posed by landslides. We can classify the use of each method/dataset depending on the size and velocity of the landslide under study (Figure 10.1). In particular, the spatial and temporal resolution of the acquisitions must be chosen in close relation with the size and velocity of the object to be detected. These new data and methods allow retrieval of displacements of slow-moving landslides from satellite remote sensing images for various objectives, including detection, monitoring, characterization and understanding of the underpinning physical processes. We review these methods and applications, with the aim of offering an overview of the topic that is as wide as possible (though likely non-exhaustive).

10.2. Image correlation applied to satellite optical images

Image correlation techniques use the recognition of patterns in two successive orthorectified images, based on their similarity and evaluated through the optimization of a cost-function calculated with a small window size (typically around 10–40 pixels; see Chapter 2). Time series of displacements can then be computed by applying the pairwise correlation to pairs of images. A first strategy consists of correlating pairs all containing the same reference image. This strategy has the advantage of providing independent measurements of the displacement at each time-step. This process is, however, influenced by the decorrelation over time of the images due to anthropogenic or natural evolution of the terrain, leading to increasing uncertainties over time. Therefore, a second strategy commonly used is the small temporal baseline approach, which allows for the generation of time series by considering only the pairs of successive images (e.g. Lacroix *et al.* 2018). This method is privileged for the acquisition of high-frequency information, but the displacement errors are accumulated from one time-step to the other, leading to biased longer-term absolute displacements. To benefit from the advantages of the two previous approaches, time-series inversion of ground displacements has been developed with specific applications for landslides (Bontemps *et al.* 2018), using the redundancy of the displacement measurements to increase the robustness of the time series and decrease the uncertainty of the individual measurements. This method uses multiple pairwise displacement measurements, followed by a least-squares inversion scheme for each pixel separately, similar to multi-temporal InSAR applications (see Chapter 5).

Among the main advantages of optical satellite images for landslide studies is the extensive availability of archives, starting as early as in the 1960s with the declassified Key-Hole spy satellite missions. Moreover, these images can be combined with older aerial photographs, and they can be interpreted relatively easily given that their acquisition is in the same wavelength as the human eye. This explains their broad use for landslide studies and in particular to study slow-moving landslides.

Image correlation has been applied in the studying and monitoring of slow-moving landslides with a wide range of satellite optical sensors: from low-resolution satellite images from Landsat 4–5 (L4–5) at 30 m resolution (Lacroix *et al.* 2020); to medium-resolution images from Landsat 7–8 (L7–8) at 15 m resolution (Caporossi *et al.* 2018; Lacroix *et al.* 2019), Sentinel-2 (S2) at 10 m resolution (e.g. Lacroix *et al.* 2018; Yang *et al.* 2019; Mulas *et al.* 2020; Dille *et al.* 2021), SPOT-1–4 at 10 m resolution (Bontemps *et al.* 2018), KH9 spy satellites at 6 m resolution (Lacroix *et al.* 2020), Planet at 3 m resolution (Bradley *et al.* 2019; Mazzanti *et al.* 2020) and SPOT-5 at 2.5 m resolution (Le Bivic *et al.* 2017; Bontemps *et al.* 2018); to high-resolution images from Pléiades at 0.7 m resolution (e.g. Stumpf *et al.* 2014; Lacroix *et al.* 2015; Desrues *et al.* 2019) and Quickbird (QB) at 1 m resolution (Delacourt *et al.* 2004). The main advantages of optical images reside in the possibility of combining different sensors to obtain long time series of landslide displacement, as has been done by Lacroix *et al.* (2020), who combined KH9, L4–5, L8 and S2 satellite images from over 40 years, or Bontemps *et al.* (2018), who combined SPOT-1–5, SPOT-6/7 and Pléiades data from over 28 years.

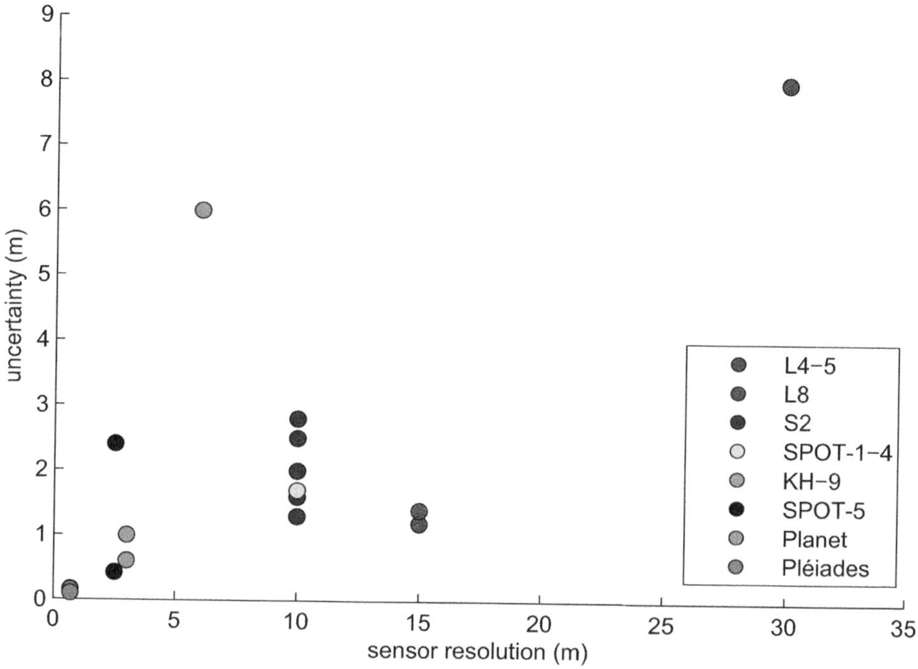

Figure 10.2. *Uncertainties of displacement fields measured from optical satellite images in landslides, as a function of the sensor type. Synthesis realized from Stumpf* et al. *(2014, 2017, 2018), Lacroix* et al. *(2015, 2018, 2019, 2020), Le Bivic* et al. *(2017), Bontemps* et al. *(2018), Caporossi* et al. *(2018), Bradley* et al. *(2019), Desrues* et al. *(2019), Yang* et al. *(2019), Mazzanti* et al. *(2020) and Mulas* et al. *(2020). For a color version of this figure, see www.iste.co.uk/cavalie/images.zip*

All these studies reveal a measurement accuracy highly dependent on the pixel size (from 0.15 m for Pléiades images to 8 m for L4–5 images; see Figure 10.2), as well as the time separation between images (the larger the time-separation between images, the higher the uncertainty, due to variations of the terrain properties over time), the difference between seasons and the type of vegetation cover (e.g. uncertainties are found to be lower in desertic areas than forested areas). The first satellite missions (L4–5, KH9, SPOT-1–4) had lower radiometric resolutions (the pixel radiometry was coded over less bytes) and a lower quality of sensor orientation information, leading to higher uncertainties. The uncertainties were also found to increase at the landslide borders, due to objects of different velocities being present in the same correlation window.

These uncertainties directly impact the range of landslide velocities that can be detected from optical satellite images. The ability to detect landslides is also dependent on the landslide size in relation to the satellite image pixel size. Smaller landslides can only be detected with higher resolution images. The different studies realized as of today have focused on landslides with a minimum size of 0.001 km^2 with Pléiades data (Stumpf *et al.* 2017), 0.08 km^2 with Planet data (Mazzanti *et al.* 2020), 0.5 km^2 with L8 data (Caporossi *et al.* 2018) and 1 km^2 with L4–5 data (Lacroix *et al.* 2020). To decrease the threshold of small object detection, correlators using a multi-resolution approach, such as, for instance, MicMac (Rupnik *et al.* 2017), are more appropriate.

Correlation techniques sometimes fail to estimate the displacement due to strong changes in the ground surface between successive images, caused, for example, by high velocities of the landslide or anthropogenic modifications. Two methods have been applied to overcome these limitations: (1) manual picking of homologous points in the satellite images (e.g. Jiang *et al.* 2016) and (2) use of a small temporal baseline approach to give more weight to pairs of images closer in time (Bontemps *et al.* 2018). Variations in the illumination (e.g. seasonal) can also result in strong displacement errors (up to 20 m in L8 images (Lacroix *et al.* 2019)) caused by variations of shadows in steep topographies. Post-processing based on decomposition of the displacement time series can be used to remove these seasonal artifacts (Lacroix *et al.* 2019).

10.2.1. *Landslide detection with optical remote sensing*

All detection algorithms applied to slow-moving landslides are based on the assumption that the landslide displacement direction is persistent over time. Indeed, as the landslide motion is driven by gravity forces, the direction is likely to be constant. Some authors even assume that the motion must be oriented in the direction of the slope that allows them to filter a large part of the incoherent motions (Desrues *et al.* 2019). The question of the scale at which the slope is calculated is therefore important, as fluctuations of the surface topography or tilted blocks due to listric faults can generate surface topography opposite to the landslide motion.

Slow-moving landslide detection has been undertaken either in pairwise processes or in time series of ground motion. In pairwise processes, the ability to detect landslides is a direct function of the uncertainty of the displacement fields produced by the correlation (Figure 10.2). Pixel-based filtering of the pairwise displacement fields can be realized, based on thresholding of geomorphological and motion indicators (slope gradient, motion orientation, motion magnitude), which makes it possible to detect slow-moving landslides, including some slow gravitational motions precursory to catastrophic failures (Lacroix *et al.* 2018; Desrues *et al.* 2019). For instance, a block of 2,000 m^2 moving at 0.7 m/yr was detected as a precursor to the Chambon landslide, France, nine months before its failure (Desrues *et al.* 2019) using Pléiades images. Also, the use of S2 data with high-frequency acquisitions allowed us to detect that the reactivation of the Harmalière landslide in June 2016 was preceded in the three days before its failure by a movement of 1 m/day of a mass of 80,000 m^2 at its headscarp (Lacroix *et al.* 2018).

These detections are however not automatic (they particularly rely on the choice of threshold parameters and also visual expertise). For this reason, and given the scientific and societal interest, some authors have developed semi-automatic detection methods for moving landslides based on time-series ground motion fields (Stumpf *et al.* 2017; Pham *et al.* 2018). Stumpf *et al.* (2017) derived a pixel-based analysis of time series of ground motion measurements that detects slow-moving landslides by thresholding an indicator of the coherence of the displacement vector directions. Using this method on high-resolution data (Pléiades images at 0.7 m resolution), Stumpf *et al.* (2017) detected 169 slow-moving landslides in an area of 200 km^2 in the French Alps active over two years. As in all the pixel-based detection methods, filtering is required due to the pixelated look of the outputs. This algorithm has proven its efficiency and is now implemented in operational platforms (e.g. ESA-GEP). In order to decrease the number of empirical parameters in this pixel-based algorithm, another method has been proposed that models and inverts the different physical properties of the signals observed in a time series of ground motion (Pham *et al.* 2018). In the proposed algorithm, the landslides are modeled as monotonous displacements over time, in opposition to noises that follow a Gaussian distribution with space and to locally sparse outliers. This method, applied to a time series of 28 years of SPOT 1–7 images at 10 m resolution, allowed the authors to detect 50% more previously unknown landslides in the Colca valley in Peru.

10.2.2. *Landslide characterization with optical remote sensing*

The detection of slow-moving landslides from previous methods allows us to easily characterize the different parameters of a landslide, such as the extent of its active part and its activity (mean and variation of velocities with time). One important parameter for characterizing landslides is volume, as it provides an indicator of their mobility (Mitchell *et al.* 2019). Assessing the landslide volume is not easy using remote sensing images alone, as these images provide direct information on the surface and not on the sub-surface. Different authors have, however, developed specific methods to estimate the volume of slow-moving landslides. The 3D landslide geometry can be obtained using the balanced cross-section method (Bishop 1999) with several profiles based on the displacements extracted from displacement maps (Aryal *et al.* 2015). Assuming that the material has elastic properties, it is possible to invert the geometries of rectangular slip surface using inverse elastic dislocation modeling (Moro *et al.* 2011; Aryal *et al.* 2015). This assumption of elasticity is often unrealistic, especially for soft slides. Therefore, Booth *et al.* (2013) determined the landslide basal surface geometry based on the equation for mass conservation, using a viscous flow model with a non-Newtonian rheology. Inversion of the resulting equations using a 3D displacement field extracted from the correlation of satellite images and from the difference between successive DEMs allowed the authors to derive a thickness of up to 163 m for the large La Clapière landslide. This method was also applied later using SAR images to characterize the landslide basal geometry of a wide sets of earthflows (Handwerger *et al.* 2021) in California.

10.2.3. *Landslide monitoring with optical remote sensing*

Landslide monitoring is based on the generation of displacement time series. The constant availability of satellite optical data since the 1970s, as well as the increased frequency of image acquisitions in recent years, have enabled scientists to monitor a wide variety of landslide activities, from transient motions at sub-weekly scales to long-term evolution over decadal scales. High-frequency S2 and L8 satellites have enabled the detection of daily variations of landslide velocities prior to a major landslide reactivation (Lacroix *et al.* 2018). Sudden acceleration of landslides was also observed thanks to S2 and L8 satellites in a Peruvian desert, with velocities reaching 1 m/day followed by a constant deceleration over three months until they reached quiescence or a steady state (Lacroix *et al.* 2019, 2020) (Figure 10.3). These motions were related to the supply of new material on the landslide body coming from regression of the headscarp. In seismically active regions, the time series of landslide displacements can also reveal the earthquake forcing the slow-moving landslide kinematics, as observed in the Colca region in Peru (Lacroix *et al.* 2015).

Over seasonal time-scales, landslides may display sinusoidal motion related to groundwater effects, as observed in the Barcelonette area, France (Stumpf *et al.* 2017). These seasonal signals in optical images must however be interpreted with caution. Indeed, seasonal differences of illumination can create an artifact of displacement, the magnitude of which is a function of the slope gradient and orientation (Lacroix *et al.* 2019). These artifacts can reach up to 20 m in steep slopes on L8 images and must therefore be corrected. To remove these artifacts, several options exist: only using images taken from the same season, removing a seasonal sinusoid from the data for the phase and amplitudes calculated over stable areas (Lacroix *et al.* 2019) or masking the shadows before the correlation step.

In the longer term, image-correlation time series have been applied in several case studies, highlighting the complex combination of different forcings (earthquake, rainfall) in the landslide kinematics of the Maca valley, Peru, over 28 years (Bontemps *et al.* 2018), for instance, or irrigation impacts on the landslides from the desertic area of the Siguas and Vitor valleys, Peru, over 40 years (Lacroix *et al.* 2020).

These different time-scales highlight different physical processes and forcings acting on landslides (Lacroix *et al.* 2020). The monitoring of landslides from optical images is therefore of significant interest for the study of the mechanics of landslides and the response to external forcings. Satellite monitoring can also be complemented with *in situ* measurements of different physical parameters to decipher the complex mechanics of landslides, although the latter depend on accessibility and feasibility.

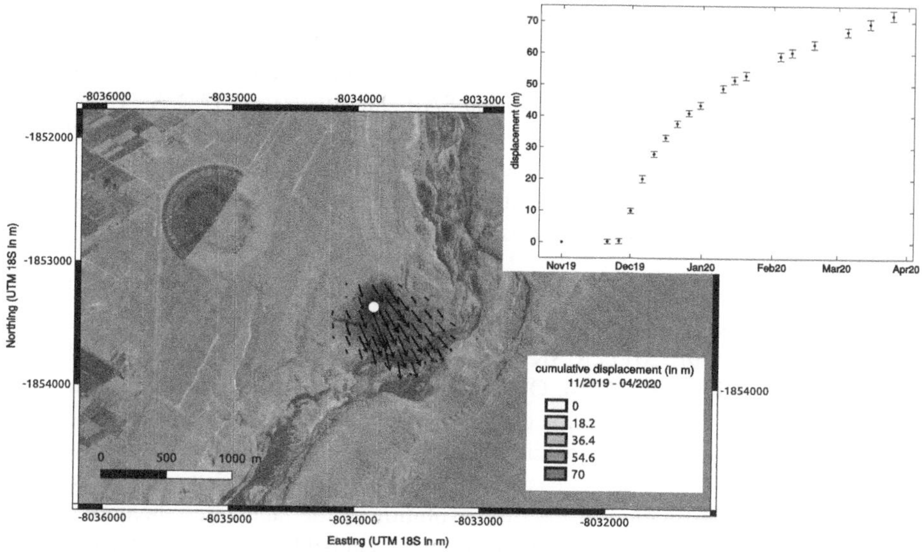

Figure 10.3. *Displacement field and time series of displacement in the Pachaqui landslide (Peru), as estimated using Sentinel-2 data (see Lacroix* et al. *(2020)). The time series shows the sudden acceleration of the landslide in December 2019, following a retrogression of the landslide headscarp and the subsequent material supply to the landslide body. For a color version of this figure, see www.iste.co.uk/cavalie/images.zip*

10.3. Offset tracking of SAR images applied to landslides

The correlation of SAR amplitude images, also known as offset tracking (see Chapter 3), has recently been applied in landslide investigations, and it efficiently complements the correlation of optical images in areas where frequent cloud cover prevents their use (Dille *et al.* 2021). This technique can be applied to landslides when patterns, due to persistent features or to correlated speckle, exist and are visible in the diachronic SAR pairs. This technique leads to the retrieval of displacement in the range and azimuth directions of the radar. Three-dimensional displacement fields can be obtained for landslides when both ascending and descending tracks are processed for the same area (e.g. Raucoules *et al.* 2013). Pre-processing of SAR images is necessary in order to obtain a multi-temporal stack coregistered to a reference SAR image (using a digital elevation model) before measuring the sub-pixel offsets from one image to another (to eliminate the topography and orbital offsets from the results). Large correlation windows are often required, as the texture patterns of the SAR amplitude are usually of larger wavelengths than the texture of objects in optical images where many ground features can be detected. Also, due to the different resolutions in the range and azimuth directions, different window sizes are sometimes chosen along these two dimensions to try to decrease the real ground pixel size in one dimension (Sun *et al.* 2017). The correlation window sizes used in past studies range between 32 and 64 pixels in the less-defined direction. Smaller window sizes are generally used for high-resolution SAR satellites (TerraSAR-X (TSX) (Sun and Muller 2016; Sun *et al.* 2017), COSMO SkyMed (CSK) (Raspini *et al.* 2015)) than for lower-resolution satellites, such as Sentinel-1 (S1) (Manconi *et al.* 2018). Large window sizes restrict the application of offset tracking techniques to only large landslides and prevent the detection of small objects.

This size limitation led most previous landslide studies to focus on high-resolution SAR data, such as TSX (Raucoules *et al.* 2013; Singleton *et al.* 2014; Sun and Muller 2016; Sun *et al.* 2017), CSK (Raspini *et al.* 2015; Caporossi *et al.* 2018) or ALOS-2 images (Jia *et al.* 2020). Other studies have, however, focused on the use of lower-resolution SAR images, such as S1 (Manconi *et al.* 2018; Li *et al.* 2020), with about 14 m pixel spacing in azimuth and 3 m in range. The latter images have the advantages of being open access and of having a high revisit rate, increasing the potential for investigation of the time evolution of the landslide kinematics. Few attempts have been made to create time series of landslide displacement with these data due to the low landslide detection ability of these middle-resolution satellites. Li *et al.* (2020) used the full archives of S1 and ALOS for the large Baige landslide in China, covering 12 years, to estimate the precursory motion of this landslide before its failure. Time series of landslide displacements are, however, more common using TSX (Raucoules *et al.* 2013; Singleton *et al.* 2014) or CSK images (Caporossi *et al.* 2018), but their cost makes their global and systematic application unfeasible.

The uncertainty of the method has barely been quantified for landslides. Some authors estimated the uncertainties by calculating the dispersion (STD) of the displacement field in stable areas and found about 0.1 m of uncertaintiy for ALOS data (Jia *et al.* 2020). Raucoules *et al.* (2013) compared the results from TSX image correlation with GNSS-2 for the La Valette landslide (France) and found 12 and 8 cm uncertainties in the horizontal and vertical motions, respectively, corresponding to about 1/10th of the pixel size. Singleton *et al.* (2014) compared the offset tracking results with 17 corner reflectors installed on a landslide in China and found RMS errors of 4 cm in range and 7 cm in azimuth using TSX stripmap data and 1 cm in range and 6 cm in azimuth for TSX spotlight data (0.8 x 0.25 m resolution images).

A limitation of the offset tracking techniques for landslides is the orientation of the landslide with respect to the radar. In order to avoid layover, the more favorably oriented landslides are those that face away from the satellite (to be covered by a sufficiently large number of radar pixels, at least several times the correlation window size) without being too steep (to avoid shadowing), or those that face towards the satellite but on gentle slopes (Figure 10.4). Furthermore, the pixels of radar images are often not squared (for instance, about 3 x 14 m in the range and azimuth directions for S1 images), which makes the accuracy of the algorithm different in range and in azimuth. This reduces the applicability of the algorithm for a specific dataset or any given landslide. All these limitations have restricted the application of this method to a few cases, despite its great potential for time-series generation with any cloud cover conditions.

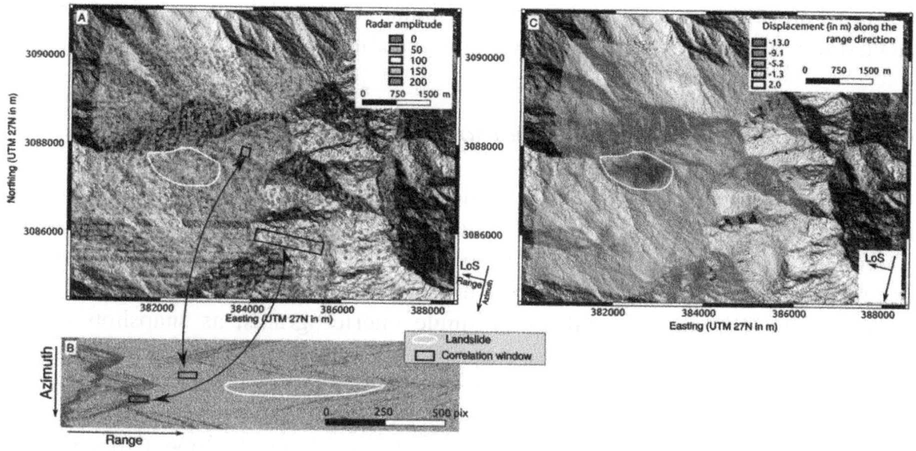

Figure 10.4. *Sentinel-1 SAR amplitude image shown projected onto a DEM (A) and in radar geometry (B). The white line shows the contour of a slow-moving landslide oriented along the range direction of the satellite. The correlation of two SAR images (C) shows a landslide displacement of –13 m along the line of sight of the satellite, that is, flowing away from the satellite. The black frames on subplots A and B are indicative of the correlation windows used at two locations and their corresponding real sizes in geographical coordinates (A). For a color version of this figure, see www.iste.co.uk/cavalie/images.zip*

10.4. InSAR for landslide studies

Since the first launches of civil SAR satellites in the 1990s (ERS, JERS, Radarsat), InSAR techniques have been applied to detect and monitor landslides (e.g. Carnec *et al.* 1996). InSAR is better suited to the study of very slow- to slow-moving landslides (Colesanti and Wasowski 2006; Handwerger *et al.* 2013; Wasowski and Bovenga 2014) with average velocities of mm/yr or cm/yr, using traditional C-band or X-band radars. However, longer wavelength (L-band and upcoming S-band) and/or high revisit rates might allow for the detection of faster displacements. The choice of the wavelength depends not only on the expected landslide velocity but also on the land cover of the area of interest. For instance, X-band signals are well suited to studying urban environments, where buildings and infrastructure offer radiometrically stable scatterers, with limited decorrelation over time. Individual, large infrastructures also benefit from monitoring with X-band due to its high sensitivity to smaller displacements and the high resolution of satellites such as the CSK and the TSX constellations (1–3 m). L-band radar is best suited for vegetated terrains, as the longer wavelength (~24 cm) is less sensitive to changes associated with growth in comparison to X- and C-band sensors (Schlögel *et al.* 2015; Nishiguchi *et al.* 2017). For example, Dini *et al.* (2019b), who carried out a study based on single interferogram analysis both in C- and L-band over a large area in Bhutan, reported that 92% of the objects detected were through L-band interferograms generated with ALOS-1 images.

10.4.1. *Standard versus multi-temporal InSAR analyses*

Standard and more advanced multi-temporal methods for the identification, classification, characterization, quantification and monitoring of displacements caused by geological and geomorphological processes of different natures in mountainous terrain are extensively documented in the literature. Standard methods generally rely on the generation and analysis of single interferograms, as snapshots of the deformation between two successive acquisitions. Examples of this application are available for the detection and inventorying of slope instabilities (Delaloye *et al.* 2007, e.g.), for the interpretation and quantification of surface displacements of different origins (e.g. Massonnet and Feigl 1998; Colesanti and Wasowski 2006; Cignetti *et al.* 2016) and for the classification of the geomorphological processes leading to the observed displacements (e.g. Dini *et al.* 2019b). Although a single interferogram is, as mentioned above, a mere snapshot of displacement occurring between two acquisitions, a detailed and in-depth analysis of a large number of interferograms over an area of interest can be quite revealing and, in fact, complementary to multi-temporal analyses in landslide applications (Figure 10.5). The analysis of interferograms of different temporal baselines can, for instance, reveal different phases of landslide activity through time.

Multi-temporal methods are based on the use of several interferograms for the extraction of velocity maps and time series. Retrieving a time-series of displacements enables us to evaluate the evolution of the kinematics underpinning the observed process. However, this is not the only advantage of multi-temporal analyses. The generation of time series can also mitigate unwanted effects (e.g. atmospheric artifacts) that may overprint the displacement signal and thus overcomes some of the limitations of standard differential SAR interferometry (Wasowski and Bovenga 2014). There are different types of approaches for InSAR time-series generation (see Chapter 5). Methods that take advantage of dominant single-point scatterers, such as the permanent scatterer approach (PS-InSAR) (Ferretti *et al.* 2001), yield good results in urbanized environments or over a large infrastructure. Other methods take advantage of the presence of distributed scatterers within each pixel. Among these, the small baseline subset (SBAS) approach (Berardino *et al.* 2002) is widely used. Distributed scatterers are those that characterize many natural land cover types, where no dominant contribution from individual features, such as debris or loose soil, exists. As opposed to urban environments, permanent scatterers are often fewer in natural landscapes, where they are generally provided by bare rock outcrops. In order to increase the number of exploitable scatterers for landslides, various methods that combine the advantages of the two previous groups have been developed, such as SqueeSAR (e.g. Ferretti *et al.* 2011; Intrieri *et al.* 2018) or StaMPS PS-InSAR (Kiseleva *et al.* 2014). The accuracy that can be obtained is in general of the order of 1–2 mm/yr for annual velocities and 5–10 mm/yr for cumulative displacements (Catani *et al.* 2014; Wasowski and Bovenga 2014).

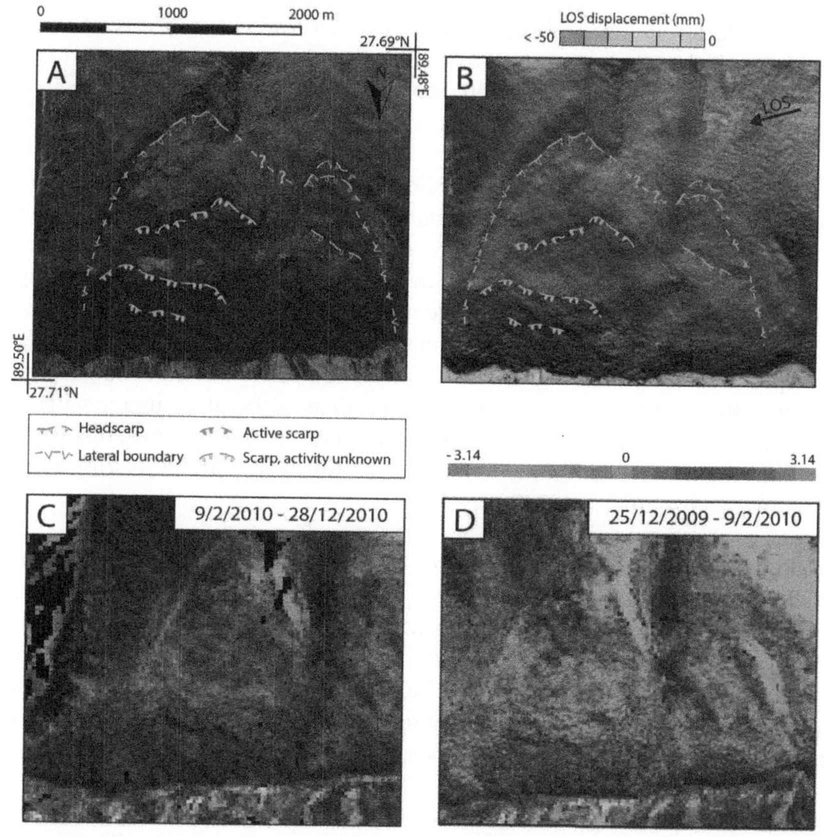

Figure 10.5. *A) Large, slow-moving landslide in NW Bhutan with mapped geomorphological features (modified from Dini* et al. *(2019b)); B) ALOS-1 cumulative displacements obtained with SBAS between February 2007 and February 2011 (modified from Dini* et al. *(2019a)); C and D) wrapped interferograms obtained with ALOS-1 data. For a color version of this figure, see www.iste.co.uk/cavalie/images.zip*

10.4.2. *Limitations in the use of SAR interferometry for landslide applications*

Despite its great potential for landslide studies, InSAR suffers from limitations that need to be carefully evaluated when conceiving new research as well as during the interpretation of results. Some of the limitations are specific to landslide applications. Landslides often occur in environmental settings that are difficult for InSAR applications; these include natural slopes on which individual, dominant scatterers are rare, dense vegetation and alpine climates entailing extensive snow cover. Land cover types such as those mentioned are unfavorable from the point of view of interferometric stability as they lead to low coherence over time (e.g. Cignetti *et al.* 2016). This, in turn, is associated with a scarcity of points for which displacement measurements can be reliably retrieved (Wasowski and Bovenga 2014), compromising the robustness and interpretation of the results.

Moreover, when working with landslides, it is not unusual to deal with areas characterized by high topographic gradients. Rough topography leads to geometrical distortions that can severely hinder the retrieval of interferometric information. This problem can, in some cases, be overcome with the use of ascending and descending geometries, but this is subject to acquisition availability and requires case-by-case evaluation. Moreover, satellites that carry SAR sensors follow orbits that are near-polar. This leads to a greater visibility of slope movements that are oriented east–west; thus, landslides that face north or south are difficult to image with InSAR as movements are only detected in the line of sight of the sensor. Thus, displacements occurring in the north–south direction cannot be resolved without making important assumptions (Samsonov *et al.* 2020).

In areas of rough topography, strong effects associated with the stratified atmosphere are often observed, affecting the entire scene. The phase changes associated with these effects are of long wavelengths and do not necessarily impair the simple detection of landslides, given the small size of the investigated area, but the retrieval of reliable time series and of reasonable average velocities can be hindered by the overprint of the atmospheric delay (e.g. Dini *et al.* 2019a) and can even cause the stable point chosen to be affected by seasonal oscillations.

An important limitation is related to an effect that is referred to as phase aliasing. The phase information is measured modulo 2π (see Chapter 6). Thus, it is not possible to unambiguously detect displacements that exceed in general a quarter of the wavelength between two successive acquisitions and two adjacent pixels (e.g. Wasowski and Bovenga 2014; Manconi 2021). When the displacement gradient exceeds this value, aliasing occurs and information on the displacement is not retrieved or severely underestimated. What this threshold corresponds to exactly is difficult to establish as it does not only depend on the displacement in the time interval between two acquisitions but also on the area affected by the displacements.

10.4.3. *Landslide detection with InSAR*

Many studies have now shown the interest of InSAR for detection and inventorying of slope movements. Delaloye *et al.* (2007), for example, used ERS data to generate a relatively small number of interferograms in which a large number of mass movements in the periglacial environment of the Swiss Alps were identified. Dini *et al.* (2019b) used a similar approach and visually inspected a large number of interferograms computed with ENVISAT (C-band) and ALOS-1 (L-band) SAR data. This allowed for the identification of a large number of previously unknown instabilities in the Himalayas of northwestern Bhutan. Moreover, Dini *et al.* (2019a) used the same dataset to generate, with the SBAS approach, average velocity and cumulative displacements maps on which more patterns associated with slope instabilities could be detected and inventoried, highlighting the complementary aspect of individual interferogram analysis and that of large-scale average velocity for landslide detection. The example in Figure 10.5 shows that, on the one hand, wrapped single interferograms may not allow the detection of slow displacement rates at shorter temporal baselines. The 46-day interferogram in Figure 10.5(D) shows only a weak signal at the bottom of the slope, likely due to a phase of deceleration during the dry season and to the short temporal baseline. The 10-month interferogram in Figure 10.5(C) shows significant activity in the lower part of the slope, but not in the upper half, hence some structures visible in the optical images and in the DEM would appear to be inactive through analysis of the interferograms. On the other hand, the difficulties that arise, for instance, the loss of coherence due to unfavorable land cover or during phase unwrapping related to aliasing, can cause loss of scatterers over large areas in multi-temporal processing approaches. For instance, the cumulative displacements obtained with a multi-temporal approach in Figure 10.5(B) show no information for the lower part of the slope, likely due to temporal decorrelation (presence of vegetation, fast displacements), while activity can be observed in the upper part of the slope where displacements are too subtle to be seen in individual interferograms. The use of PS-InSAR results at the regional scale, and in combination with ancillary data, has been attempted for the interpretation of velocity ranges, aiming at improving existing landslide inventories (e.g. Rosi *et al.* 2018).

The availability of InSAR-based inventories thus requires careful interpretation of the possible processes causing the detected displacements. Geomorphological analyses carried out visually with optical images are again commonly associated with this task. This does not only require expertise in the recognition of geomorphological features as descriptors of the investigated mass movements but also a sound knowledge of the main pitfalls and possible artifacts related to InSAR results. For example, Cignetti *et al.* (2016) performed a regional multi-temporal analysis based on the SBAS approach and compared the average velocity results with existing large, slow-moving landslide and rock glacier inventories in order to steer the interpretation of the InSAR results.

10.4.4. *Landslide characterization with InSAR*

At the site scale, some authors have demonstrated methods that combine the analysis of the spatial distribution of both surface geomorphological features and SAR phase values (Schlögel *et al.* 2015). Morphological features such as scarps, fractures, lobes and gullies are mapped on optical images or DEMs and are then used to guide the interpretation of wrapped and unwrapped interferograms. Although this type of methodology requires detailed geomorphological mapping and geophysical knowledge of the landslide for the interpretation of the InSAR deformation fields, it provides a basis for the investigation landslide mechanisms for rotational, translational and complex movements. At the regional scale, recent studies have shown the potential of InSAR in combination with geomorphological analyses for the association of the displacements obtained with specific landforms and with the geological material types involved, in order to give an interpretation to the processes (e.g. Dini *et al.* 2019a). The authors of these studies have used different approaches in order to unravel reversible seasonal and irreversible gravitational displacements, including those occurring in periglacial environments. They have also demonstrated the use of ascending and descending geometries in the decomposition between vertical and east–west components in obtaining 2D InSAR displacements (Rouyet *et al.* 2019; Daout *et al.* 2020). Moreover, just the analysis of the aspect dependence of the signal in the two acquisition geometries can reveal whether the largest component of the displacement is vertical or horizontal. With this type of analysis, insights can be gained regarding the dominant processes occurring in understudied slopes, as well as in the interannual variability of freeze–thaw cycles, in long-term permafrost variations and in groundwater table-induced displacements below the permafrost region. Methods involving the use of PS-InSAR and SqueeSAR data, as well as single interferograms, have been developed to characterize the kinematics of large, slow-moving landslides in the Italian Alps (Crippa *et al.* 2021). Such methods focus on the classification of observed average velocities, as well as on the presence of nested sectors within a landslide and the style of activity. This type of analyses is able to reveal, even at the regional scale, composite slope instabilities with nested sectors that might undergo different evolutionary trends towards failure.

10.4.5. *Landslide monitoring with InSAR*

InSAR-derived time-series can be used to evaluate external forcing on landslides of natural or anthropogenic origin. Handwerger *et al.* (2013) were able to show the annual velocity variations of slow-moving landslides in northern California with ALOS data and the delayed response of landslide accelerations to rainfall. The accuracy and coverage obtained with L-band SAR data made it possible to unravel the dependency between landslide kinematics and minor pore pressure changes. Coseismic reactivations of large landslides in Italy were imaged with the generation of interferograms, and the displacements observed were used to validate the displacements obtained through modeling of the rupture surface and the instantaneous deformation postulated (Moro *et al.* 2011). The availability of multiple sensors enabled the generation of long time-series ($\approx$20 yrs) and thus established causal effects between landslide velocities and triggers. For example, Strozzi *et al.* (2010) used InSAR in combination with other data to show the long-term evolution of a landslide in the Swiss Alps that was activated as a consequence of glacier retreat and presented acceleration phases related to snow melt and intense rainfall events. The quantification of displacements for very slow landslides using multi-temporal and multi-sensor InSAR analyses has also been shown to reveal evidence of recent activity in response to natural and anthropogenic factors (e.g. fluvial erosion, excavation, overloading, enhanced infiltration) (Herrera *et al.* 2013). Furthermore, excavations and stabilization measures have effects on natural slopes and the quantification of velocities through InSAR time-series has been shown to be a valuable tool to enhance the understanding of the interactions (Dini *et al.* 2020).

In the context of hazard assessment, landslide damage maps can be produced based on thresholds of deformation calculated over InSAR-derived LOS velocities or its projection along the line of maximum gradient of a landslide (e.g. Herrera *et al.* 2013; Barra *et al.* 2017; Solari *et al.* 2019). This type of analysis can be used with predictive aims, if thresholds are to be set above which damage might be expected. Barra *et al.* (2017) and Solari *et al.* (2019) used multi-temporal InSAR analyses based on PS-InSAR and SqueeSAR to retrieve "hot spots" of deformation over large regions. The aim was identifying and monitoring, through regular updates of the time series, areas characterized by active deformation. As the spatio-temporal noise of the InSAR results and velocity maps can contain a large number of points, this type of methodology strives to improve the readability of the results to be used in an operational context. This type of work is essential to bridge the gap that exists between retrieving the interferometric information and feeding this data into hazard and risk assessments in an easily readable, reliable and quantitative way.

The launch of satellites with high revisit rates (up to six days revisit time) is now allowing scientists to begin to consider the viability of closely monitoring displacements occurring on slopes and landslide velocity through time, as opposed to carrying out retrospective studies using archived data. This is appealing because effective monitoring can be used to establish clear triggering mechanisms and potentially provide alerts on thresholds. Timely monitoring would make it possible to apply inverse velocity methods with predictive abilities to InSAR velocity data. This possibility is still limited by unfavorable combinations of wavelengths and revisit rates with regard to landslide velocity, with phase aliasing potentially causing severe underestimations of the displacement rates which would prevent accurate failure forecasts (e.g. Manconi 2021). However, more authors are beginning to point towards the possibility of establishing regional-scale early-warning systems based on the continuous processing of S1 data due to its regularity of acquisition. Updated displacement time-series can be used to systematically feed landslide failure forecast models that require the timely identification of accelerations (Raspini *et al.* 2018). Recently, some notable case studies have been analyzed by means of back-analyses with InSAR multi-temporal techniques (e.g. Intrieri *et al.* 2018), in the attempt to detect precursory signals of failure and to establish whether such information could have been obtained pre-emptively and converted into an early warning. Although these studies are carried out post-failure, thus with plenty of available time for processing and analysis and with previous knowledge of the studied landslide, the findings are encouraging and seem to indicate that the better temporal resolution of S1 and the future launch of L-band satellites with high revisit rates (NISAR) might allow scientists to better analyze the evolution of a landslide as it unfolds, implying that some viable lead time for alerts might be reached in the future with this technique.

10.5. Conclusion

Satellite images have proven their efficiency in detecting, characterizing and monitoring slow-moving landslides with velocities from mm/yr to m/day, often situated in remote and/or steep areas. The specificities of landslides (small size, rough topography, high variability of velocity through time, variable terrain cover such as vegetation or snow) make the use of a combination of various sensors onboard satellites key for all these applications.

The availability of an increasingly large pool of sensors and the development of new methodologies allows for expert combination of different types of satellite data and methods. This represents a key factor in many landslide studies based on remote sensing, maximizing the potential for displacement retrieval and making use of the advantages and the complementarity of different techniques.

The significant time depth now covered by satellite images (up to 60 years) allows for the generation of long time-series of slow-moving landslide displacements over wide areas. This is of specific interest for the study of landslide processes in a changing world, where both climate change and anthropogenic stress on the natural environment are prominent. The study of these two forcings on landslide processes can therefore start to be inferred thoroughly.

The launch of high-resolution satellites and high-frequency satellites provides new opportunities to better study the physical processes of landslides in relation to their forcing factors, to map them more accurately and to estimate the related hazard. In particular, some back-analyses of catastrophic landslides have shown the existence of coherent motion before failure that can be measured with both SAR and optical images. The combination of different high-frequency satellites will certainly be of interest for such objectives. This rather operational prospective must not hide the large challenges research still has to face, related, for instance, to big data processing, transient motion detection abilities over large areas and integration of different sources of data.

10.6. References

Aryal, A., Brooks, B.A., Reid, M.E. (2015). Landslide subsurface slip geometry inferred from 3-D surface displacement fields. *Geophysical Research Letters*, 42(5), 1411–1417.

Barra, A., Solari, L., Béjar-Pizarro, M., Monserrat, O., Bianchini, S., Herrera, G., Crosetto, M., Sarro, R., González-Alonso, E., Mateos, R.M., Ligüerzana, S., López, C., Moretti, S. (2017). A methodology to detect and update active deformation areas based on Sentinel-1 SAR images. *Remote Sensing*, 9(10), 1002.

Berardino, P., Fornaro, G., Lanari, R., Sansosti, E. (2002). A new algorithm for surface deformation monitoring based on small baseline differential SAR interferograms. *IEEE Transactions on Geoscience and Remote Sensing*, 40(11), 2375–2383.

Bishop, K.M. (1999). Determination of translational landslide slip surface depth using balanced cross sections. *Environmental and Engineering Geoscience*, 2, 147–156.

Bontemps, N., Lacroix, P., Doin, M.-P. (2018). Inversion of deformation fields time-series from optical images, and application to the long term kinematics of slow-moving landslides in Peru. *Remote Sensing of Environment*, 210, 144–158.

Booth, A.M., Lamb, M.P., Avouac, J.-P., Delacourt, C. (2013). Landslide velocity, thickness, and rheology from remote sensing: La Clapière landslide, France. *Geophysical Research Letters*, 40(16), 4299–4304.

Bradley, K., Mallick, R., Andikagumi, H., Hubbard, J., Meilianda, E., Switzer, A., Du, N., Brocard, G., Alfian, D., Benazir, B. *et al.* (2019). Earthquake-triggered 2018 Palu Valley landslides enabled by wet rice cultivation. *Nature Geoscience*, 12(11), 935–939.

Caporossi, P., Mazzanti, P., Bozzano, F. (2018). Digital image correlation (DIC) analysis of the 3 December 2013 Montescaglioso Landslide (Basilicata, Southern Italy): Results from a multi-dataset investigation. *ISPRS International Journal of Geo-Information*, 7(9), 372.

Carnec, C., Massonnet, D., King, C. (1996). Two examples of the use of SAR interferometry on displacement fields of small spatial extent. *Geophysical Research Letters*, 23(24), 3579–3582.

Catani, F., Canuti, P., Casagli, N. (2014). The use of radar interferometry in landslide monitoring. In *Landslides in Cold Regions in the Context of Climate Change*, Shan, W., Guo, Y., Wang, F., Marui, H., Strom, A. (eds). Springer, Cham.

Cignetti, M., Manconi, A., Manunta, M., Giordan, D., De Luca, C., Allasia, P., Ardizzone, F. (2016). Taking advantage of the ESA G-POD service to study ground deformation processes in high mountain areas: A Valle D'Aosta case study, Northern Italy. *Remote Sensing*, 8(10), 852.

Colesanti, C. and Wasowski, J. (2006). Investigating landslides with space-borne synthetic aperture radar (SAR) interferometry. *Engineering Geology*, 88(3–4), 173–199.

Crippa, C., Valbuzzi, E., Frattini, P., Crosta, G.B., Spreafico, M.C., Agliardi, F. (2021). Semi-automated regional classification of the style of activity of slow rock-slope deformations using PS InSAR and SqueeSAR velocity data. *Landslides*, 1–19.

Daout, S., Dini, B., Haeberli, W., Doin, M.-P., Parsons, B. (2020). Ice loss in the northeastern Tibetan Plateau permafrost as seen by 16 yr of ESA SAR missions. *Earth and Planetary Science Letters*, 545, 116404.

Delacourt, C., Allemand, P., Casson, B., Vadon, H. (2004). Velocity field of the "La Clapière" landslide measured by the correlation of aerial and QuickBird satellite images. *Geophysical Research Letters*, 31(15).

Delaloye, R., Lambiel, C., Lugon, R., Raetzo, H., Strozzi, T. (2007). Typical ERS InSAR signature of slope movements in a periglacial mountain environment (Swiss Alps). *Envisat Symposium*, Montreux, Switzerland, 23–27 April.

Desrues, M., Lacroix, P., Brenguier, O. (2019). Satellite pre-failure detection and *in situ* monitoring of the landslide of the Tunnel du Chambon, French Alps. *Geosciences*, 9(7), 313.

Dille, A., Kervyn, F., Handwerger, A.L., d'Oreye, N., Derauw, D., Bibentyo, T.M., Samsonov, S., Malet, J.-P., Kervyn, M., Dewitte, O. (2021). When image correlation is needed: Unravelling the complex dynamics of a slow-moving landslide in the tropics with dense radar and optical time series. *Remote Sensing of Environment*, 258, 112402.

Dini, B., Daout, S., Manconi, A., Loew, S. (2019a). Classification of slope processes based on multitemporal DInSAR analyses in the Himalaya of NW Bhutan. *Remote Sensing of Environment*, 233, 111408.

Dini, B., Manconi, A., Loew, S. (2019b). Investigation of slope instabilities in NW Bhutan as derived from systematic DInSAR analyses. *Engineering Geology*, 259, 105111.

Dini, B., Manconi, A., Loew, S., Chophel, J. (2020). The Punatsangchhu-I dam landslide illuminated by InSAR multitemporal analyses. *Scientific Reports*, 10(1), 1–10.

Ferretti, A., Prati, C., Rocca, F. (2001). Permanent scatterers in SAR interferometry. *IEEE Transactions on Geoscience and Remote Sensing*, 39(1), 8–20.

Ferretti, A., Fumagalli, A., Novali, F., Prati, C., Rocca, F., Rucci, A. (2011). A new algorithm for processing interferometric data-stacks: SqueeSAR. *IEEE Transactions on Geoscience and Remote Sensing*, 49(9), 3460–3470.

Froude, M.J. and Petley, D.N. (2018). Global fatal landslide occurrence from 2004 to 2016. *Natural Hazards and Earth System Sciences*, 18(8), 2161–2181.

Guzzetti, F., Mondini, A.C., Cardinali, M., Fiorucci, F., Santangelo, M., Chang, K.-T. (2012). Landslide inventory maps: New tools for an old problem. *Earth-Science Reviews*, 112(1–2), 42–66.

Handwerger, A.L., Roering, J.J., Schmidt, D.A. (2013). Controls on the seasonal deformation of slow-moving landslides. *Earth and Planetary Science Letters*, 377, 239–247.

Handwerger, A.L., Booth, A.M., Huang, M.-H., Fielding, E.J. (2021). Inferring the subsurface geometry and strength of slow-moving landslides using 3-D velocity measurements from the NASA/JPL UAVSAR. *Journal of Geophysical Research: Earth Surface*, 126(3), e2020JF005898.

Herrera, G., Gutiérrez, F., García-Davalillo, J., Guerrero, J., Notti, D., Galve, J., Fernández-Merodo, J., Cooksley, G. (2013). Multi-sensor advanced DInSAR monitoring of very slow landslides: The Tena Valley case study (Central Spanish Pyrenees). *Remote Sensing of Environment*, 128, 31–43.

Hungr, O., Leroueil, S., Picarelli, L. (2014). The Varnes classification of landslide types, an update. *Landslides*, 11(2), 167–194.

Intrieri, E., Raspini, F., Fumagalli, A., Lu, P., Del Conte, S., Farina, P., Allievi, J., Ferretti, A., Casagli, N. (2018). The Maoxian landslide as seen from space: Detecting precursors of failure with Sentinel-1 data. *Landslides*, 15(1), 123–133.

Jia, H., Wang, Y., Ge, D., Deng, Y., Wang, R. (2020). Improved offset tracking for predisaster deformation monitoring of the 2018 Jinsha River landslide (Tibet, China). *Remote Sensing of Environment*, 247, 111899.

Jiang, S., Wen, B.-P., Zhao, C., Li, R.-D., Li, Z.-H. (2016). Kinematics of a giant slow-moving landslide in northwest China: Constraints from high resolution remote sensing imagery and GPS monitoring. *Journal of Asian Earth Sciences*, 123, 34–46.

Kiseleva, E., Mikhailov, V., Smolyaninova, E., Dmitriev, P., Golubev, V., Timoshkina, E., Hooper, A., Samiei-Esfahany, S., Hanssen, R. (2014). PS-InSAR monitoring of landslide activity in the Black Sea coast of the Caucasus. *Procedia Technology*, 16, 404–413.

Lacroix, P., Berthier, E., Maquerhua, E.T. (2015). Earthquake-driven acceleration of slow-moving landslides in the Colca Valley, Peru, detected from Pléiades images. *Remote Sensing of Environment*, 165, 148–158.

Lacroix, P., Bièvre, G., Pathier, E., Kniess, U., Jongmans, D. (2018). Use of Sentinel-2 images for the detection of precursory motions before landslide failures. *Remote Sensing of Environment*, 215, 507–516.

Lacroix, P., Araujo, G., Hollingsworth, J., Taipe, E. (2019). Self entrainment motion of a slow-moving landslide inferred from Landsat-8 time-series. *Journal of Geophysical Research – Earth Surface*, 124, 1201–1216.

Lacroix, P., Dehecq, A., Taipe, E. (2020a). Irrigation-triggered landslides in a Peruvian desert caused by modern intensive farming. *Nature Geoscience*, 13(1), 56–60.

Lacroix, P., Handwerger, A.L., Bièvre, G. (2020b). Life and death of slow-moving landslides. *Nature Reviews Earth & Environment*, 1(8), 404–419.

Le Bivic, R., Allemand, P., Quiquerez, A., Delacourt, C. (2017). Potential and limitation of SPOT-5 ortho-image correlation to investigate the cinematics of landslides: The example of "Mare à Poule d'Eau" (Réunion, France). *Remote Sensing*, 9(2), 106.

Li, M., Zhang, L., Ding, C., Li, W., Luo, H., Liao, M., Xu, Q. (2020). Retrieval of historical surface displacements of the Baige landslide from time-series SAR observations for retrospective analysis of the collapse event. *Remote Sensing of Environment*, 240, 111695.

Manconi, A. (2021). How phase aliasing limits systematic space-borne DInSAR monitoring and failure forecast of alpine landslides. *Engineering Geology*, 287, 106094.

Manconi, A., Kourkouli, P., Caduff, R., Strozzi, T., Loew, S. (2018). Monitoring surface deformation over a failing rock slope with the ESA sentinels: Insights from Moosfluh instability, Swiss Alps. *Remote Sensing*, 10(5), 672.

Massonnet, D. and Feigl, K.L. (1998). Radar interferometry and its application to changes in the earth's surface. *Reviews of Geophysics*, 36(4), 441–500.

Mazzanti, P., Caporossi, P., Muzi, R. (2020). Sliding time master digital image correlation analyses of CubeSat images for landslide monitoring: The Rattlesnake Hills landslide (USA). *Remote Sensing*, 12(4), 592.

Mitchell, A., McDougall, S., Nolde, N., Brideau, M.-A., Whittall, J., Aaron, J. (2019). Rock avalanche runout prediction using stochastic analysis of a regional dataset. *Landslides*, 17, 777–792.

Moro, M., Chini, M., Saroli, M., Atzori, S., Stramondo, S., Salvi, S. (2011). Analysis of large, seismically induced, gravitational deformations imaged by high-resolution COSMO-SkyMed synthetic aperture radar. *Geology*, 39(6), 527–530.

Mulas, M., Ciccarese, G., Truffelli, G., Corsini, A. (2020). Integration of digital image correlation of Sentinel-2 data and continuous GNSS for long-term slope movements monitoring in moderately rapid landslides. *Remote Sensing*, 12(16), 2605.

Nishiguchi, T., Tsuchiya, S., Imaizumi, F. (2017). Detection and accuracy of landslide movement by InSAR analysis using PALSAR-2 data. *Landslides*, 14(4), 1483–1490.

Pham, M.Q., Lacroix, P., Doin, M.P. (2018). Sparsity optimization method for slow-moving landslides detection in satellite image time-series. *IEEE Transactions on Geoscience and Remote Sensing*, 57(4), 2133–2144.

Raspini, F., Ciampalini, A., Del Conte, S., Lombardi, L., Nocentini, M., Gigli, G., Ferretti, A., Casagli, N. (2015). Exploitation of amplitude and phase of satellite SAR images for landslide mapping: The case of Montescaglioso (South Italy). *Remote Sensing*, 7(11), 14576–14596.

Raspini, F., Bianchini, S., Ciampalini, A., Del Soldato, M., Solari, L., Novali, F., Del Conte, S., Rucci, A., Ferretti, A., Casagli, N. (2018). Continuous, semi-automatic monitoring of ground deformation using Sentinel-1 satellites. *Scientific Reports*, 8(1), 1–11.

Raucoules, D., de Michele, M., Malet, J.-P., Ulrich, P. (2013). Time-variable 3D ground displacements from high-resolution synthetic aperture radar (SAR). Application to La Valette landslide (South French Alps). *Remote Sensing of Environment*, 139, 198–204.

Rosi, A., Tofani, V., Tanteri, L., Stefanelli, C.T., Agostini, A., Catani, F., Casagli, N. (2018). The new landslide inventory of Tuscany (Italy) updated with PS-InSAR: Geomorphological features and landslide distribution. *Landslides*, 15(1), 5–19.

Rouyet, L., Lauknes, T.R., Christiansen, H.H., Strand, S.M., Larsen, Y. (2019). Seasonal dynamics of a permafrost landscape, Adventdalen, Svalbard, investigated by InSAR. *Remote Sensing of Environment*, 231, 111236.

Rupnik, E., Daakir, M., Deseilligny, M.P. (2017). MicMac – A free, open-source solution for photogrammetry. *Open Geospatial Data, Software and Standards*, 2(1), 1–9.

Samsonov, S., Dille, A., Dewitte, O., Kervyn, F., d'Oreye, N. (2020). Satellite interferometry for mapping surface deformation time series in one, two and three dimensions: A new method illustrated on a slow-moving landslide. *Engineering Geology*, 266, 105471.

Schlögel, R., Doubre, C., Malet, J.-P., Masson, F. (2015). Landslide deformation monitoring with ALOS/PALSAR imagery: A D-InSAR geomorphological interpretation method. *Geomorphology*, 231, 314–330.

Schulz, W.H., McKenna, J.P., Kibler, J.D., Biavati, G. (2009). Relations between hydrology and velocity of a continuously moving landslide – Evidence of pore-pressure feedback regulating landslide motion? *Landslides*, 6(3), 181–190.

Singleton, A., Li, Z., Hoey, T., Muller, J.-P. (2014). Evaluating sub-pixel offset techniques as an alternative to D-InSAR for monitoring episodic landslide movements in vegetated terrain. *Remote Sensing of Environment*, 147, 133–144.

Solari, L., Del Soldato, M., Montalti, R., Bianchini, S., Raspini, F., Thuegaz, P., Bertolo, D., Tofani, V., Casagli, N. (2019). A Sentinel-1 based hot-spot analysis: Landslide mapping in north-western Italy. *International Journal of Remote Sensing*, 40(20), 7898–7921.

Strozzi, T., Delaloye, R., Kääb, A., Ambrosi, C., Perruchoud, E., Wegmüller, U. (2010). Combined observations of rock mass movements using satellite SAR interferometry, differential GPS, airborne digital photogrammetry, and airborne photography interpretation. *Journal of Geophysical Research: Earth Surface*, 115(F1).

Stumpf, A., Malet, J.P., Allemand, P., Ulrich, P. (2014). Surface reconstruction and landslide displacement measurements with Pléiades satellite images. *ISPRS Journal of Photogrammetry and Remote Sensing*, 95, 1–12.

Stumpf, A., Malet, J.-P., Delacourt, C. (2017). Correlation of satellite image time-series for the detection and monitoring of slow-moving landslides. *Remote Sensing of Environment*, 189, 40–55.

Stumpf, A., Michéa, D., Malet, J.-P. (2018). Improved co-registration of Sentinel-2 and Landsat-8 imagery for earth surface motion measurements. *Remote Sensing*, 10(2), 160.

Sun, L. and Muller, J.-P. (2016). Evaluation of the use of sub-pixel offset tracking techniques to monitor landslides in densely vegetated steeply sloped areas. *Remote Sensing*, 8(8), 659.

Sun, L., Muller, J.-P., Chen, J. (2017). Time series analysis of very slow landslides in the three Gorges region through small baseline SAR offset tracking. *Remote Sensing*, 9(12), 1314.

Wasowski, J. and Bovenga, F. (2014). Investigating landslides and unstable slopes with satellite multi temporal interferometry: Current issues and future perspectives. *Engineering Geology*, 174, 103–138.

Yang, W., Wang, Y., Sun, S., Wang, Y., Ma, C. (2019). Using Sentinel-2 time series to detect slope movement before the Jinsha River landslide. *Landslides*, 16(7), 1313–1324.

11

Remote Sensing of Glacier Motion

Amaury DEHECQ[1,2], Bas ALTENA[3], Alex S. GARDNER[4], Emmanuel TROUVÉ[5] and Silvan LEINSS[5,6]

[1] *Laboratory of Hydraulics, Hydrology and Glaciology (VAW), ETH Zurich, Switzerland*

[2] *IGE, CNRS, IRD, INP, Grenoble Alpes University, France*

[3] *IMAU, Utrecht University, The Netherlands*

[4] *NASA Jet Propulsion Laboratory, California Institute of Technology, Pasadena, USA*

[5] *LISTIC, Savoie Mont Blanc University, Annecy, France*

[6] *Institute of Environmental Engineering, ETH Zurich, Switzerland*

11.1. Introduction

The Earth's frozen water, composed of seasonal snow, sea ice, lake ice, river ice, glaciers and permafrost, makes up the cryosphere. This chapter focuses on the observation of the Earth's glacier ice, which is found in the large Antarctic and Greenlandic ice sheets, in ice caps, in icefields and in valley glaciers.

Glaciers cover about 10% of Earth's land area and make up two-thirds of the world's freshwater resources (International Panel on Climate Change 2019). They play a key role in the Earth energy balance by reflecting solar radiation and through melting, which in turn cools the ocean and atmosphere. Glaciers are an important part of the water cycle, storing water in the form of snow and ice that is then released to streams,

Surface Displacement Measurement from Remote Sensing Images,
coordinated by Olivier CAVALIÉ and Emmanuel TROUVÉ.
© ISTE Ltd 2022.

groundwater and the oceans during warm periods. As glaciers are highly sensitive to changes in climate, they are an iconic representation of human-induced climate warming, with impacts reaching far beyond their regional extent. Indeed, glaciers contribute to about half of the modern (1993–2015) observed sea level rise, which threatens nearly 10% of the world's population living in low-lying areas (International Panel on Climate Change 2019, Chapter 4). Another 10% of the world's population living in high mountainous areas is directly affected by changes in stream flow and seasonality, which is partially linked to glacier retreat (International Panel on Climate Change 2019, Chapter 2). For these reasons, monitoring glacier changes is of great societal importance, and remote sensing has recently offered a unique opportunity to study these changes at a global scale.

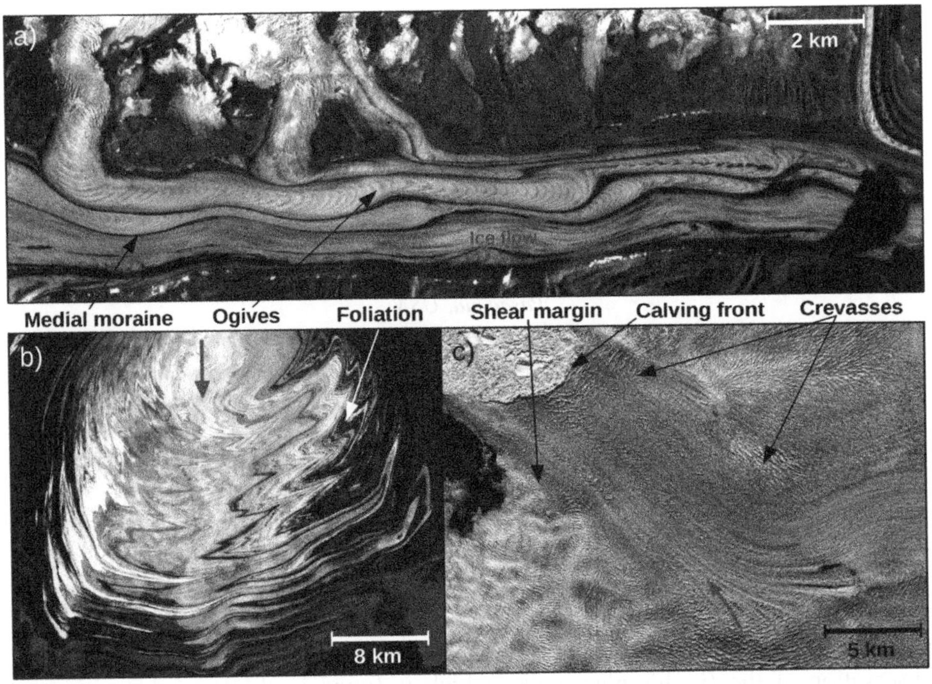

Figure 11.1. *Features of ice flow visible in Sentinel-2 images. a) Art Lewis glacier, Alaska, on September 15, 2019, showing medial moraines, ogives and crevasses; b) Matanuska glacier, Alaska, on September 15, 2019, showing foliations; and c) Sermeq Kujalleq, Greenland, on August 22, 2020, depicting shear margins, crevasses and calving front. Red arrows indicate the general direction of ice flow. For a color version of this figure, see www.iste.co.uk/cavalie/images.zip*

Two properties distinguish glaciers from other elements of the cryosphere. First, unlike sea, lake and river ice or permafrost, which form from the refreezing of water, glaciers form from an accumulation of snow. Second, as opposed to passively drifting sea ice or river ice, glaciers flow under the pull of gravity, shaping valleys and transporting fresh water into the oceans. Many features express the flow of ice on the surface (Figure 11.1). Crevasses are a sign of deformation of the brittle ice when local stresses overcome the tensile strength of ice (Vaughan 1993). Medial moraines, which are rock debris trapped at glacier convergence zones and clearly visible from satellite images, draw streaklines as the ice moves down valley. Ogives or Forbes bands, alternations of light and dark bands on the glacier surface that form below icefalls, are timelines and their U shape highlights the impact of the drag on valley walls in counteracting flow. These surface features directed the attention of scientists in the 18th and 19th centuries to understanding ice flow, which was initially demonstrated by following the displacement of painted stones placed on the glacier surface (Rémy and Testut 2006). The displacement of surface features was later measured more precisely using theodolites and later yet using GPS stations (Cuffey and Paterson 2010, section 8.1.4). The extensive availability of remote sensing imagery has recently opened a wide range of opportunities to observe spatially distributed ice flow at local to regional scales. This chapter first highlights the main characteristics of ice flow compared to other kinds of geophysical motion. It then describes methodologies specifically developed to measure ice flow from remote sensing imagery. Finally, it provides an overview of glaciological advances made available by remote sensing and future perspectives.

11.2. What is glacier motion?

11.2.1. *Physical basis of ice flow*

Ice flows, under the pull of gravity, follow two main processes (Cuffey and Paterson 2010). The first is the slow and continuous deformation of ice under an applied stress, which is called creep. The rate of creep depends on the local ice thickness and surface slope, as well as the viscosity of the ice, which in turn depends on ice temperature, water content and crystal size and orientation. Ice deformation is maximum at the glacier bed, where vertical shear stresses are highest, and zero at the surface. Since motion is the sum of all ice deformation occurring below a given depth, the motion is minimum at the base, where it is equal to the rate of slip, and maximum at the surface (Figure 11.2). The second process explaining glacier flow is basal motion, the sum of bed deformation (in the case of a soft bed, for example, till), and sliding on the bed . Basal sliding only occurs when the ice is at or near its melting temperature. It depends on the bed roughness and rheological properties and on basal water pressure, which can raise the ice off the bed and reduce friction.

Spatial heterogeneity in ice flow leads to zones of mass convergence or divergence. Due to the incompressibility of ice, this causes vertical motion. This vertical motion is essentially downward in the accumulation zones, where accumulated snow compacts into ice, and upward in the ablation zone, where ice melt exceeds snowfall (Figure 11.2). Ice flow thus enables a glacier to reach a geometry that is in equilibrium with the climate, by redistributing the mass from accumulation zones to ablation zones and by compensating for local mass gain and loss.

To date, remote sensing only enables the observation of surface motion (or near- surface for ice-penetrating radar images). However, these observations, together with ice mechanics theory, can provide valuable information on depth properties such as ice thickness or basal sliding, which are otherwise extremely difficult to observe (section 11.4). Offset tracking only measures the two-dimensional displacement of the surface, and InSAR measures the displacement along the satellite's line-of-sight, although the combination of several acquisitions can enable the estimation of the 3D displacement (section 11.3.3). It is important to note that this surface motion, either observed in the horizontal plane or projected parallel to the slope, can differ from the actual three-dimensional displacement of the ice (Figure 11.2). Finally, remotely sensed ice flow is generally described in an Eulerian framework, in which the movement of ice at a fixed location is known through time, as opposed to a Lagrangian framework in which the position and velocity of individual features are known (despite the commonly used term "feature tracking"). The conversion from one framework to another is, however, straightforward (Altena and Kääb 2017b).

11.2.2. *Scales and processes of displacement in glaciology*

Ice flow occurs at different temporal and spatial scales. Stick-slip events cause centimeter-scale displacements of the glacier on the bedrock within minutes (Fischer and Clarke 1997). Changes in basal water pressure cause fluctuations in basal sliding at hourly, daily and seasonal scales that translate into the cumulative displacement of tens or hundreds of meters (e.g. Bartholomew *et al.* 2010; Allstadt *et al.* 2015). GPS and remote sensing observations have demonstrated that ice motion is typically faster during the melt season and slower in the cold season (e.g. Armstrong *et al.* 2017). The fastest recorded velocities, for example, at Sermeq Kujalleq glacier, Greenland, lead to displacements of up to 15 km over the course of a year (Figure 11.6). Changes in climate also impact glacier accumulation and surface and subaqueous (for ocean-terminating glaciers) melt, which in turn influence the glacier's mass and velocity at decadal scales (Joughin *et al.* 2018; Dehecq *et al.* 2019). For the Greenland and Antarctic ice sheets, velocity fluctuations can take millennia to propagate over the several thousand kilometers of ice and reach an equilibrium (Alley and Whillans 1984). Observing and understanding the different processes that lead to ice flow from the millisecond to millennial scale, from the centimeter to the continental scale, are some of the most exciting challenges of glaciology. Remote sensing has been, and will continue to be, a powerful tool to tackle these challenges.

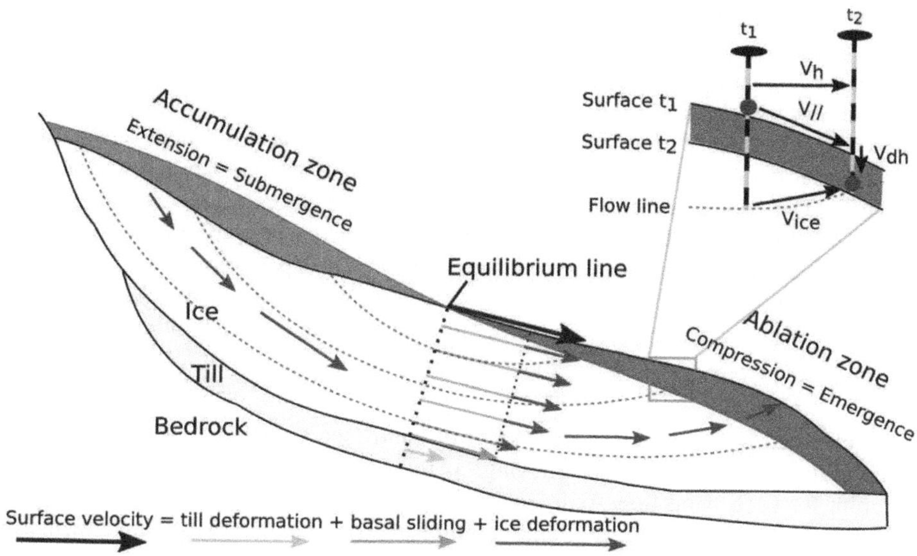

Figure 11.2. *Longitudinal sketch of ice velocity in a glacier. Velocity varies with depth (green arrows), and surface velocity is the sum of till deformation, sliding of the ice on its bed and deformation of the ice. Velocity varies longitudinally, with a maximum near the equilibrium line: zones of extension lead to submergence of the surface, compensated by annual snow accumulation (blue); zones of compression lead to emergence of the surface, compensated by annual ablation (red). Remote sensing typically estimates horizontal velocity (V_h) or slope-parallel velocity ($V_{//}$), which may differ from the three-dimensional ice velocity (V_{ice}) measured, for example, by GPS, which again differs from the velocity of an object at the surface (gray circle) given by the sum of the slope-parallel velocity and surface lowering (V_{dh}). For a color version of this figure, see www.iste.co.uk/cavalie/images.zip*

11.3. Measuring glacier displacement from satellite data

Many characteristics of glaciers make the measurement of surface displacement challenging: lack of image contrast, rapidly changing surface properties, strong topographic relief and spatially and temporally varying ice flow. Many approaches have been developed to overcome these issues. Here, we summarize most of these approaches and discuss best practices in a glaciological context, while the general principles of optical offset tracking, SAR offset tracking and InSAR are detailed in Chapters 2, 3 and 4, respectively. This section presents the methodologies needed at each step to determine displacements from satellite imagery; from the choice of the image preprocessing and displacement measurement method to the post-processing filtering and displacement field fusion strategies.

11.3.1. *Image preprocessing*

This section primarily focuses on optical images, as SAR-specific preprocessing has been extensively covered in Chapter 3. The highly reflective snow-covered glacier surfaces can lead to saturation of imaging sensors, particularly for older sensors with lower dynamic range and which stored information with an 8-bit depth (Landsat 1–7, ASTER). The lack of contrast or surface features in glaciers is also common when fresh snow cover is prevalent, be it in the high altitude accumulation zones or during the cold season. Two prepossessing steps are typically performed to mitigate the impact of low contrast and saturated imagery: a selection of the best spectral band or band combination and a filtering of the images to enhance local contrast.

11.3.1.1. *Spectral band selection*

The selection of an appropriate spectral band (see Chapter 1, section 1.1) can be critical for successful offset-tracking, as the reflectance of snow and ice will be different for each band (e.g. Tedesco 2015; Figure 7.1). Dowdeswell and McIntyre (1986) have shown that most Landsat multispectral scanner (MSS) and thematic mapper (TM) bands are subject to saturation for high sun angles in the polar regions. Consequently, near-infrared bands are usually preferable over ice as they are less subject to saturation (Dehecq *et al.* 2015; Rosenau *et al.* 2015) and can show features resulting from different snow grain sizes. However, the numerous types of glacier surface (bare ice, snow, rock debris) make it difficult to select one spectral band that performs well in all conditions (Redpath *et al.* 2013). Therefore, a principal component analysis (PCA) of all or a selection of the bands can be performed (Scambos *et al.* 1992; Ahn and Howat 2011; Dehecq *et al.* 2015), which maximizes the variance of the intensity into a single band and can improve the image correlation (Figure 11.3).

11.3.1.2. *High-pass filtering*

To reduce the impact of larger-scale features that do not move with the glacier surface, such as shadows or topographic shading, it is recommended to apply a high-pass filter to the images (Figure 11.3, Scambos *et al.* 1992; Ahn and Howat 2011). Additional filters, such as edge detection filters (Sobel filter) used in so-called orientation correlation (Heid and Kääb 2012; Dehecq *et al.* 2015) or Wallis filters (Gardner *et al.* 2018), have been proposed, while band-pass filtering is used in geodetic imaging software like CosiCorr (Leprince *et al.* 2007). The choice of the optimal filter depends on the quality of the imagery and sensor, but filter window sizes of 3–7 pixels are typical.

11.3.1.3. *Image striping*

Early optical images often suffer from varying degrees of striping (sometimes also referred to as banding), which can be accentuated over low-contrast surfaces such as glaciers. Striping is caused by small differences in the radiometric calibration of individual sensors. Such striping can be highly problematic for measuring surface displacements as it often represents regularly spaced artifacts in image intensity that produce high correlations between images. Striping can be partially mitigated by decomposing the image using Fourier image decomposition and reconstruction. In this approach, the image is decomposed into Fourier series, components with high-power around the typical stripe frequency are removed (power set to zero) and the image is reconstructed (Rosenau *et al.* 2015).

11.3.1.3.1. Missing data

Some images may present gaps due to malfunctioning of the sensors or downlink segment, especially for early satellite missions (the failure of the Scan Line Corrector onboard Landsat 7 on May 31, 2003, is a prominent example). Many cross-correlation algorithms use discrete Fourier transforms that can make explicit handling of missing data challenging. One approach to resolve this is to use a high-pass filter that normalizes the local mean and variance (e.g. a Wallis filter) and then fills missing data with white noise with the same mean and variance as the filtered image (Warner and Roberts 2013; Gardner *et al.* 2018). This allows for Fourier decomposition without affecting the correlation score between images.

11.3.2. *Offset-tracking algorithms*

Offset tracking, also frequently referred to as "feature tracking" or "image correlation", is a routine that repeatedly finds the best-matching location of a subset of image 2 (*I2* or template) in image 1 (*I1*). This is done by "sliding" the templates over *I1*, comparing overlapping patches and identifying the location that maximizes the similarity (see Chapter 2, Figure 2.1). Similarity is typically calculated using normalized cross-correlation or variants of it (see Chapter 2, section 2.2.3.2 and Chapter 3, section 3.2.2). The area over which the template is "slid" in *I1* is referred to as the search window. The values of similarity within that search window are referred to as the "correlation score" in this chapter, given the predominant use of the normalized cross-correlation. Several considerations (e.g. template size, search window) should be taken into account to optimize spatial coverage and displacement accuracy.

One such consideration is the size of the template. Selection of a template size is a trade-off between improved signal (large template) and improved spatial resolution (small template). An increased spatial resolution is particularly critical in areas of complex deformation, such as glacier icefalls or shear margins, while large templates are required to resolve ice motion in areas with low contrast and sparse surface features (e.g. accumulation areas). Several methods can be implemented to optimize the selection of template size. A first method is to work with adaptive template sizes. The idea is to adaptively select a template size that achieves a good correlation score (Debella-Gilo and Kääb 2011) or good "coherence" (Gardner *et al.* 2018). In cases of more complex deformation, an irregular or rotated template can be considered, or rotation-invariant correlation methods (Liu *et al.* 2018). Alternatively, the combined estimation of displacement, rotation and strain rate can provide highly accurate solutions (Debella-Gilo and Kääb 2012). Methods that account for template distortion allow for the use of larger templates but the increase in the degrees of freedom in the template matching can lead to an increase in spurious matches and computational cost, both of which often favor simpler approaches.

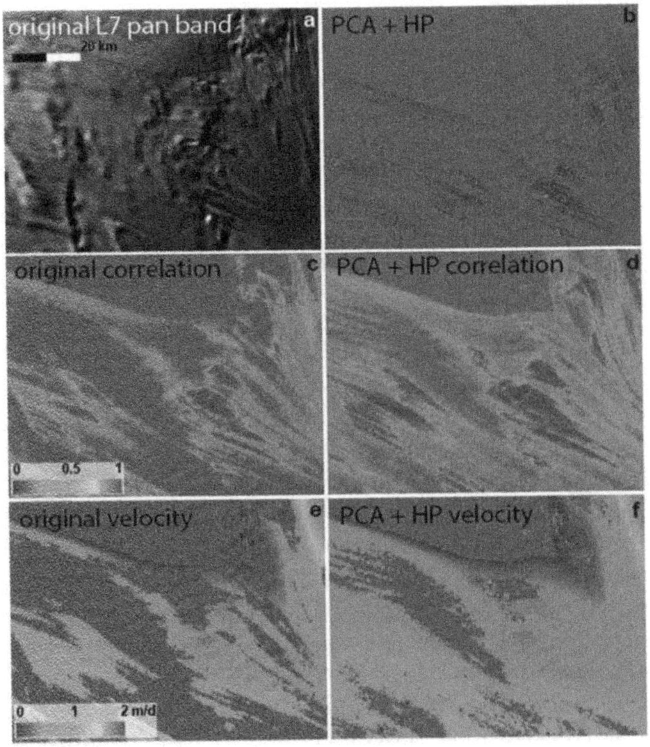

Figure 11.3. *Benefits of the preprocessing steps illustrated with a Landsat 7 image pair (year 2001) of West Antarctica. High-pass filter and principal component analysis (PCA) of bands 2, 3, 4 and 8 (b) of the original image (a) lead to enhanced glacier features. This produces a higher correlation score (d) and more spatially complete filtered velocity fields (f) compared to the original image (c and e). Figure reproduced from Fahnestock* et al. *(2016), copyright Elsevier. For a color version of this figure, see www.iste.co.uk/cavalie/images.zip*

A second consideration is the size of the search window, which can also influence the quality of offset-tracking results. Small search windows can be used when the expected displacement between repeat images is small (e.g. short time separation and/or slow moving surface) but large search windows are required when the location of the template within *I1* is highly uncertain or the expected displacement is large. Large search windows, however, are not desirable, as they increase the number of computations and the probability of identifying a spurious "best" match. A solution to reduce the size of the search window is by centering the search window on the expected offset between the template and *I1*. This can be done using a pyramidal approach to iteratively refine the displacement and search window (Leprince *et al.* 2007; Gardner *et al.* 2018, section 2.3.3). The template and search window are initially set at a large value to estimate the displacement on a sparse grid. This prior information can then be used to estimate an average displacement and narrow the search window for a second iteration on a denser grid. The sparse search can be done either by degrading the image resolution or by estimating the displacement with a sparse grid on the full-resolution imagery. Over glacier surfaces, the latter is more advantageous as it ensures that the small-scale features that are used to identify matches between images are not lost. Alternatively, search windows can be centered using an *a priori* estimate of glacier motion. This approach is often referred to as a downstream search (e.g. Gardner *et al.* 2018).

Offset-tracking results can also be improved using a combined estimation of displacements from a triplet of images (Altena and Kääb 2017a). This approach can be performed in a computationally efficient manner in the frequency domain, where displacement is calculated from a simple multiplication of the image's discrete Fourier transforms (Altena and Kääb 2017a). This methodology is highly robust to outliers, though if a single image is corrupted, the estimation fails. Another approach to make tracking more robust consists of using the temporal redundancy in time series of cross-correlations. Averaging the correlation functions from different image pairs (referred to as "ensemble matching") increases the signal-to-noise ratio, or peakiness, of the correlation at the location of the matched feature. Ensemble matching reduces the template size and increases the coverage ratio of successful displacement estimates (Altena and Kääb 2020; Chapter 3, section 3.2.3).

Optical flow is an alternative to offset tracking, in which velocity is computed from spatiotemporal derivatives of the image intensity (Fleet and Weiss 2006). These methods usually make the assumption that a given feature has a constant illumination throughout the images, referred to as the brightness constancy assumption (Vogel *et al.* 2012). This assumption is generally violated for remote sensing images covering glaciers due to varying illumination and surface conditions, which hampers the use of optical flow in glaciology. Additionally, the system of equations in differential optical flow methods is under-determined (one relationship for two velocity components) and requires an assumption about the smoothness of the velocity, an assumption that can be either local or global (Vogel *et al.* 2012). However, these methods present several advantages over offset tracking, such as substantially faster computation and high resolution (e.g. Charrier *et al.* 2020), which is the reason why several studies have implemented these methods in a glaciological context. Vogel *et al.* (2012) showed that a variation of optical flow allowing for smooth illumination variations can provide better matches than manual tracking except in areas of large surface changes. Altena and Kääb (2017b) demonstrated that by extracting pixel intensities along a flow line in a series of images, pixel displacement can be manually identified as lines in a space–time graph, enabling the velocity evolution to be resolved with great temporal detail. The brightness constancy issue can also be solved by applying a rank filter to the images, with promising results obtained both for SAR and optical images (Charrier *et al.* 2020), but implementations have been tested only at small spatial scales so far.

All of these approaches estimate a sub-pixel displacement, with a theoretical accuracy of typically 1/10 of a pixel (see Chapter 2, section 2.4). However, many factors influence the actual accuracy of the retrieved offsets. Environmental factors such as snowfall, melt, ice deformation, ionospheric propagation delay (for SAR data; see Chapter 4, section 4.3.2) and clouds (for optical data) make the matching of features between consecutive images more difficult and lead to spurious matches. The methods used to filter spurious matches are discussed in section 11.3.4.1. Image processing factors such as georeferencing and orthorectification (section 11.3.4.2) or pixel locking (Chapter 3, section 3.2.1.1) may also influence the measurement accuracy.

11.3.3. *InSAR*

SAR differential interferometry (DInSAR) has been used to measure glacier displacement since ERS-1 made this technique possible from repeat-pass spaceborne SAR images. DInSAR has been applied to measure ice sheet motion since 1993 (Goldstein *et al.* 1993) and now allows continent-wide mapping of Antarctic ice velocity by mosaicking DInSAR results from different SAR missions (Mouginot *et al.* 2019). Over the polar regions, DInSAR is also used to separate the ice flow velocity from the tidal motion in order to locate grounded versus floating ice (Rignot *et al.* 2011a). Over mountain glaciers, DInSAR measurements are affected by several, constraints described in Chapter 4, and remain limited by the availability of appropriate acquisitions. The three-day repeat-pass phase at the beginning of the ERS-1 mission and the one-day repeat-pass Tandem phase of the ERS-1 and ERS-2 missions were particularly well suited for obtaining high-quality interferograms over mountain glaciers (Rabus and Fatland 2000; Luckman *et al.* 2007). In this section, we present the main constraints and processing issues specific to InSAR measurement of glacier surface displacement.

11.3.3.1. *Temporal decorrelation*

One of the main challenges is the temporal decorrelation caused by changing dielectric properties of glacier surfaces. This is caused in particular by melt, snow/firn transformation, water runoff, debris motion, snowfall and wind redistribution. On low latitude glaciers, useful interferograms can be obtained only during winter with short temporal baselines. In the European Alps for instance, in C-band (see Chapter 1, section 1.1.2 for a list of radar bands), a series of one-day interferograms were successfully obtained during the ERS-Tandem mission (Trouvé *et al.* 2007) but six-day interferograms from Sentinel-1 are rarely exploitable (Jauvin *et al.* 2018). In X-band, the 11-day pairs from TerraSAR-X and PAZ data (see Chapter 1, section 1.2.5) acquired in winter over this site rarely provide a coherence level sufficient to monitor displacement by DInSAR. InSAR results in L-band are rare for such glaciers due to the 46-day and 14-day repeat cycles of ALOS and ALOS-2, respectively. For glaciers that provide sufficient coherence, time series can be used to implement multi-temporal InSAR techniques, such as SBAS (see Chapter 5), or to apply multi-temporal filters to improve the overall quality of the measurements (Prébet *et al.* 2019).

11.3.3.2. *Phase unwrapping*

Another major challenge is the unwrapping of the InSAR phase (see Chapter 6), which is even more complex for fast-moving temperate glaciers. The following factors can increase the difficulty of phase unwrapping:

– the presence of disconnected areas due to shadow or layover created by steep topography or due to low coherence of fast-changing structures such as crevasses;

– strong velocity gradients of ice flow, both across and along glacier, which result in variable fringe patterns or aliased fringes when the spatial sampling does not respect the Nyquist criterion (the phase varies more than π between adjacent pixels);

– the sparse spatial distribution of fast flowing glaciers amidst stationary mountains, which can cause the InSAR coregistration to fail for fast-moving areas, resulting in a complete loss of coherence (see Chapter 4, section 4.2.4) or discontinuities in displacement near glacier boundaries.

Several techniques have been proposed to help overcome difficulties in phase unwrapping. First, adaptive phase and coherence filters (e.g. Goldstein and Werner 1998) can be applied to reduce phase noise and reveal low-coherence areas where unwrapping cannot be performed. Second, if the unwrapping is based on a conventional minimum cost flow approach, use of accurate masks is recommended to avoid paths going through hidden discontinuities. Third, an alternative approach based on weighted least squares can also be used with appropriate weights to avoid the connection of the glacier surface with motion-free areas. Such an approach provides a global result where low-coherence areas are interpolated and errors visible as remaining fringes in the difference between the wrapped phase and rewrapped result (Trouvé *et al.* 2007). With any of these approaches, if the glacier does not include a motion-free area, the result is obtained with an unknown multiple of 2π due to the lack of a "zero" reference. This additional constant can be found by using a punctual *in situ* measurement or prior knowledge of the range of the expected velocity derived, for instance, from other displacement measurement techniques, such as offset tracking.

However, solving this problem is not necessary if only the spatial derivative of the displacement field is needed, for example, for strain rate computations.

11.3.3.3. *Reprojection of the line-of-sight displacement vector*

Another constraint is the geometry of the DInSAR technique, which only measures a single component of the displacement vector, i.e. its projection on the radar line-of-sight (LOS). Deriving 3D displacement vectors requires either combining complementary measurements or hypothesizing flow direction. Complementary measurements may come from different orbits (ascending and descending) and from different techniques, such as along-track interferometry (Gourmelen *et al.* 2011) or offset-tracking results (Joughin 2002; see also Chapter 3). When at least three projection directions of the displacement vector are known, the east, north and up components can be computed by inverting a linear system, as described in Chapter 3, section 3.2.4.3. If no other observation is available, the hypothesis of a "surface parallel flow", i.e. along the main slope direction, can be used in glaciology (Reeh *et al.* 2003) as an approximation to transform the measured DInSAR 1D velocity v_{LOS} into a 3D velocity vector $\vec{v}_{3d}$ by retro-projection:

$$\vec{v}_{3d} = \frac{v_{\text{LOS}}}{\vec{l}_{\text{LOS}} \cdot \vec{d}_m} . \vec{d}_m \qquad [11.1]$$

where $\vec{l}_{\text{LOS}}$ denotes the LOS direction and $\vec{d}_m$ the flow direction estimated from a low-pass-filtered DEM (if possible about 10 times the ice thickness to account for longitudinal stress coupling (Kamb and Echelmeyer 1986)).

11.3.3.4. *Sources of uncertainty*

Finally, different sources of error (see Chapter 6) should be considered in the estimation of the InSAR result uncertainty:

– the presence of tropospheric delays (see Chapter 4, section 4.3.1) can often be neglected due to the low temperature and low humidity in glacier environments. If required, interferograms can be corrected by referencing them to the surrounding non-moving areas, since atmospheric fringes usually have wavelengths much longer than the length scales of glaciers themselves;

– residual topographic phase errors due either to glacier elevation change with respect to the available DEM or interferometric bias in the presence of snow/ice penetration (e.g. Dehecq *et al.* 2016). As the topographic phase scales with the perpendicular baselines, such errors are reduced for recent missions such as Sentinel-1, which have a relatively small perpendicular baseline;

– the "surface parallel flow" hypothesis accounts for the vertical displacement associated with down slope flow, but ignores the vertical component of the ice deformation (see Figure 11.2). However, this vertical component is in the order of a few centimeters to meters per year (Vincent *et al.* 2021) and is generally negligible compared to the horizontal velocity.

11.3.4. *Filtering and corrections of displacement fields*

11.3.4.1. *Filtering*

Changing surface conditions, complex terrain or peculiar glacier features can influence offset tracking and lead to mismatches that need to be filtered out to isolate the true displacement of the ice surface. Specific challenges include 1) the lack of image contrast in areas covered with fresh snow, glacier accumulation zones or mountain shadows; 2) the presence of features that do not move with ice flow such as nunataks, supraglacial lakes, glacier fronts, bedrock bumps and firn or snow lines; and 3) the presence of periodic features, such as ogives and crevasses, or one-dimensional features (e.g. longitudinal moraines) for which no unique match exists (Figures 11.1 and 11.4). Filtering mismatches from displacement measurements is the most important and challenging step in generating good estimates of glacier motion. Since most offset-tracking algorithms simply identify the offset that maximizes the similarity between two images, a displacement estimate is always provided, regardless of the confidence in the match.

Many methodologies have been implemented to detect and remove mismatches in a glaciological context. The most straightforward filtering step is to select matches based on their correlation score or signal-to-noise ratio (see Chapter 3, equation [3.6]; Scambos *et al.* 1992; Strozzi *et al.* 2002). More complex indicators, such as the width of the correlation peak or its height relative to a second peak, can also be used (Fahnestock *et al.* 2016). However, a high correlation score or signal-to-noise ratio is not a guarantee of an accurate match and additional filters must be used. When the maximum velocity or general direction is known, these characteristics can be used to remove outliers. Many studies have taken into account the relative spatial smoothness of ice flow velocity to filter out outliers based on their deviation from neighbor values. This can be done using either the velocity direction (Burgess *et al.* 2012) or its magnitude by employing a low-pass (mean, median) filter, generally of sizes of 3×3 and 5×5 pixels (Heid and Kääb 2012; Fahnestock *et al.* 2016), or by calculating the strain rate at each pixel (Rosenau *et al.* 2015). However, the choice of the deviation threshold largely depends on the search window size used and hence on the maximum expected velocity. A solution is to apply the filtering to search-normalized displacements, i.e. displacements divided by the search window size (Gardner *et al.* 2018). Another effective, but costly, solution is to estimate the displacement using several preprocessing and template sizes, which can help exclude mismatches while reducing the impact of user-defined parameters (Ahn and Howat 2011).

All of the above approaches are targeted at removing random mismatches. Removal of spatially coherent blunders is much more challenging and requires ancillary information (*a priori* displacement or multiple overlapping displacement fields). When a time series of images is available, and the velocity can be assumed constant over the study period, the redundancy of the observations can be used to filter outliers through averaging (Dehecq *et al.* 2015) or voting (Altena *et al.* 2019). To reduce the influence of non-glacial objects, it is recommended to mask areas outside of the glacier during offset tracking, assuming that an accurate glacier outline is available. Finally, to reduce the impact of periodic features that can appear artificially static or move in a direction opposite to ice flow (locking, Figure 11.4), a solution can be to process image pairs with many different time spans.

Figure 11.4. *Example of locking due to a localized topographic feature (icefall, top row) and time-invariant radiance (curved moraine, bottom) (Gardner* et al. *2019). For a color version of this figure, see www.iste.co.uk/cavalie/images.zip*

11.3.4.2. *Coregistration, terrain and sensor corrections*

The problem of image coregistration and orthorectification is common to all remote sensing disciplines. It is, however, more pronounced for glaciers because 1) the surface elevation evolves rapidly over time while the digital elevation models used for the orthorectification are generally from a single date or a compilation of dates, resulting in inaccurate orthorectification and an artificial horizontal displacement; and 2) for large ice caps and ice sheets, tie points for the horizontal georeferencing of the images may be lacking.

The effects of orthorectification errors are minimal for nadir-looking sensors and repeat orbits (e.g. Landsat images from the same path/row), but geolocation or relative geometric properties in the image might be off by tens of meters (Kääb *et al.* 2016). Orthorectification errors due to outdated or inaccurate DEMs and apparent offsets due to variable radar penetration depth are especially crucial for

SAR sensors that image scenes with incidence angles of 20–50°. Such errors can be avoided when a simultaneously acquired DEM is used for orthorectification (Leinss and Bernhard 2021). When deriving displacements from imagery that is sensitive to orthorectification errors (e.g. non-nadir and/or non-repeat), it is recommended to work with raw images in the sensor geometry when available and to apply the orthorectification with a known elevation source. When this is not possible, but the DEM used for orthorectification is known, a solution is to correct (undo) the orthorectification and to perform the orthorectification again with a DEM which is more accurate or closer in time to the images (Rosenau *et al.* 2015). Complete reverse engineering is also possible, since orthorectification for pushbroom sensors is only applied in the across-track direction, and elevation offsets can thus be estimated directly when a minimum of three images are matched (see, for example, Figure 11.5; Altena and Kääb 2017a). In this example, the angle between the two viewing directions is 15° and the DEM used for orthorectification is outdated. The displacements calculated by offset tracking (red arrows on the left panel) are the sum of the offsets due to ice flow (blue arrows on the left panel) and elevation differences between the date of the DEM used for orthorectification and the date of the images (right panel). A lowering of the glacier tongue can be seen, which would otherwise contaminate the velocity estimate. The surrounding terrain displays vertical offsets that can be explained by a coarser resolution and a horizontal offset of the DEM used in the original orthorectification.

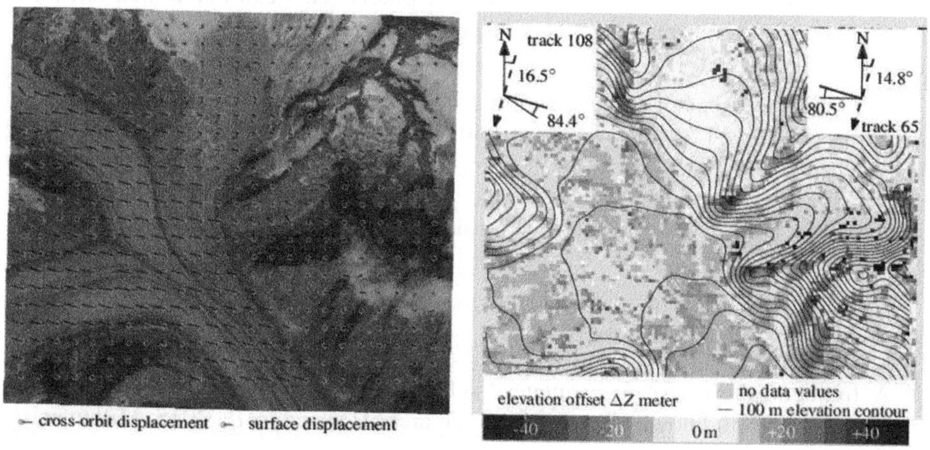

Figure 11.5. *Left: Sentinel-2 displacement over Concordiaplatz, Switzerland (see also Chapter 3, Figure 3.10). Cross-track displacements are shown in blue, surface displacements in red. Right: estimated vertical offset (ΔZ) between surface elevation and DEM. For a color version of this figure, see www.iste.co.uk/cavalie/images.zip*

Planimetric errors, due to absolute platform positioning and orientation, are typically corrected by removing a horizontal shift, estimated as the mean or median displacement in each direction on stable grounds. This approach is challenged over large glacierized terrain where little stable terrain is available. One solution is to apply a low-pass filter to the images to remove small-scale features that move with the ice flow and leave only large topographic features that remain static (Scambos *et al.* 1992). Offset tracking can then be applied to the low-pass-filtered images to estimate a horizontal shift or more complex transformations. A second approach is to identify areas of low velocities in an ancillary velocity field and to adjust the derived velocities to these known velocities (Fahnestock *et al.* 2016). Finally, a third solution is to tie the image to neighbor images, by constraining the velocity to be equal between the two scenes. By building up a system of overlapping displacement fields, it is possible to obtain sufficient terrain to tie the displacement field to a known stable terrain (e.g. as in section 3.4 in Bekaert *et al.* (2019)).

11.3.4.3. *Comparison with ancillary observations and validation*

For validation purposes, remote sensing-derived velocity fields can be compared to ground-based measurements, for example those obtained through precise GPS techniques. There are, however, limitations to such comparisons (Paul *et al.* 2017):

– the diurnal and seasonal variability of ice flow (e.g. Allstadt *et al.* 2015) requires a comparison over similar time periods;

– remote sensing-derived velocities are obtained for relatively large templates, with dimensions of a few tens to hundreds of pixels. This can make them difficult to compare with point measurements, especially in areas of large spatial gradients (e.g. shear margins);

– the displacement of stakes anchored in the ice is typically measured in three directions (Figure 11.2). In comparison, offset tracking measures a two-dimensional displacement (horizontal for optical, slant-range/azimuth for radar), while InSAR measures only the displacement in the LOS. The comparison between these observations requires a careful transformation between the different coordinate systems (e.g. section 11.3.3);

– glaciers are situated in hostile environments (crevasses, avalanche risk) and consequently validation data are sparse.

A frequent alternative to estimate displacement measurement uncertainty is to analyze residuals in known or assumed stable areas (Dehecq *et al.* 2015; Millan *et al.* 2019).

11.3.5. *Working with time series*

The temporal resolution at which velocity can be estimated evolves in line with the number of repeat images. While the sporadic coverage of glacier regions was a common limitation at the start of this century, today satellite systems with increased sensing capabilities have opened up the possibility of capturing short-lived glacier dynamics from space. Figure 11.6 illustrates this data availability clearly: gaps exist at the start of the time series, but it progresses into high temporal coverage and greater detail. This section focuses on the temporal signal and is ordered according to complexity, starting with simple procedures of general flow estimation (see also Chapter 3, section 3.2.3) followed by velocity change and advanced time-series generation.

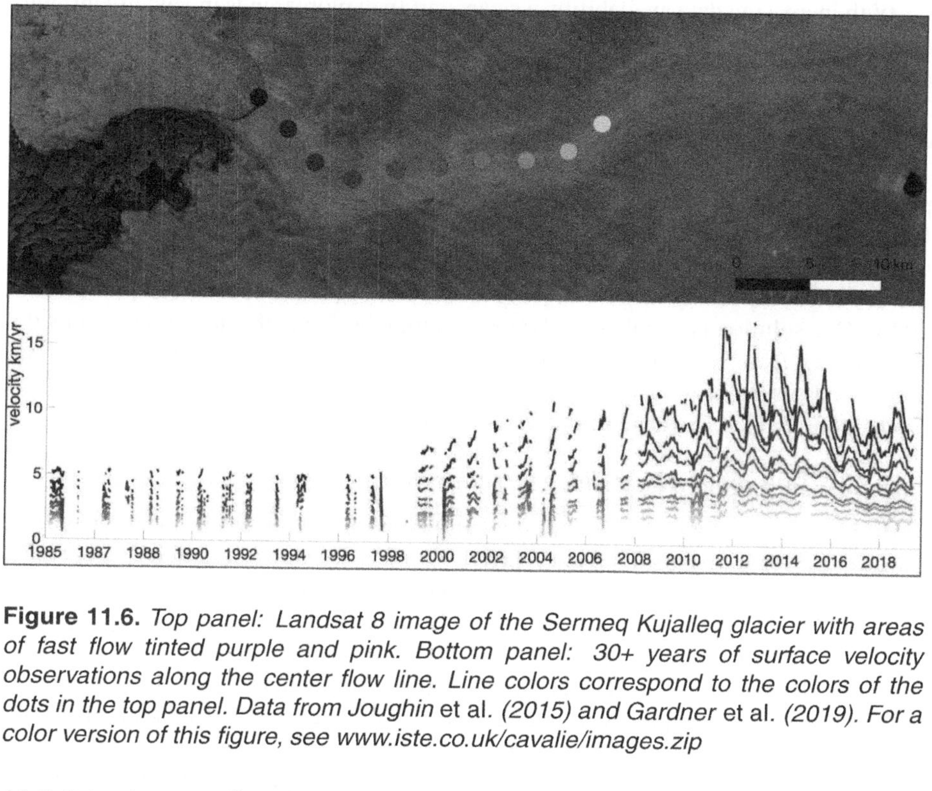

Figure 11.6. *Top panel: Landsat 8 image of the Sermeq Kujalleq glacier with areas of fast flow tinted purple and pink. Bottom panel: 30+ years of surface velocity observations along the center flow line. Line colors correspond to the colors of the dots in the top panel. Data from Joughin* et al. *(2015) and Gardner* et al. *(2019). For a color version of this figure, see www.iste.co.uk/cavalie/images.zip*

11.3.5.1. *Average flow (degree 0)*

The starting point of many post-processing algorithms is a conversion from displacement to velocity (i.e. displacement rate). Many of these algorithms are based upon an Eulerian viewpoint, where velocity estimates at the same location, but covering different time spans, are lumped together. Then, an estimate of the mean velocity, and its uncertainty, is extracted through sample statistics. This can be either done on a grid, to get a more spatially complete overview, or along a flowline. From this collection of velocities, a selection can be made to remove outliers from the sample set. This can be done by looking at group characteristics to assess the agreement in flow direction and/or magnitude (coherence (Dehecq *et al.* 2015), compatibility (Evans 2000), isotropy and interest (Maksymiuk *et al.* 2016)). Another option is to include weights, which can depend on the similarity score or the spatial distance from the spot location or the central flowline. Given that image matching has long tails in its distribution, robust statistics (median, median absolute deviation) are typically favored over standard statistics that assume a Gaussian distribution (mean, standard deviation).

11.3.5.2. *Pluri-annual trends (degree 1)*

With increasing data availability, we can start to explore trends in velocity through time. Depending on the geophysical process of interest, variability in velocity can be examined at inter- (Dehecq *et al.* 2019) or intra-annual periodicity (Armstrong *et al.* 2017; Lemos *et al.* 2018). To help understand underlying drivers of trends, results can be clustered based on glacier characteristics such as glacier type, for example ocean- versus land-terminating (Millan *et al.* 2017). When estimating trends, it is preferable to work directly with velocity components, rather than velocity magnitudes. Estimates of the velocity magnitude indeed have a non-symmetric distribution, since magnitude values cannot be negative, and have a central estimate (e.g. mean, median) that is biased, especially in slow-moving areas (Dehecq *et al.* 2019; Derkacheva *et al.* 2020). One solution is to calculate "velocity anomalies", i.e. the deviation in velocity components from a reference velocity (e.g. the median of the entire record). An unbiased estimate of a change in velocity can then be determined by taking the dot product of the anomalies and reference velocities (Dehecq *et al.* 2019).

11.3.5.3. *Sub-annual fluctuations (higher degree)*

It may also be of interest to measure fluctuations in flow at sub-annual time scales, for example, to understand seasonal flow dynamics. To do so, ice flow is generally formulated as a combination of multi-annual and sub-annual fluctuations (Greene *et al.* 2020). The multi-annual velocity can be described through a moving mean or median (Moon *et al.* 2014; Derkacheva *et al.* 2020), splines (Greene *et al.* 2020) or B-splines (Riel *et al.* 2021). The ability to measure sub-annual fluctuations, however, is dependent on the frequency of image acquisition. Indeed, offset tracking measures the cumulative displacement that occurs between two consecutive image acquisitions, which precludes the direct measurement of instantaneous velocity at any given time. Estimating sub-annual fluctuations is challenged in particular by temporal gaps in observations, due to polar night (for optical data) or summer melt (for SAR data), irregular image acquisitions and the presence of outliers (Derkacheva *et al.* 2020; Greene *et al.* 2020). To overcome these issues and retrieve seasonal fluctuations, data reduction and smoothing is performed using kernel smoothing (Derkacheva *et al.* 2020) or by fitting sinusoidal functions (Greene *et al.* 2020) or B-splines (Riel *et al.* 2021). When displacement is measured over various time separations with a certain degree of overlap, the redundancy of observations can be exploited by minimizing a temporal closure error (see Chapter 5, section 5.4; Altena and Kääb 2017a; Hadhri *et al.* 2019); for example, through least-squares. For intra-annual time scales, the periodic nature of certain natural forcings can be taken into account. For example, formulating image acquisitions relative to the tidal or seasonal cycle helps to fill in the observational gaps in imagery (Minchew *et al.* 2017). Through a time-dependent co-variance function, it is possible to control the behavior of intra-annual effects in the time series (Riel *et al.* 2021).

11.3.6. *Summary of benefits and drawbacks of the different methods*

Although it is difficult to compare the merits of the three described displacement methodologies directly against each other, different limitations and advantages can be identified for each sensor and methodology in a glaciological setting. A non-exhaustive list is proposed in Table 11.1.

Property/Phenomenon	InSAR	SAR tracking	Optical tracking
Power consumption	▼ Active	▼ Active	▲ Passive
Spatial resolution	▲ Full resolution	▼ Large template size	▲▼ Small template size
3D	▲▼ 1D but 3D possible	▲ 2D but 3D possible	▼ Only 2D
Clouds	▲ Insensitive	▲ Insensitive	▼ Sensitive
Atmosphere	▼ Propagation delays	▲ Insensitive	▲ Insensitive
Steep topography	▼ Shadow/layover	▼ Shadow/layover	▲ Near-nadir view
Orthorectification	▲ DInSAR	▼ Critical	▼ Important
Works with snow/firns	▲ Yes	▲ Yes in winter	▼ Difficult
Slow flow	▲ Very accurate	▼ Inaccurate	▲▼ Less accurate
Fast flow	▼ Aliasing	▲ Accurate	▲ Accurate
Melting	▼ Decorrelates in days	▼ Decorrelates in weeks	▲ Enhances features

Table 11.1. *Summary of pros (▲) and cons (▼) of the different methods. For a color version of this figure, see www.iste.co.uk/cavalie/images.zip*

11.4. What can we learn from glacier displacement?

Remote sensing has opened up possibilities for the observation of ice flow at regional scales with increasing accuracy and repetition over the past 30 years. The first ice-sheet-wide maps of velocity were develped in the early 2010s (e.g. Rignot *et al.* 2011b) and are now available at annual time steps for the past decades (NASA Inter-mission Time Series of Land Ice Velocity and Elevation (ITS_LIVE)[1], ESA Climate Change Initiative (CCI)[2]). Although mountain glaciers represent an additional challenge due to their small size, observations of ice flow for glaciers are available with improving spatial and temporal resolutions (Millan *et al.* 2019; Gardner *et al.* 2019). What can we learn from such observations?

Glaciological research spans many different disciplines with different interests. The sections below are organized around the different perspectives that attempt to extract information and products from glacier surface kinematics.

1. https://its-live.jpl.nasa.gov/.
2. https://climate.esa.int/en/projects/glaciers/.

11.4.1. *From a topological perspective*

One of the simplest pieces of information to extract from velocity fields is glacier topological characteristics. Areas of coherent flow can help to identify ice-covered areas, especially under debris cover (Dehecq *et al.* 2015), where glacier mapping from images can be challenging. Flow velocities can help map ice divides[3], especially in flat accumulation areas where topographic gradients are not sufficient (Mouginot and Rignot 2015). At the glacier terminus, ice flow direction points outward of the glacier outlines. This property can be used for automatic calving-front detection.

11.4.2. *From a climatological perspective*

Glaciers and ice sheets have experienced periods of advance and retreat, particularly pronounced over the past decades. These volume fluctuations are a direct response to changes in climate forcing, but their timing and amplitude cannot be explained without an understanding of ice flow. When a glacier is in equilibrium with climate, it reaches a state where surface elevation and flow velocity are in balance. The mass gained in the accumulation zone is conveyed in the ablation zone where it compensates for the mass loss. If ice thickness is known, the flux of ice through a glacier cross-section provides information on the mass gained upstream or lost downstream.

Ice flux can be calculated, for example, at the equilibrium line altitude, i.e. the altitude where the ice flow changes from submergence towards emergence (Figure 11.2), to estimate the total glacier accumulation (Miles *et al.* 2021). For glaciers flowing into the ocean, the same method can be used to estimate the ice discharge into the ocean, the sum of iceberg calving and subaqueous melting (McNabb *et al.* 2015). When the surface mass balance is known or can be modeled, the total glacier mass balance can be calculated using the "input–output" method (e.g. Millan *et al.* 2017). Ultimately, when the thickness and velocity are known, it is possible to estimate the ice flux divergence, either between glacier cross-sections (Berthier and Vincent 2012) or at individual points (Vincent *et al.* 2021), and to estimate the local mass balance. This can, for example, provide information on glacier imbalance (Miles *et al.* 2021) or factors influencing ice melt, such as debris cover (Bisset *et al.* 2020).

11.4.3. *From a mechanical perspective*

Many of the rheological parameters of clean ice are extracted from laboratory experiments, but some mechanical properties are scale-dependent. In this regard, ice shelves represent a great opportunity to study ice mechanical properties since friction can be assumed to be zero and ice thickness, typically of a few hundred meters,

3. Boundary of an ice sheet, ice cap or glacier separating opposing ice-flow directions.

varies slowly. Maps of ice flow can be used to look at mechanical processes such as crevasse and rift formation, compression (Minchew *et al.* 2017), flexure and bending properties (Wild *et al.* 2018) or shear heating (Minchew *et al.* 2018b).

On the other hand, when the ice is grounded, friction at the glacier base is important and has an effect on surface velocity that can be exploited. Surface velocity is frequently used to estimate strain rates and ice rheology (Bindschadler *et al.* 1996), basal friction and bed roughness (Joughin *et al.* 2004; Minchew *et al.* 2016). This can be used to estimate weaknesses in the bed, such as unconsolidated sediments (till) that could cause future destabilization of ice flow (Joughin *et al.* 2009) and an acceleration of sea level rise, or help understand the impact of localized grounded points (pinning points) on this stabilization (Minchew *et al.* 2018a). Surface velocity can also be used to constrain models used to infer a glacier thermal regime (Gilbert *et al.* 2018) or date the age of the ice, particularly in ice cores (Urbini *et al.* 2008).

Another cause of basal sliding is the reduction of basal friction by meltwater, which can cause large seasonal velocity fluctuations (Bartholomew *et al.* 2010). The increased repeat-time of satellite acquisitions has enabled observation of these inter-annual fluctuations with increased accuracy and an understanding of its impact on dynamic mass loss (Moon *et al.* 2014; Tedstone *et al.* 2014; Armstrong *et al.* 2017). On the other hand, multi-decadal satellite archives have enabled estimation of long-term trends in velocity (Rignot 2006; Mouginot and Rignot 2015) and in particular the study of the impact of basal sliding on its evolution (Joughin *et al.* 2018; Dehecq *et al.* 2019).

11.4.4. *From a hydrological perspective*

Glaciers and ice sheets are important contributors to sea level rise and freshwater resources (see section 11.1). It is therefore of interest to estimate their total volume and understand their past, current and future evolution. From this perspective, velocity fields can help in the estimation of ice thickness, a pre-requisite to model glacier evolution and contribution to the water cycle.

Many of the ice thickness inversion models are based on mass and momentum conservation, along with a constitutive relation for ice (usually Glen's law) and boundary conditions such as basal sliding (Farinotti *et al.* 2017). While several models rely on assumptions on mass balance gradients, several take advantage of surface velocity observations. Direct inversion of these constitutive equations on a grid-cell basis can provide a first-order estimate of ice thickness (Gantayat *et al.* 2014); however, this heavily relies on accurate velocity data and ice thickness is not influenced by neighboring estimates. Other methods apply the principle of mass conservation either to glacier cross-sections (Rabatel *et al.* 2018), flowlines (McNabb *et al.* 2012) or a grid (Gudmundsson and Bauder 1999). However, these methods also rely on auxiliary data such as an elevation model and information about the mass balance.

11.4.5. *From a geomorphological perspective*

From a geomorphological perspective, glaciers function as agents for erosion or conveyor belts for glacier debris. The mean ice velocity of a glacier and its altitudinal distribution can help uncover information about glacier change and influence on the landscape. Scherler *et al.* (2011) showed, using satellite-derived velocities of glaciers in high-mountain Asia, that cold-based, clean ice glaciers have little erosional power, limited to their lower part, but can evolve into steeper and faster glaciers with a larger erosional power. Similarly, knowing the full velocity profile and thus the transportation rate of debris, the accumulated debris thickness at the terminus can then be linked to erosion rates of high-mountain headwalls (Banerjee and Wani 2018). In another example, velocity observations have shown that ice falls seem to be a barrier for summer speed-up migrating upwards, leading to minimal erosion above the ice fall. This would mean that small topographic perturbations can establish a self-promoting barrier between two different erosion regimes that can enhance relief (Armstrong *et al.* 2017).

11.4.6. *From a hazards perspective*

As glaciers are dynamic landforms, they can represent a natural hazard through, for example, ice detachment, calving, glacier collapse, lake formation or surges. In order to assess hazardous glaciers or surrounding landforms, remote sensing and offset tracking can be valuable tools. Ice velocity speed-up can be used to predict the break-off of hanging glaciers (Faillettaz *et al.* 2016) or precursors to whole glacier collapses (Gilbert *et al.* 2018). Ice velocity can also be used to detect areas prone to supraglacial lake formation and potentially hazardous glaciers (Quincey *et al.* 2007). Glacier surges are sudden increases in the ice flow of up to 10 or 100 times the normal velocity within a short period of time (weeks to months) that cause mass redistribution and front advance. Surges can cause damage to infrastructures, block a valley or create ice-dammed lakes that can cause potential outburst floods (Round *et al.* 2017). Sub-weekly repeat imagery and offset tracking provide a great opportunity to study the onset, duration and periodicity of surges and help understand the factors responsible for surges (e.g. Burgess *et al.* 2012; Round *et al.* 2017).

11.5. Perspectives and future directions

11.5.1. *What will we learn from future missions?*

The increasing spatial resolution of satellite sensors will make it possible to not only monitor the bulk flow of ice but also more detailed surface changes.

Processes currently observed with aerial images, drones or terrestrial images provide an indication of what such aspects could look like. Very high resolution enables the study of individual objects moving with the ice, in a Lagrangian framework, such as the back-wasting of ice cliffs, debris dynamics and their impact on local ablation rates (e.g. Brun *et al.* 2018), and helps quantify processes at the ice front, such as calving, ice fracturing and rift mechanics (Jouvet *et al.* 2019). It can help resolve shear margins and grounding line evolution at an increased resolution (Moon *et al.* 2020). To complement such data, the recent developments in laser altimetry (ICESat-2) open new possibilities for tracking individual features such as crevasses or rifts and studying their evolution over time.

The ever increasing number of Earth-observation satellites and image acquisitions will make it possible to study processes at a much finer temporal scale, such as weekly or daily time scales. In this regard, it will be all the more important to develop methods to combine data from different missions and to optimize for displacement time series (section 11.3.5.3). Such advances will help better understand processes generally observed by GPS, such as the influence of subglacial water on ice flow (Armstrong and Anderson 2020), rift and calving mechanisms or tidal forcing (Riel *et al.* 2021). Agile satellites and satellite constellations are starting to provide overpasses separated by a time span of a few minutes only and can be used to map fast changing movements. The tracking of icebergs drifting in a fjord can be used to infer water circulation within a fjord system (Figure 11.7), which in turn helps in obtaining a better understanding of the complicated dynamics of ice–ocean interaction.

Figure 11.7. *Streamlines of fjord circulation in Leconte Bay, Alaska. On the left are results adopted from terrestrial imagery by Kienholz* et al. *(2019). On the right are results from near-simultaneous overpasses from PlanetScope imagery. For a color version of this figure, see www.iste.co.uk/cavalie/images.zip*

Future SAR satellite missions such as NISAR (see Chapter 1, section 1.3.3), which will carry two L-band and S-band SAR instruments on a 12-day repeat orbit, will provide increased possibilities for the derivation of phase- and feature-based displacement measurements. NISAR observations are expected to be superior to higher-frequency observations thanks to increased repeat-image coherence and will provide a significant coverage increase of the Antarctic Ice Sheet thanks to a left-looking configuration.

11.5.2. *What future sensors are needed for glaciology?*

Despite incredible progress in satellite data availability and capabilities, two main features of glaciers essentially remain out of reach with current missions: sufficient repeat topographic information to account for the changing glacier surfaces and subsurface information.

The former information is particularly crucial given the increased frequency of satellite acquisitions and the accelerating glacier mass loss. One way to achieve this is through systematic stereoscopic acquisitions. Sources of systematic stereoscopic images are currently limited to the ASTER sensor (to be decommissioned in 2023), which offers a lower spatial resolution and radiometric quality compared to current systematic monoscopic satellite missions (Landsat, Sentinel-2), as well as operational elevation products that are generally of insufficient quality (Girod *et al.* 2017). Very high-resolution images, on the other hand, are available for purchase and without a systematic acquisition plan, for example, from SPOT, Pléiades or WorldView missions (see Chapter 1, section 1.4.1). High repeat photogrammetry is possible with monoscopic sensors (Sentinel-2 (Altena and Kääb 2017a), Planet (Aati and Avouac 2020)), but all configurations have small baselines resulting in inferior products. Future missions should therefore aim at improving the stereoscopic abilities of optical satellite missions to provide elevation information along the images. Additionally, providing raw, unorthorectified images and precise ephemerides is crucial to allow users to generate the orthorectified images based on the best available topographic information for a given area. A second way to retrieve topographic information is from single-pass SAR satellites such as TanDEM-X, but such global mapping missions are rare. The use of receive-only radar satellites may provide a cost-effective way of increasing the availability of such data. Harmony, a selected mission for the ESA's Earth Explorers program, envisions the launch of two companion satellites of Sentinel-1D to provide precise and frequent elevation measurements (Kleinherenbrink *et al.* 2021).

Subsurface information is a much more challenging goal to reach, but recent developments in radar missions could help fill the gap in the near future. Simultaneous acquisitions in different radar bands could allow imaging of snow and ice with different penetration depths and the derivation of the displacement field at the subsurface. SAR tomography has been applied sporadically for airborne radar data and showed promising results in measuring ice thickness (Morlighem *et al.* 2013) and imaging the interior of glaciers (Tebaldini *et al.* 2016). Biomass, a planned SAR mission with tomography abilities (see Chapter 1, section 1.3.4), will provide more data to develop methodology but will certainly represent a challenge in a glaciological context due to the frequent presence of liquid water or the mountainous terrain. However, this may open the door to methodological developments in SAR tomography, which in turn will provide opportunities to better document subsurface processes at large scales, such as rates of ice deformation and basal sliding or the development of englacial and subglacial water channels.

11.6. Acknowledgments

We thank Twila Moon and Brent Minchew for their careful reviews and comments that helped to improve this chapter.

11.7. References

Aati, S. and Avouac, J.-P. (2020). Optimization of optical image geometric modeling, application to topography extraction and topographic change measurements using PlanetScope and SkySat imagery. *Remote Sensing*, 12(20), 3418.

Ahn, Y. and Howat, I.M. (2011). Efficient automated glacier surface velocity measurement from repeat images using multi-image/multichip and null exclusion feature tracking. *IEEE Transactions on Geoscience and Remote Sensing,* 49(8), 2838–2846.

Alley, R.B. and Whillans, I.M. (1984). Response of the East Antarctica ice sheet to sea-level rise. *Journal of Geophysical Research: Oceans*, 89(C4), 6487–6493.

Allstadt, K., Shean, D., Campbell, A., Fahnestock, M., Malone, S. (2015). Observations of seasonal and diurnal glacier velocities at Mount Rainier, Washington, using terrestrial radar interferometry. *Cryosphere*, 9(6), 2219–2235.

Altena, B. and Kääb, A. (2017a). Elevation change and improved velocity retrieval using orthorectified optical satellite data from different orbits. *Remote Sensing*, 9(3), 300.

Altena, B. and Kääb, A. (2017b). Weekly glacier flow estimation from dense satellite time series using adapted optical flow technology. *Frontiers in Earth Science*, 5, 53.

Altena, B. and Kääb, A. (2020). Ensemble matching of repeat satellite images applied to measure fast-changing ice flow, verified with mountain climber trajectories on Khumbu icefall, Mount Everest. *Journal of Glaciology*, 66(260), 905–915.

Altena, B., Scambos, T., Fahnestock, M., Kääb, A. (2019). Extracting recent short-term glacier velocity evolution over southern Alaska and the Yukon from a large collection of Landsat data. *The Cryosphere*, 13, 795–814.

Armstrong, W.H. and Anderson, R.S. (2020). Ice-marginal lake hydrology and the seasonal dynamical evolution of Kennicott Glacier, Alaska. *Journal of Glaciology*, 66(259), 699–713.

Armstrong, W.H., Anderson, R.S., Fahnestock, M. (2017). Spatial patterns of summer speedup on south central Alaska glaciers. *Geophysical Research Letters*, 44(18), 9379–9388.

Banerjee, A. and Wani, B. (2018). Exponentially decreasing erosion rates protect the high-elevation crests of the Himalaya. *Earth and Planetary Science Letters*, 497, 22–28.

Bartholomew, I., Nienow, P., Mair, D., Hubbard, A., King, M.A., Sole, A. (2010). Seasonal evolution of subglacial drainage and acceleration in a Greenland outlet glacier. *Nature Geoscience*, 3(6), 408–411.

Bekaert, D.P.S., Jones, C.E., An, K., Huang, M.-H. (2019). Exploiting UAVSAR for a comprehensive analysis of subsidence in the Sacramento Delta. *Remote Sensing of Environment*, 220, 124–134.

Berthier, E. and Vincent, C. (2012). Relative contribution of surface mass-balance and ice-flux changes to the accelerated thinning of Mer de Glace, French Alps, over 1979–2008. *Journal of Glaciology*, 58(209), 501–512.

Bindschadler, R., Vornberger, P., Blankenship, D., Scambos, T., Jacobel, R. (1996). Surface velocity and mass balance of Ice Streams D and E, West Antarctica. *Journal of Glaciology*, 42(142), 461–475.

Bisset, R.R., Dehecq, A., Goldberg, D.N., Huss, M., Bingham, R.G., Gourmelen, N. (2020). Reversed surface-mass-balance gradients on Himalayan Debris-covered glaciers inferred from remote sensing. *Remote Sensing*, 12(10), 1563.

Brun, F., Wagnon, P., Berthier, E., Shea, J.M., Immerzeel, W.W., Kraaijenbrink, P.D.A., Vincent, C., Reverchon, C., Shrestha, D., Arnaud, Y. (2018). Ice cliff contribution to the tongue-wide ablation of Changri Nup Glacier, Nepal, central Himalaya. *The Cryosphere*, 12(11), 3439–3457.

Burgess, E.W., Forster, R.R., Larsen, C.F., Braun, M. (2012). Surge dynamics on Bering Glacier, Alaska, in 2008–2011. *The Cryosphere*, 6(6), 1251–1262.

Charrier, L., Godet, P., Rambour, C., Weissgerber, F., Erdmann, S., Koeniguer, E.C. (2020). Analysis of dense coregistration methods applied to optical and SAR time-series for ice flow estimations. *IEEE Radar Conference*, Florence.

Cuffey, K. and Paterson, W. (2010). *The Physics of Glaciers*, 4th edition. Academic Press. Cambridge.

Debella-Gilo, M. and Kääb, A. (2011). Sub-pixel precision image matching for measuring surface displacements on mass movements using normalized cross-correlation. *Remote Sensing of Environment*, 115(1), 130–142.

Debella-Gilo, M. and Kääb, A. (2012). Locally adaptive template sizes for matching repeat images of Earth surface mass movements. *ISPRS Journal of Photogrammetry and Remote Sensing*, 69, 10–28.

Dehecq, A., Gourmelen, N., Trouvé, E. (2015). Deriving large-scale glacier velocities from a complete satellite archive: Application to the Pamir–Karakoram–Himalaya. *Remote Sensing of Environment*, 162, 55–66.

Dehecq, A., Millan, R., Berthier, E., Gourmelen, N., Trouvé, E., Vionnet, V. (2016). Elevation changes inferred from TanDEM-X data over the Mont-Blanc area: Impact of the X-band interferometric bias. *IEEE Journal of Selected Topics in Applied Earth Observations and Remote Sensing*, 9(8), 3870–3882.

Dehecq, A., Gourmelen, N., Gardner, A., Brun, F., Goldberg, D., Nienow, P., Berthier, E., Vincent, C., Wagnon, P., Trouvé, E. (2019). Twenty-first century glacier slowdown driven by mass loss in high mountain Asia. *Nature Geoscience*, 12(1), 22–27.

Derkacheva, A., Mouginot, J., Millan, R., Maier, N., Gillet-Chaulet, F. (2020). Data reduction using statistical and regression approaches for ice velocity derived by Landsat8, Sentinel-1 and Sentinel-2. *Remote Sensing*, 12(12), 1935.

Dowdeswell, J.A. and McIntyre, N.F. (1986). The saturation of LANDSAT MSS detectors over large ice masses. *International Journal of Remote Sensing*, 7(1), 151–164.

Evans, A. (2000). Glacier surface motion computation from digital image sequences. *IEEE Transactions on Geoscience and Remote Sensing*, 38(2), 1064–1072.

Fahnestock, M., Scambos, T., Moon, T., Gardner, A., Haran, T., Klinger, M. (2016). Rapid large-area mapping of ice flow using Landsat 8. *Remote Sensing of Environment*, 185, 84–94.

Faillettaz, J., Funk, M., Vagliasindi, M. (2016). Time forecast of a break-off event from a hanging glacier. *The Cryosphere*, 10, 1191–1200.

Farinotti, D., Brinkerhoff, D., Clarke, G., Fürst, J., Frey, H., Gantayat, P., Gillet-Chaulet, F., Girard, C., Huss, M., Leclercq, P.W. *et al.* (2017). How accurate are estimates of glacier ice thickness? Results from ITMIX, the ice thickness models intercomparison experiment. *The Cryosphere*, 11(2), 949–970.

Fischer, U.H. and Clarke, G.K.C. (1997). Stickslip sliding behaviour at the base of a glacier. *Annals of Glaciology*, 24, 390–396.

Fleet, D. and Weiss, Y. (2006). Optical flow estimation. In *Handbook of Mathematical Models in Computer Vision*, Paragios, N., Chen, Y., Faugeras, O. (eds). Springer, New York.

Gantayat, P., Kulkarni, A., Srinivasan, J. (2014). Estimation of ice thickness using surface velocities and slope: Case study at Gangotri Glacier, India. *Journal of Glaciology*, 60(220), 277–282.

Gardner, A.S., Moholdt, G., Scambos, T., Fahnstock, M., Ligtenberg, S., van den Broeke, M., Nilsson, J. (2018). Increased West Antarctic and unchanged East Antarctic ice discharge over the last 7 years. *The Cryosphere*, 12(2), 521–547.

Gardner, A., Fahnestock, M., Scambos, T. (2019). ITS_LIVE regional glacier and ice sheet surface velocities. NSIDC.

Gilbert, A., Leinss, S., Kargel, J., Kääb, A., Gascoin, S., Leonard, G., Berthier, E., Karki, A., Yao, T. (2018). Mechanisms leading to the 2016 giant twin glacier collapses, Aru range, Tibet. *The Cryosphere*, 12(9), 2883–2900.

Girod, L., Nuth, C., Kääb, A., McNabb, R., Galland, O. (2017). MMASTER: Improved ASTER DEMs for elevation change monitoring. *Remote Sensing*, 9(7), 704.

Goldstein, R.M. and Werner, C.L. (1998). Radar interferogram filtering for geophysical applications. *Geophysical Research Letters*, 25(21), 4035–4038.

Goldstein, R.M., Engelhardt, H., Kamb, B., Frolich, R.M. (1993). Satellite radar interferometry for monitoring ice sheet motion: Application to an Antarctic ice stream. *Science*, 262, 1525–1530.

Gourmelen, N., Kim, S., Shepherd, A., Park, J., Sundal, A., Bjornsson, H., Palsson, F. (2011). Ice velocity determined using conventional and multiple-aperture InSAR. *Earth and Planetary Science Letters*, 307(1), 156–160.

Greene, C.A., Gardner, A.S., Andrews, L.C. (2020). Detecting seasonal ice dynamics in satellite images. *The Cryosphere*, 14(12), 4365–4378.

Gudmundsson, G. and Bauder, A. (1999). Towards an indirect determination of the mass-balance distribution of glaciers using the kinematic boundary condition. *Geografiska Annaler: Series A, Physical Geography*, 81(4), 575–583.

Hadhri, H., Vernier, F., Atto, A., Trouvé, E. (2019). Time-lapse optical flow regularization for geophysical complex phenomena monitoring. *ISPRS Journal of Photogrammetry and Remote Sensing*, 150, 135–156.

Heid, T. and Kääb, A. (2012). Evaluation of existing image matching methods for deriving glacier surface displacements globally from optical satellite imagery. *Remote Sensing of Environment*, 118, 339–355.

Intergovernmental Panel on Climate Change (2019). Special Report on the Ocean and Cryosphere in a Changing Climate. Monaco.

Jauvin, M., Yan, Y., Trouvé, E., Fruneau, B. (2018). Potential and limits of Sentinel-1 data for small alpine glaciers monitoring. *IEEE International Geoscience and Remote Sensing Symposium*, Valencia.

Joughin, I. (2002). Ice-sheet velocity mapping: A combined interferometric and speckle-tracking approach. *Annals of Glaciology*, 34, 195–201.

Joughin, I., MacAyeal, D., Tulaczyk, S. (2004). Basal shear stress of the Ross ice streams from control method inversions. *Journal of Geophysical Research: Solid Earth*, 109(B9).

Joughin, I., Tulaczyk, S., Bamber, J.L., Blankenship, D., Holt, J.W., Scambos, T., Vaughan, D.G. (2009). Basal conditions for Pine Island and Thwaites Glaciers, West Antarctica, determined using satellite and airborne data. *Journal of Glaciology*, 55(190), 245–257.

Joughin, I., Smith, B., Howat, I., Scambos, T. (2015). MEaSUREs Greenland ice sheet velocity map from InSAR data, version 2. NASA National Snow and Ice Data Center, Boulder, CO.

Joughin, I., Smith, B., Howat, I. (2018). A complete map of Greenland ice velocity derived from satellite data collected over 20 years. *Journal of Glaciology*, 64(243), 1–11.

Jouvet, G., Weidmann, Y., van Dongen, E., Lüthi, M.P., Vieli, A., Ryan, J.C. (2019). High-endurance UAV for monitoring calving glaciers: Application to the Inglefield Bredning and Eqip Sermia, Greenland. *Frontiers in Earth Science*, 7.

Kääb, A., Winsvold, S., Altena, B., Nuth, C., Nagler, T., Wuite, J. (2016). Glacier remote sensing using Sentinel-2. Part I: Radiometric and geometric performance, and application to ice velocity. *Remote Sensing*, 8(7), 2072–4292.

Kamb, B. and Echelmeyer, K. (1986). Stress-gradient coupling in glacier flow I: Longitudinal averaging of the influence of ice thickness and surface slope. *Journal of Glaciology*, 32(111), 267–284.

Kienholz, C., Amundson, J., Motyka, R., Jackson, R., Mickett, J., Sutherland, D., Nash, J., Winters, D., Dryer, W., Truffer, M. (2019). Tracking icebergs with time-lapse photography and sparse optical flow, LeConte Bay, Alaska, 2016–2017. *Journal of Glaciology*, 65(250), 195–211.

Kleinherenbrink, M., Korosov, A., Newman, T., Theodosiou, A., Komarov, A.S., Li, Y., Mulder, G., Rampal, P., Stroeve, J., Lopez-Dekker, P. (2021). Estimating instantaneous sea-ice dynamics from space using the bi-static radar measurements of Earth Explorer 10 candidate Harmony. *The Cryosphere*, 15(7), 3101–3118.

Leinss, S. and Bernhard, P. (2021). TanDEM-X: Deriving InSAR height changes and velocity dynamics of Great Aletsch Glacier. *IEEE Journal of Selected Topics in Applied Earth Observations and Remote Sensing*, 14, 1–18.

Lemos, A., Shepherd, A., McMillan, M., Hogg, A. (2018). Seasonal variations in the flow of land-terminating glaciers in central-west Greenland using Sentinel-1 imagery. *Remote Sensing*, 10(12), 1878.

Leprince, S., Barbot, S., Ayoub, F., Avouac, J.-P. (2007). Automatic and precise orthorectification, coregistration, and subpixel correlation of satellite images, application to ground deformation measurements. *IEEE Transactions on Geoscience and Remote Sensing*, 45(6), 1529–1558.

Liu, T., Niu, M., Yang, Y. (2018). Ice velocity variations of the polar record glacier (East Antarctica) using a rotation-invariant feature-tracking approach. *Remote Sensing*, 10(1), 42.

Luckman, A., Quincey, D., Bevan, S. (2007). The potential of satellite radar interferometry and feature tracking for monitoring flow rates of Himalayan glaciers. *Remote Sensing of Environment*, 111(2), 172–181.

Maksymiuk, O., Mayer, C., Stilla, U. (2016). Velocity estimation of glaciers with physically-based spatial regularization – Experiments using satellite SAR intensity images. *Remote Sensing of Environment*, 172, 190–204.

McNabb, R.W., Hock, R., O'Neel, S., Rasmussen, L., Ahn, Y., Braun, M., Conway, H., Herreid, S., Joughin, I., Pfeffer, W., Smith, B., Truffer, M. (2012). Using surface velocities to calculate ice thickness and bed topography: A case study at Columbia Glacier, Alaska, USA. *Journal of Glaciology*, 58(212), 1151–1164.

McNabb, R.W., Hock, R., Huss, M. (2015). Variations in Alaska tidewater glacier frontal ablation, 1985–2013. *Journal of Geophysical Research: Earth Surface*, 120(1), 120–136.

Miles, E., McCarthy, M., Dehecq, A., Kneib, M., Fugger, S., Pellicciotti, F. (2021). Health and sustainability of glaciers in High Mountain Asia, *Nature Communications*, 12(1), 2868.

Millan, R., Mouginot, J., Rignot, E. (2017). Mass budget of the glaciers and ice caps of the Queen Elizabeth Islands, Canada, from 1991 to 2015. *Environmental Research Letters*, 12(2), 024016.

Millan, R., Mouginot, J., Rabatel, A., Jeong, S., Cusicanqui, D., Derkacheva, A., Chekki, M. (2019). Mapping surface flow velocity of glaciers at regional scale using a multiple sensors approach. *Remote Sensing*, 11(21), 2498.

Minchew, B., Simons, M., Björnsson, H., Pálsson, F., Morlighem, M., Seroussi, H., Larour, E., Hensley, S. (2016). Plastic bed beneath Hofsjökull Ice Cap, central Iceland, and the sensitivity of ice flow to surface meltwater flux. *Journal of Glaciology*, 62(231), 147–158.

Minchew, B., Simons, M., Riel, B., Milillo, P. (2017). Tidally induced variations in vertical and horizontal motion on Rutford ice stream, West Antarctica, inferred from remotely sensed observations. *Journal of Geophysical Research: Earth Surface*, 122(1), 167–190.

Minchew, B., Gudmundsson, H., Gardner, A.S., Paolo, F.S., Fricker, H.A. (2018a). Modeling the dynamic response of outlet glaciers to observed ice-shelf thinning in the Bellingshausen Sea Sector, West Antarctica. *Journal of Glaciology*, 64(244), 333–342.

Minchew, B., Meyer, C., Robel, A., Gudmundsson, G., Simons, M. (2018b). Processes controlling the downstream evolution of ice rheology in glacier shear margins: Case study on Rutford Ice Stream, West Antarctica, *Journal of Glaciology*, 64(246), 583–594.

Moon, T.A., Joughin, I., Smith, B., van den Broeke, M., Berg, W., Noël, B., Usher, M. (2014). Distinct patterns of seasonal Greenland glacier velocity. *Geophysical Research Letters*, 41(20), 7209–7216.

Moon, T.A., Gardner, A.S., Csatho, B., Parmuzin, I., Fahnestock, M.A. (2020). Rapid reconfiguration of the Greenland ice sheet coastal margin. *Journal of Geophysical Research: Earth Surface*, 125(11), e2020JF005585.

Morlighem, M., Rignot, E., Mouginot, J., Wu, X., Seroussi, H., Larour, E., Paden, J. (2013). High-resolution bed topography mapping of Russell Glacier, Greenland, inferred from operation IceBridge data. *Journal of Glaciology*, 59(218), 1015–1023.

Mouginot, J. and Rignot, E. (2015). Ice motion of the Patagonian icefields of South America: 1984–2014. *Geophysical Research Letters*, 42(5), 1441–1449.

Mouginot, J., Rignot, E., Scheuchl, B. (2019). Continent-wide, interferometric SAR phase, mapping of Antarctic ice velocity. *Geophysical Research Letters*, 46(16), 9710–9718.

Paul, F., Bolch, T., Briggs, K., Kääb, A., McMillan, M., McNabb, R., Nagler, T., Nuth, C., Rastner, P., Strozzi, T., Wuite, J. (2017). Error sources and guidelines for quality assessment of glacier area, elevation change, and velocity products derived from satellite data in the Glaciers_CCI project. *Remote Sensing of Environment*, 203, 256–275.

Prébet, R., Yan, Y., Jauvin, M., Trouvé, E. (2019). A data-adaptive EOF-based method for displacement signal retrieval from InSAR displacement measurement time series for decorrelating targets. *IEEE Transactions on Geoscience and Remote Sensing*, 57(8), 5829–5852.

Quincey, D., Richardson, S., Luckman, A., Lucas, R., Reynolds, J., Hambrey, M., Glasser, N. (2007). Early recognition of glacial lake hazards in the Himalaya using remote sensing datasets. *Global and Planetary Change*, 56(1–2), 137–152.

Rabatel, A., Sanchez, O., Vincent, C., Six, D. (2018). Estimation of glacier thickness from surface mass balance and ice flow velocities: A case study on Argentière Glacier, France. *Frontiers in Earth Science*, 6, 112.

Rabus, B.T. and Fatland, D.R. (2000). Comparison of SAR-interferometric and surveyed velocities on a mountain glacier: Black rapids glacier, Alaska, U.S.A. *Journal of Glaciology*, 46(152), 119–128.

Redpath, T.A.N., Sirguey, P., Fitzsimons, S.J., Kääb, A. (2013). Accuracy assessment for mapping glacier flow velocity and detecting flow dynamics from ASTER satellite imagery: Tasman Glacier, New Zealand. *Remote Sensing of Environment*, 133, 90–101.

Reeh, N., Mohr, J.J., Madsen, S.N., Oerter, H., Gundestrup, N. (2003). Three-dimensional surface velocities of Storstrommen glacier, Greenland, derived from radar interferometry and ice-sounding radar measurements. *Journal of Glaciology*, 49(165), 201–209.

Rémy, F. and Testut, L. (2006). But how a glacier can flow? Historical outlines. *Comptes Rendus Geoscience*, 338(5), 368–385.

Riel, B., Minchew, B., Joughin, I. (2021). Observing traveling waves in glaciers with remote sensing: New flexible time series methods and application to Sermeq Kujalleq (Jakobshavn Isbræ), Greenland. *The Cryosphere*, 15(1), 407–429.

Rignot, E. (2006). Changes in ice dynamics and mass balance of the Antarctic ice sheet. *Philosophical Transactions of the Royal Society A: Mathematical, Physical and Engineering Sciences*, 364(1844), 1637–1655.

Rignot, E., Mouginot, J., Scheuchl, B. (2011a). Antarctic grounding line mapping from differential satellite radar interferometry. *Geophysical Research Letters*, 38(10).

Rignot, E., Mouginot, J., Scheuchl, B. (2011b). Ice flow of the Antarctic ice sheet. *Science*, 333(6048), 1427–1430.

Rosenau, R., Scheinert, M., Dietrich, R. (2015). A processing system to monitor Greenland outlet glacier velocity variations at decadal and seasonal time scales utilizing the Landsat imagery. *Remote Sensing of Environment*, 169, 1–19.

Round, V., Leinss, S., Huss, M., Haemmig, C., Hajnsek, I. (2017). Surge dynamics and lake outbursts of Kyagar Glacier, Karakoram. *The Cryosphere*, 11(2), 723–739.

Scambos, T., Dutkiewicz, M., Wilson, J., Bindschadler, R. (1992). Application of image cross-correlation to the measurement of glacier velocity using satellite image data. *Remote Sensing of Environment*, 42(3), 177–186.

Scherler, D., Bookhagen, B., Strecker, M. (2011). Hillslope-glacier coupling: The interplay of topography and glacial dynamics in High Asia. *Journal of Geophysical Research: Earth Surface*, 116(F2).

Strozzi, T., Luckman, A., Murray, T., Wegmuller, U., Werner, C.L. (2002). Glacier motion estimation using SAR offset-tracking procedures. *IEEE Transactions on Geoscience and Remote Sensing*, 40(11), 2384–2391.

Tebaldini, S., Nagler, T., Rott, H., Heilig, A. (2016). Imaging the internal structure of an alpine glacier via L-band airborne SAR tomography. *IEEE Transactions on Geoscience and Remote Sensing*, 54(12), 7197–7209.

Tedesco, M. (ed.) (2015). *Remote Sensing of the Cryosphere.* Wiley Blackwell, Hoboken, NJ.

Tedstone, A.J., Nienow, P.W., Gourmelen, N., Sole, A.J. (2014). Greenland ice sheet annual motion insensitive to spatial variations in subglacial hydraulic structure. *Geophysical Research Letters*, 41(24), 8910–8917.

Trouvé, E., Vasile, G., Gay, M., Bombrun, L., Grussenmeyer, P., Landes, T., Nicolas, J.-M., Bolon, P., Petillot, I., Julea, A. (2007). Combining airborne photographs and spaceborne SAR data to monitor temperate glaciers: Potentials and limits. *IEEE Transactions on Geoscience and Remote Sensing*, 45(4), 905–924.

Urbini, S., Frezzotti, M., Gandolfi, S., Vincent, C., Scarchilli, C., Vittuari, L., Fily, M. (2008). Historical behaviour of Dome C and Talos Dome (East Antarctica) as investigated by snow accumulation and ice velocity measurements. *Global and Planetary Change*, 60(3), 576–588.

Vaughan, D. (1993). Relating the occurrence of crevasses to surface strain rates. *Journal of Glaciology*, 39(132), 255–266.

Vincent, C., Cusicanqui, D., Jourdain, B., Laarman, O., Six, D., Gilbert, A., Walpersdorf, A., Rabatel, A., Piard, L., Gimbert, F., Gagliardini, O., Peyaud, V., Arnaud, L., Thibert, E., Brun, F., Nanni, U. (2021). Geodetic point surface mass balances: A new approach to determine point surface mass balances on glaciers from remote sensing measurements. *The Cryosphere*, 15(3), 1259–1276.

Vogel, C., Bauder, A., Schindler, K. (2012). Optical flow for glacier motion estimation. *22nd ISPRS Congress*, Melbourne.

Warner, R.C. and Roberts, J.L. (2013). Pine Island Glacier (Antarctica) velocities from Landsat7 images between 2001 and 2011: FFT-based image correlation for images with data gaps. *Journal of Glaciology*, 59(215), 571–582.

Wild, C., Marsh, O., Rack, W. (2018). Unraveling InSAR observed Antarctic ice-shelf flexure using 2-D elastic and viscoelastic modeling. *Frontiers in Earth Science*, 6, 28.

12

New Applications of Spaceborne Optical Image Cross-Correlation: Digital Elevation Models of Volcanic Clouds and Shallow Bathymetry from Space

Marcello DE MICHELE and Daniel RAUCOULES
BRGM, Orléans, France

12.1. Introduction

12.1.1. *General introduction*

In this chapter, we describe how we create offset maps by using cross-correlation between two satellite images in an unconventional way. Instead of using images acquired at multiples dates, as we do in SAR/optical offset tracking or SAR interferometry, we use two images from the same acquisition date and from the same dataset. The reader needs to know that within the pushbroom sensor geometry, the physical separation of panchromatic and multispectral charge-coupled device (CCD) bands on the focal plane of the instrument yields a baseline and a temporal lag between the two images. The baseline is associated with a base-to-height (B/H) ratio that is

Surface Displacement Measurement from Remote Sensing Images,
coordinated by Olivier CAVALIÉ and Emmanuel TROUVÉ.
© ISTE Ltd 2022.

sufficient to extract the elevation models of a tall object such as a volcanic ash cloud. The time lag also yields pixel offsets. These offsets, perpendicular to the epipolar direction, are proportional to the ash cloud velocity or, if the dataset is acquired over the sea, to the wave celerity. In the following section, we concentrate on how this physical separation and time lag can be used as an instrument itself. We will see how we can exploit it to create a digital elevation model of the volcanic ash cloud. Moreover, we will see how we can measure wave celerity from space and how this parameter is related to water depth, that is, bathymetry.

12.2. Digital elevation models of volcanic ash clouds

12.2.1. *Introduction: can we precisely measure the height of a volcanic ash cloud and what physical process controls the injection height and the speed of a volcanic ash plume?*

The routine retrieval of both the height and velocity of volcanic clouds is a crucial point in volcanology. The injection altitude is an important parameter that controls the ash and gas dispersion in the atmosphere and their impact and lifetime around the Earth. As is known today, information on the volcanic cloud height is critical for ash dispersion modeling and air traffic security. The volcanic cloud height related to explosive volcanism is indeed the primary parameter for estimating the mass eruption rate. A number of other reasons also make the volcanic cloud altitude a key parameter to measure: for example, the retrieval of the SO_2 concentration from dedicated spaceborne spectrometers (Corradini *et al.* 2009) and the distribution of ash deposits on the ground during a given eruption. Furthermore, the understanding of a large part of the volcanic system itself strongly relies on knowledge of the volcanic cloud height since the physical characteristics of the cloud are related to key parameters of the volcanic system beneath the surface. For example, the ash and gas emission rates are related to the state of pressurization of the magmatic chamber at depth. The height reached by a volcanic cloud, together with the atmospheric wind regime, controls the dispersal of tephra, the fragmental material of any size and composition produced by a volcanic eruption. The eruption column is itself a function of quite a few important physical parameters, such as the vent radius, gas exit velocity, gas content of eruption products and efficiency of conversion of thermal energy contained in juvenile material to potential and kinetic energy during the entrainment of atmospheric air (Wilson *et al.* 1978). Conventional photogrammetric restitution based on satellite stereo imagery cannot precisely retrieve a spatially detailed digital elevation model of volcanic plumes or plume elevation models (PEMs), as the plume's own velocity – mostly due to the wind – induces an apparent parallax that adds up to the standard parallax given by the stereoscopic view. Therefore, measurements based on standard satellite photogrammetric restitution do not apply here, as there is an ambiguity in the measurement of the plume position. Other methods exist, which we do not treat in detail here but we wish to mention for the curious reader. The earliest use of satellite data to estimate ash cloud heights with photogrammetric methods relied on manually measuring the length of the shadow cast by the cloud under known illumination conditions (Prata and Grant 2001; Spinetti *et al.* 2013). This technique is still used today (Pailot *et al.* 2020). In the literature, we can find some approaches to cloud photogrammetry using instruments with multi-angle observation capabilities; for example, Prata and Turner (1997) used the forward and nadir views of the along-track scanning radiometer (ATSR) to determine volcanic cloud top height (VCTH) for the 1996 Mt. Ruapehu eruption. The multi-angle imaging spectroradiometer (MISR) has been used to retrieve the VCTH, optical depth, type and shape of the finest particles for several eruptions (Scollo *et al.* 2010; Nelson *et al.* 2013; Flower and Kahn 2017).

12.2.2. *Principles*

The method that we describe here is a geometric technique. Consequently, height estimates are not sensitive at all to radiometric calibration uncertainties or to the emissivity of the plume material. In this method, we use optical sensors with a very high spatial resolution in the range of 30–10 m (e.g. Landsat 8, Sentinel 2) and 0.5 m (e.g. Pléiades). This method only works with non-transparent volcanic plumes. It restitutes high-resolution maps of volcanic cloud top height and horizontal velocities, as described by de Michele *et al.* (2016).

The general concept for extracting a plume elevation model is that the panchromatic (PAN) and the multispectral (MS) sensors onboard a pushbroom satellite platform cannot occupy the same position in the focal plane of the instrument (Figure 12.1). Therefore, there is a physical separation between the two CCDs. This geometry is key and relies on the concept of the sensor itself. We remind the reader that this sensor was not built on purpose to extract PEMs; we actually exploit a default of the sensor geometry to prove a concept. This CCD separation yields a baseline and a time lag between the PAN and MS image acquisitions. On the one hand, the small baseline has already been successfully exploited for retrieving digital elevation models (DEMs) of still surfaces such as topography or building heights (e.g. Massonnet *et al.* 1997; Vadon 2003; Mai and Latry 2009). On the other hand, the time lag has been successfully exploited to measure the velocity field of moving surfaces, such as ocean waves and arctic river discharges (de Michele *et al.* 2012; Kääb and Leprince 2014).

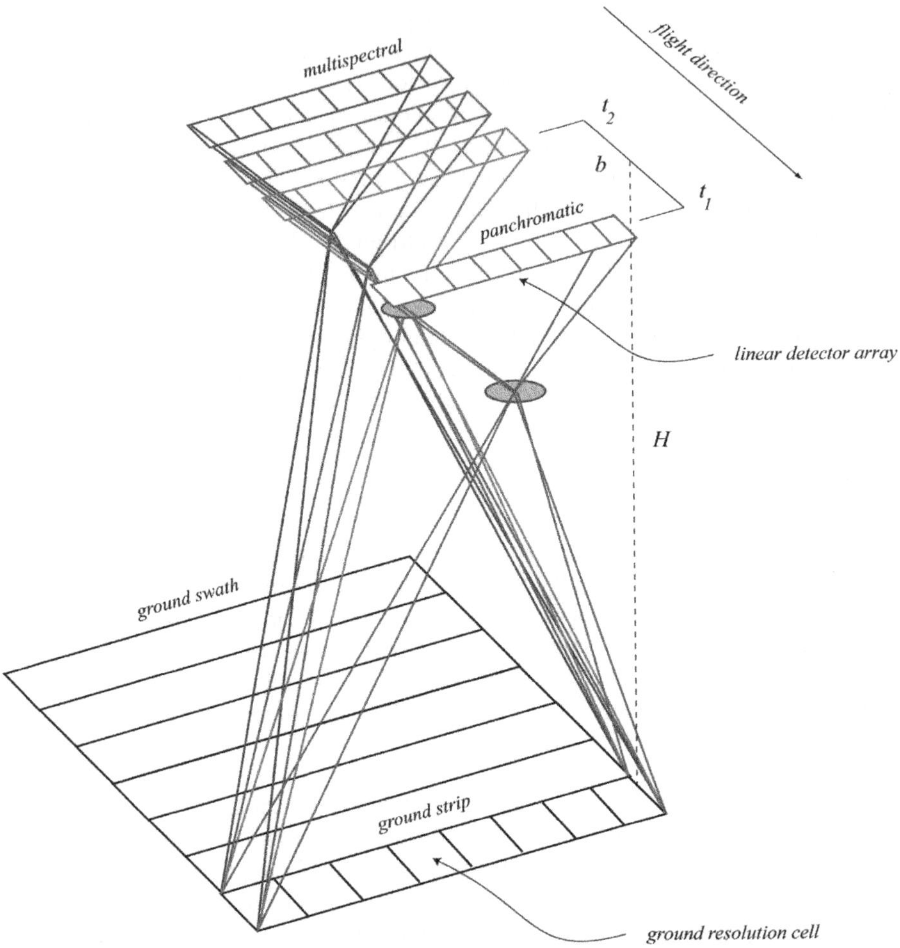

Figure 12.1. *Simplified geometry of the focal plane of a pushbroom sensor. This illustration stretches out the separation between the panchromatic CCD and the multispectral CCDs. This separation yields a time lag (t_1 and t_2) as well as a baseline (distance b). These two values can be used to measure the heights and velocities of features in the scenes. Illustration by M. de Michele, ©BRGM. For a color version of this figure, see www.iste.co.uk/cavalie/images.zip*

We consider two main directions: the epipolar (EP) direction, i.e. the azimuth direction of the satellite (i.e. the flight direction), and the perpendicular to the EP (P2E) direction. The pixel offset between PAN and MS sensors in the EP direction is proportional to the height of the plume plus the pixel offset contribution induced by the motion of the plume itself in between the two acquisitions. The pixel offset in the P2E direction, also controlled by the time lag, is proportional to the plume motion only, as there is no parallax in the P2E direction by definition. In principle, the offset in the P2E direction is proportional to the movements of every feature in the imaged scene (e.g. meteorological clouds, lahars, river flow, ocean waves, vehicles). In our case study, we are only interested in the volcanic cloud motion. We use this latter information to measure the volcanic cloud velocity and to compensate for the apparent parallax recorded in the EP offset. Sometimes things become complicated because the focal plane of the instrument is not made of long CCDs, as in the Centre National d'Etudes Spatiales (CNES) SPOT family, but of small focal plane modules (FPMs), such as the National Aeronautics and Space Administration/United States Geological Survey (NASA/USGS) Landsat 8 or the European Union/ESA (European Space Agency) Copernicus program Sentinel 2 satellite. For instance, the Landsat 8 operational land imager (OLI) instrument is a pushbroom (linear array) imaging system that collects visible, near infra-red and short-wave infra-red spectral band imagery at 30 m multispectral and 15 m panchromatic ground sample distances. It collects 190 km wide image swaths from 705 km orbital altitude. The OLI detectors are distributed across 14 separate FPMs, each of which covers a portion of the 15° OLI cross-track field of view. The internal layout of all the 14 FPMs is the same for all the FPMs, with alternate FPMs being rotated by 180° to keep the active detector areas as close together as possible. This feature has the effect of inverting the along-track order of the spectral bands in adjacent FPMs. Consequently, this has the effect of inverting the signs of the cross-correlation measurements when calculating pixel offsets between PAN and MS bands, which is related to the volcanic cloud velocity and parallax. The Copernicus Sentinel 2 focal plane hosts a similar geometry, called a "staggered" geometry. A correlator is used to perform pixel (or sub-pixel) offset measurements. If we consider the OLI image as a matrix made of lines and columns, offsets among the lines are the EP offsets O_e, while offsets among columns are the P2E offsets (O_{p2e}). In the offset map, there may exist a ramp resulting from band misregistration; the ramp, if found, is removed. The direction of the plume (θ) is measured with respect to the azimuth direction, with the convention depicted in Figure 12.2. The absolute value of the pixel offsets due to the VCTH is calculated as O_h. We have to take into account that O_e has the same sign throughout the image. However, the wind direction does not. If we make the assumption that the plume direction is driven by the wind, there are cases where $|O_{p2e}||tan\theta|$ has to be subtracted and there are cases where it has to be added to O_e, depending on whether the wind-generated offset is the same sign with respect to Oe or not. In the first case, the wind generates an apparent parallax that adds up to the one generated by the plume height; therefore, $|O_p2e||tan\theta|$ needs to be subtracted. In the latter case, the wind generates an apparent parallax that subtracts from the one generated by the plume height; therefore, $|O_p2e||tan\theta|$ needs to be added

(Figure 12.2). To sum up, generally, O_h from staggered sensors should be calculated as follows:

$$|O_h| = \begin{cases} |O_e| - |O_{p2e}||tan\theta| & \text{for } 0^\circ \leq \theta \leq 180^\circ \\ |O_e| + |O_{p2e}||tan\theta| & \text{for } 180^\circ < \theta \leq 360^\circ \end{cases} \qquad [12.1]$$

For each pixel, O_h is then converted into a VCTH using the formula:

$$h = |O_h|\frac{sH}{Vt}, \qquad [12.2]$$

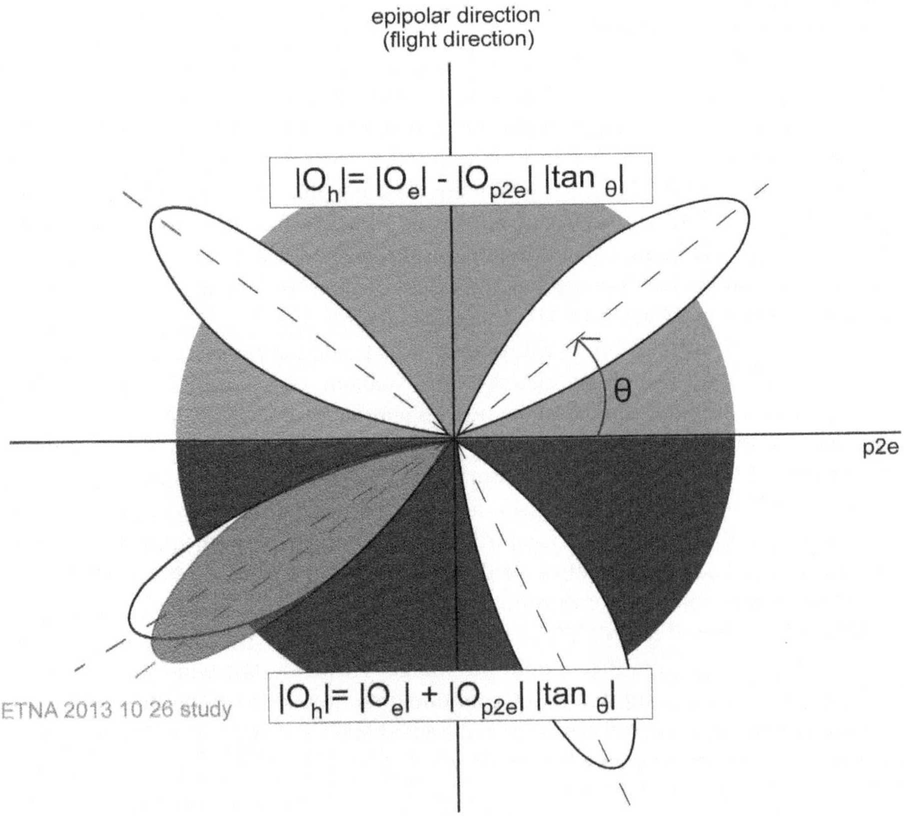

Figure 12.2. *Conventions for RAW images when the spacecraft is flying positive along the Y axis: a simplified illustration of plume directions, θ, the epipolar direction, the P2E direction and the formulas to use if the plume is in the top or bottom quadrants. The plume is represented by the white lobe, with an example from Etna, October 26, 2016 (blue lobe). For a color version of this figure, see www.iste.co.uk/cavalie/images.zip*

The result is a digital elevation model of the volcanic ash cloud. Here, h is the plume height (m), s is the pixel size (m), V is the platform velocity (m/s), t is the temporal lag between the two image bands (s) and H is the platform height (m). Peculiar cases are: $\theta = 15°$ and $\theta = 180°$. In these cases, the system is no longer sensitive to the plume velocity. Therefore,

$$|O_h| = |O_e| \qquad [12.3]$$

If $\theta = 90°$ and $\theta = 270°$, the methodology is no longer applicable as the plume velocity yields an apparent parallax that is non-separable from the height-related parallax. In addition to the height of the ash cloud, this method provides a spatially dense measure of the ash cloud velocity at the PEM pixel resolution:

$$v_p = \frac{sO_{p2e}}{t\cos\theta} \qquad [12.4]$$

12.2.3. *Applications*

The extraction of digital elevation models of volcanic ash plumes, and their velocities, from pushbroom sensors is based on a rather new methodology. Therefore, examples are limited to only a few published studies (de Michele *et al.* 2016, 2019). In the following sections, we present some examples found in the literature using sensors from the Landsat 8 mission and the Pléiades mission.

12.2.3.1. *Holuhraun (Landsat 8)*

The 2014–2015 Holuhraun eruption in the Bárðarbunga volcanic system was the largest fissure eruption in Iceland since the 1783 Laki eruption. It started at the end of August 2014 and lasted six months, until late February 2015. It was characterized by large degassing processes and emission of SO_2 into the atmosphere. The eruption steam and gas column were nicely captured by Landsat 8 on September 6, 2014 at 12:25 UTC (Figure 12.3). The nominal PAN/red time lag for Landsat 8 is 0.52 seconds, while the nominal angular separation is 0.3 degrees. The platform flies at a nominal altitude of 705 km and a nominal speed of 7.5 km/s. Therefore, the base-to-height ratio (b/h) is 0.0055. Alternatively, we can use the green channel instead of the red (the time lag of which is 0.65 s), or both. We chose the Holuhraun eruption as it represents a challenging test case for us, as its plume was rapidly moving and reached low altitudes. Therefore, if our method works on the Holuhraun test case, then it should apply to other types of volcanic plumes (higher and slower).

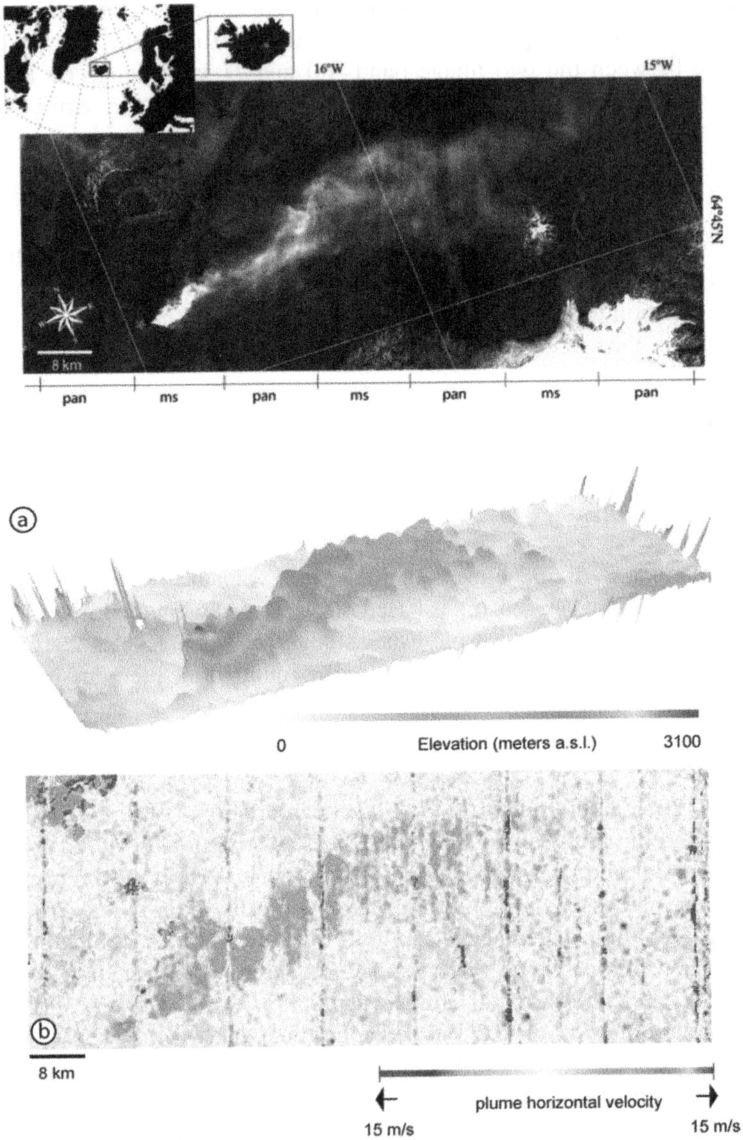

Figure 12.3. *Top: the Holuhraun eruption seen in the reconstructed image from Landsat, which is made up of alternating panchromatic (pan) and multispectral (ms) bands. (a) The digital elevation model of the volcanic cloud, visualized in 3D, was obtained by applying the proposed method. (b) A map of the resulting horizontal velocities associated with the volcanic cloud. For a color version of this figure, see www.iste.co.uk/cavalie/images.zip*

Figure 12.4. *Top left: Ubinas volcano eruption as seen from the Pléiades satellite. Top right: the horizontal velocity of the volcanic cloud measured from the cross-correlation of Pléiades images. Bottom: a 3D visualization of the volcanic plume elevation model extracted from Pléiades images. Courtesy of BRGM; Pléiades data provided by CNES; processing done in the framework of the French CIEST2 initiative. For a color version of this figure, see www.iste.co.uk/cavalie/images.zip*

12.2.3.2. *Ubinas 2019 (Pléiades)*

Ubinas is Peru's most active volcano. It is the northernmost of three young volcanoes located along a regional structural lineament east of the main volcanic front of Peru. It has had 10 eruptions since 1900, with the last being reported in November 2016. On July 19, 2019, the Ubinas volcano started to erupt (Figure 12.4). An ash plume developed and the ashes fell on several departments of southern Peru. The International Charter "Space and Major Disasters" was activated following Peru declaring a state of emergency for the areas surrounding the Ubinas volcano. Via the International Charter "Space and Major Disasters", the CNES programmed several acquisitions from the Pléiades satellite, the very high resolution panchromatic and multispectral imager. A clear image was acquired on July 19th, with an ash plume clearly visible in the dataset. We used the lag between the panchromatic and the multispectral CCDs in the Pléiades data to extract a digital elevation model of the volcanic ash plume and a map of the plume horizontal velocity. This information, along with ash plume roughness, is crucial for atmospheric physicists to update ash dispersion models. In this exercise, we proved the concept and tested the potential of Pléiades data for ash plume height monitoring.

12.2.3.3. *Etna (SPOT-1)*

The methodology can be applied to past volcanic eruptions, which opens the way to the investigation of ash cloud injection heights in the past. Mount Etna is one of the most active volcanoes in Europe. The years 2001 and 2002 were marked by intense episodes of flank eruptions. These eruptions generated a large ash clouds. The SPOT-2 satellite captured the eruption on July 27, 2001, which was used in this exercise to retrieve a digital elevation model of this particular ash cloud (Figure 12.5).

Figure 12.5. *Left: Etna flank eruption as seen by SPOT-2 on July 21, 2001. Right: the volcanic cloud digital elevation model extracted from SPOT-2. For a color version of this figure, see www.iste.co.uk/cavalie/images.zip*

12.3. Shallow bathymetry: measuring wave characteristics from space

12.3.1. *Introduction: can we map tectonic fault motion under shallow water and can we measure differential bathymetry?*

As 75% of the Earth's crust lies under water, it is thus unobservable with the electromagnetic energy used by visible remote sensing techniques. However, since coastal bathymetry controls swell celerities and swell wavelengths, if it is possible to measure these parameters from quasi-synchronous optical satellite images, we can potentially retrieve coastal bathymetry from space indirectly. Differential bathymetry consists of subtracting bathymetries acquired at different dates in order to identify changes. Therefore, potentially, the answer to this challenge is "yes". After the discovery of a measurable baseline and a time lag between the panchromatic and the multispectral bands of virtually every optical (multispectral) spaceborne pushbroom system, several scientific teams (e.g. Poupardin *et al.* 2016; Bergsma *et al.* 2019; Almar *et al.* 2019) have exploited this lag to measure swell velocity from space. This has led to a potential new technique for retrieving bathymetry from space, at up to 100 meters depth. Many applications can be foreseen; the one that is most related to the topic of this book is shallow seafloor geodesy (or differential bathymetry) applied to seismotectonics at shallow depth. Our vision is that one day we will be able to map bathymetric changes from space with a great deal of precision, which is of major importance as many geological phenomena occur under water.

12.3.2. *Principles*

In section 12.1, we saw how certain pushbroom sensors present a peculiar focal plane geometry. This peculiarity results from the fact that the panchromatic band and the multispectral bands cannot sit in the same position in the focal plane. For technical reasons, they are adjacent to one another. This translates into image offsets due to a tiny time lag between the band acquisitions in a single dataset or satellite pass. For instance, there are 2.05 seconds between the acquisition of the panchromatic band and the acquisition of the green band in a single SPOT-5 dataset, for one single acquisition date. This, combined with the very high resolution of SPOT-5 (up to 2.5 meters), means that we are potentially able to measure the kinematics of a moving "object" at the surface, at a high spatial resolution. The "object" could be a natural phenomenon, such as ocean waves. It could be a human-made machine, such as a car, a plane or a ship. The attentive reader will notice that if the object is high on the ground, like a plane, there will be an ambiguity in resolving its speed since the offset in the position of the object in the panchromatic image and the multispectral image will also be due to the parallax induced by the elevation of the object itself. This is exactly what we saw in the above section of this chapter, but the problem between speed and elevation is reversed. Fortunately, here we focus on the speed of the ocean or sea waves, the height of which we can neglect with respect to their speed and with respect to the satellite altitude. To be precise, it is the phase velocity not the group velocity that we want to measure here. It is called the celerity. Returning to the example of SPOT-5, during the 2.05 seconds between the panchromatic and the multispectral image acquisition, the ocean or sea waves move. Therefore, using cross-correlation in a Fourier domain, for instance, we can calculate pixel offsets between the panchromatic and multispectral images, proportional to wave celerities and directions. As a result, we obtain a map of wave celerities and directions at a high spatial resolution (Figure 12.6).

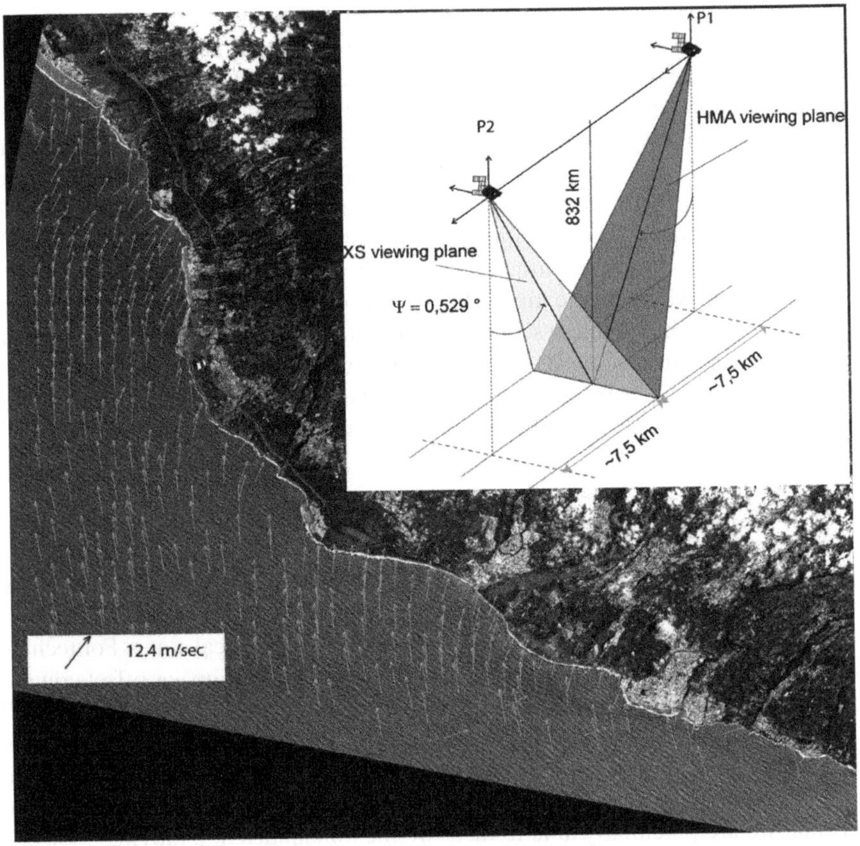

Figure 12.6. *Celerities and directions of ocean waves from SPOT-5 at La Réunion Island, France. Directions here are the combinations of the ocean swell directions and the wind-generated wave directions. Inset: illustration of SPOT-5 acquisition geometry. P1 and P2 represent the satellite position at time 1 and time 1 + 2.05 seconds. XMA is the panchromatic image and XS is the multispectral image (redrawn by de Michele (2020), following the SPOT-5 user guide). For a color version of this figure, see www.iste.co.uk/cavalie/images.zip*

Why it is important to map the celerity of ocean waves at high resolution? For a number of reasons. Here, we are particularly interested in a law that relates ocean wave kinematics to water depth. In other words, according to the linear wave theory, when water depth is less than about half the dominant wavelength (λ), wave celerity (c) and λ decrease towards the coast due to the decrease in water depth. It is then possible, for free surface waves, to assess the water depth by knowing the dominant wavelengths and their associated celerities using the linear dispersion relation [12.5]:

$$h = \frac{\lambda}{2\pi} \tanh^{-1}\left(\frac{2\pi c^2}{g\lambda}\right) \qquad [12.5]$$

where h is the water depth, c is the celerity, λ is the wavelength and g is the universal gravitational constant. Equation [12.5] is referred to as the dispersion relation. This method was first used in a practice during World War II (Williams 1947).

We provide below some history on this particular satellite-based method. The aim of what follows is to offer an introductory overview of existing studies that have used this same approach (i.e. the time lag due to the satellite sensor geometry). We note that we cannot pretend to be perfect in this bibliographic exercise. Therefore, we do apologize to the authors who might feel excluded.

The use of wave propagation to retrieve shallow bathymetry works either with two sequential images acquired with a short time interval, shorter than a wave period, or with one image, which requires *in situ* measurement for estimating the wave celerity, as explained by Danilo and Melgani (2016). The retrieval of wave celerity from the satellite is thus a key element and is still challenging today. Access to at least two consecutive images with a few seconds of time delay is crucial to extract bathymetry. As we discussed earlier, this peculiar image acquisition geometry exists onboard specific satellites with pushbroom sensors. With these sensors, the retrieval of c relies on the non-simultaneous stereo acquisition capability of platforms such as Ikonos. To our knowledge, Abileah (2006) was the first to exploit this capability to retrieve coastal bathymetry from Ikonos. Ikonos presents an 11 second time lag between the PAN and the MS sensors, so there may be ambiguity when tracking a wave, depending on its speed. However, the retrieval of c can rely on the fact that two multispectral bands cannot coexist on the focal plane of the instrument, such as in the SPOT family, Pléiades, Sentinel 2 and Landsat 8, yielding *de facto* a small time lag, sufficient to measure c directly (de Michele *et al.* 2012). With the joint measurement of c and λ, h becomes accessible from the satellite, without the need for ground calibration. To provide some history of previously published studies, Abileah (2006) suggested a method to measure c from optical satellite data using Ikonos stereo pairs. De Michele *et al.* (2012) improved the method by using the small time lag between two bands of the SPOT-5 dataset. Danilo and Binet (2013) have detailed the concept and limiting factors. On these bases, Poupardin *et al.* (2016) proposed local spectral decomposition before the celerity estimates in order to produce local $c-\lambda$ pairs to derive water depth based on the dispersion relation. Bergsma *et al.* (2019, 2021) proposed a method based on multiple Sentinel 2 bands and the Fourier slice theorem. Almar *et al.* (2019) proposed a method to extract $c-\lambda$-based bathymetry from Pléiades "persistent" mode. Most of the aforementioned studies make use of only one $c-\lambda$ couple – the most energetic – to extract bathymetry for a resolution cell, while Poupardin *et al.* (2016) and de Michele *et al.* (2021) used multiple c and λ pairs to improve water depth retrieval.

Practically, we consider a sub-section of two bands, normalized by the mean of each sub-section amplitude, resulting in two matrices [A] and [B]. We call this a "window" or resolution cell. With i and j, we can identify the column/line indices in the window. Over a given window, we compute the two-dimensional discrete fast Fourier transform (DFFT) per detector band (matrices $[F_A]$ and $[F_B]$). ν and μ are the column/line indices in the Fourier domain. Then, we compute the matrix [R] whose coefficients are given by:

$$R_{\nu,\mu} = F_{A_{\nu,\mu}} \times F^*_{B_{\nu,\mu}} = Am_{\nu,\mu}e^{i\theta(\nu,\mu)} \qquad [12.6]$$

where * represents the complex conjugate and Am is the amplitude of the spectrum. On the module of $R_{\nu,\mu}$, we identify multiple peaks that can be considered to be representative of the most significant waves. Then, we can derive their associated wavelengths $\lambda_{\nu,\mu}$ and the phases $\phi_{\nu,\mu}$. [A] and [B], being separated in time by 1.05 s, are assumed to have the same spatial frequency content. For this reason, the positions of peaks in |[R]|, $[|F_A|]$ and $[|F_B|]$ coincide. Therefore, we can identify significant wavelengths in [R]. Finally, we determine c by measuring the pixel offset in the wave motion direction ($\delta r_{\nu,\mu}$) between the two bands, which is directly related to the phase of $R_{\nu,\mu}$ by

$$\delta r_{\nu,\mu} = \frac{\lambda_{\nu,\mu}\theta_{\nu,\mu}}{2\pi} \qquad [12.7]$$

Then, we obtain c by dividing $\delta r_{\nu,\mu}$ by $\delta t_{k,l}$, where $\delta t_{k,l}$ is the time span between band k and band l. When the pixel offset is not wavelength-dependent (i.e. $r_{\nu,\mu} = r, \forall(\nu, \mu)$), this formula (equation [12.7]) is equivalent to the phase correlation algorithm used for earthquake-induced ground-offset measurements (e.g. Puymbroeck *et al.* 2000). Now, knowing δt, we aim for $((\lambda_{\nu,\mu}), (\theta_{\nu,\mu}))$ pairs that can be converted to $((\lambda_{\nu,\mu}), (\theta_{\nu,\mu}))$. We use the spectral amplitude of R (*Am* in [12.6]) as a criterion for selecting the most significant waves. This particular method was developed by de Michele *et al.* (2021) under the framework of the ESA/BRGM-funded BathySent project (2017–2020).

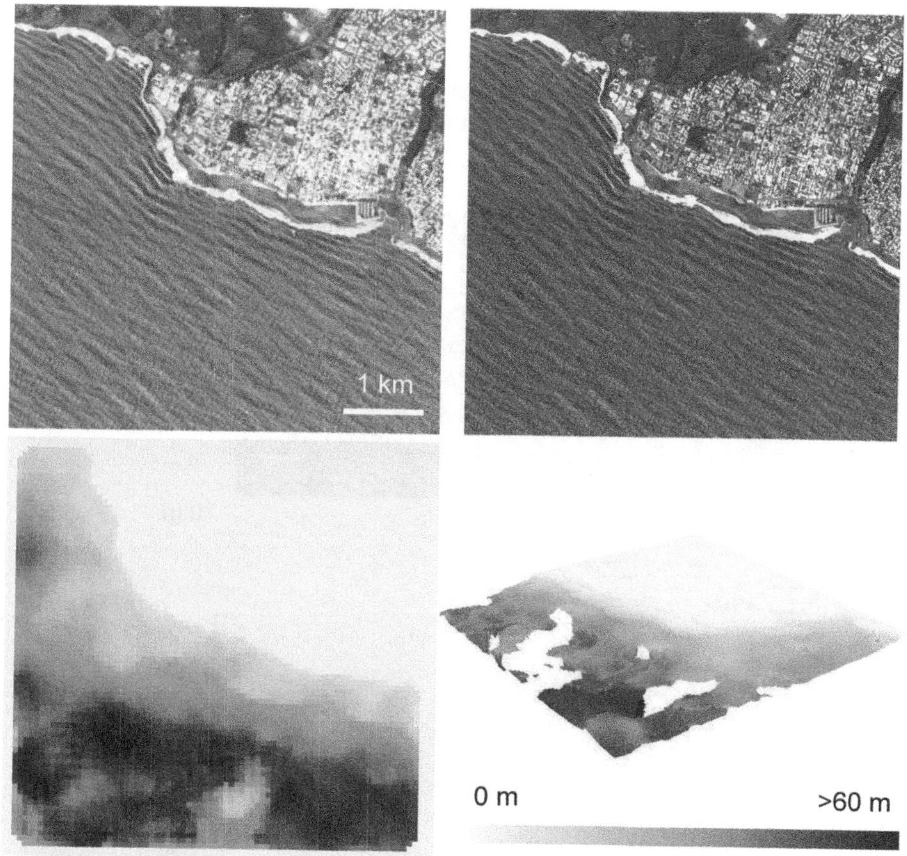

Figure 12.7. *Coastal bathymetry from quasi-synchronous SPOT-5 images and linear wave theory. Top: a subset of two SPOT-5 images belonging to the same dataset at 2.05 seconds apart, multispectral on the left and panchromatic on the right. Bottom: the bathymetry retrieved with the BathySent method with SPOT-5 data and a 3D visualization. ©BRGM 2020. For a color version of this figure, see www.iste.co.uk/cavalie/images.zip*

12.3.3. *Applications*

In this chapter, we provide two examples. The first case is based on the CNES SPOT-5 images of La Réunion Island. During a flight, SPOT-5 acquires panchromatic and multispectral images separated by 2.05 seconds, as shown in Figure 12.7. Note how the ocean waves are easily observable in both the images. Also, note how their wavelength decreases, as one approaches the coast. Their celerities also decrease; we could see this here if we could switch back and forth from one image to the other.

In the second case (Figure 12.8), we use the EU Copernicus Sentinel 2 to calculate bathymetry. Sentinel 2 focal plane modules are arranged in a staggered geometry. They present useful time lags between multispectral bands for the aims of bathymetry retrieval. Below we show the application of the BathySent method to Sentinel 2 images for the Gulf of Lion (France).

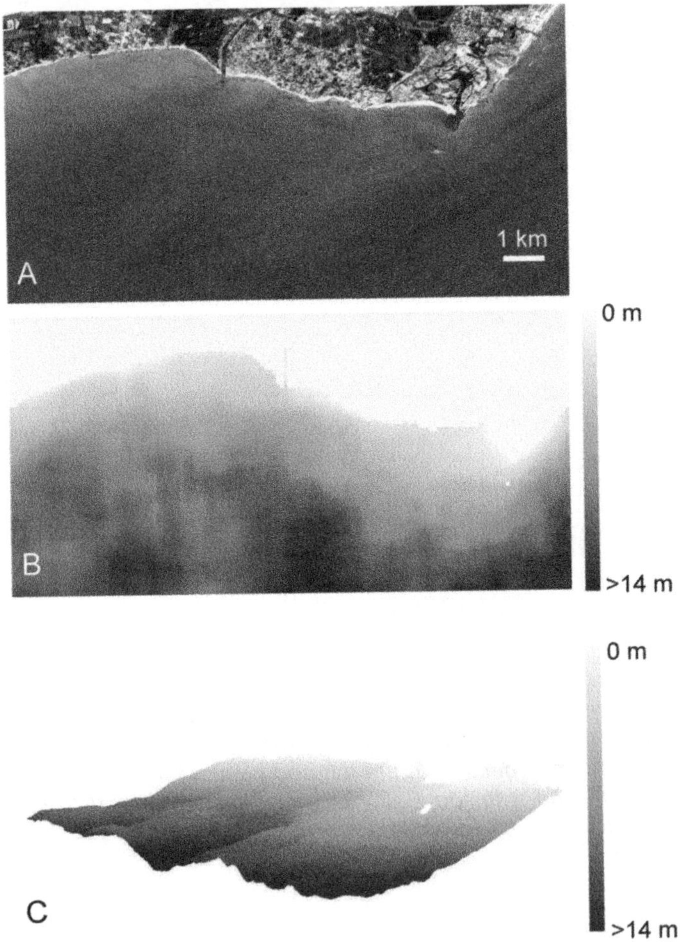

Figure 12.8. *Coastal bathymetry derived from Sentinel 2 and wave theory. Computation undertaken at BRGM (2019) under the ESA/BRGM-funded BathySent project. A) Sentinel 2 data acquired for the Gulf of Lion, France. B) Coastal bathymetry retrieved between 0 and 14 meters depth. C) 3D visualization. Figure contains modified Copernicus Sentinel data. ©BRGM 2020. For a color version of this figure, see www.iste.co.uk/cavalie/images.zip*

12.4. Concluding remarks

As occurred in the case of SAR and InSAR, sometimes a sensor can be used to extract information in a different manner from the main scope of the mission the sensor was built for. In the field of scientific research, most often, a measurement error (or a bias in a sensor) for one scientist is an invaluable source of information for another scientist. In this chapter, we showed how it is possible to exploit a peculiar characteristic of the focal plane of a high-resolution pushbroom instrument to extract crucial geophysical parameters via conventional cross-correlation techniques based on FFTs. Here, these parameters provide a detailed map of elevation and velocity of a volcanic ash cloud and coastal bathymetry. Frequent image acquisitions, enabled by the joint use of multiple platforms, help to multiply the number of measurements. On the one hand, frequent acquisitions are key to performing multi-temporal analysis of ash plume evolution and ground deformation monitoring under water. On the other hand, they are crucial to improving the precision of measurements through data stacking. Future spaceborne missions will assure the continuity with actual missions and encourage scientists to invest energy in creative research and method development in this field. On the one hand, new high-resolution sensors, such as the Co3D and Pléiades NEO from the CNES, among others, will help in the routine monitoring of the Earth's surface. On the other hand, the EU Copernicus program, with the Sentinels, will ensure long-term mission continuity, thus providing a solid framework for research and development.

12.5. Acknowledgments

We would like to thank BRGM for supporting these research initiatives. We thank Deborah Idier, Adrien Poupardin, Michael Foumelis (BRGM, France), Þórður Arason (IMO, Iceland), Claudia Spinetti, Stefano Corradini, Luca Merucci (INGV, Italy) and Vivi Drakapoulou (ECMR, Greece) for the scientific contributions. We are thankful to NASA/USGS, Catherine Proy, Claire Tinel (CNES), Espen Volden (ESA) and the Copernicus program (EU) for releasing the satellite data. We thank Céline Danilo and Jean-Marc Delvit for their careful reviews and comments that helped to improve this chapter.

12.6. References

Abileah, R. (2006). Mapping shallow water depth from satellite. *Proceedings of the ASPRS Annual Conference*, 1–7.

Almar, R., Bergsma, E.W., Maisongrande, P., de Almeida, L.P.M. (2019). Wave-derived coastal bathymetry from satellite video imagery: A showcase with Pleiades persistent mode. *Remote Sensing of Environment*, 231(111263).

Bergsma, E.W., Almar, R., Maisongrande, P. (2019). Radon-augmented Sentinel 2 satellite imagery to derive wave-patterns and regional bathymetry. *Remote Sensing*, 11(1918).

Bergsma, E.W., Almar, R., Rolland, A., Binet, R., Brodie, K.L., Bak, A.S. (2021). Coastal morphology from space: A showcase of monitoring the topography–bathymetry continuum. *Remote Sensing of Environment*, 261, 112469.

Corradini, S., Merucci, L., Prata, A.J. (2009). Retrieval of SO2 from thermal infrared satellite measurements: Correction procedures for the effects of volcanic ash. *Atmospheric Measurement Techniques*, 2, 177–191.

Danilo, C. and Binet, R. (2013). Bathymetry estimation from wave motion with optical imagery: Influence of acquisition parameters. *MTS/IEEE OCEANS Bergen*, 10–14 June.

Danilo, C. and Melgani, F. (2016). Wave period and coastal bathymetry using wave propagation on optical images. *IEEE Transactions on Geoscience and Remote Sensing*, 54(11), 1–13.

Flower, V.J. and Kahn, R.A. (2017). Assessing the altitude and dispersion of volcanic plumes using MISR multi-angle imaging from space: Sixteen years of volcanic activity in the Kamchatka Peninsula, Russia. *Journal of Volcanology and Geothermal Research*, 337, 1–15.

Kääb, A. and Leprince, S. (2014). Motion detection using near-simultaneous satellite acquisitions. *Remote Sensing of Environment*, 1154, 164–179.

Mai, S. and Latry, C. (2009). Digital elevation model computation with SPOT5 panchromatic and multispectral images using low stereoscopic angle and geometric model refinement. *IEEE International Geoscience and Remote Sensing Symposium*, 12–17 July, Cape Town, South Africa.

Massonnet, D., Giros, A., Breton, E. (1997). Forming digital elevation models from single pass spot data: Results on a test site in the Indian ocean. *IEEE International Geoscience and Remote Sensing Symposium*, 3–8 August, Singapore.

de Michele, M., Leprince, S., Thiébot, J., Raucoules, D., Binet, R. (2012). Direct measurement of ocean waves velocity field from a single SPOT-5 dataset. *Remote Sensing of Environment*, 119, 266–271.

de Michele, M., Raucoules, D., Arason, P. (2016). Volcanic plume elevation model and its velocity derived from Landsat 8. *Remote Sensing of Environment*, 176, 219–224.

de Michele, M., Raucoules, D., Corradini, S., Merucci, L., Salerno, G., Sellitto, P., Carboni, E. (2019). Volcanic cloud top height estimation using the plume elevation model procedure applied to orthorectified Landsat 8 data. Test case: 26 October 2013 mt. Etna eruption. *Remote Sensing*, 11(785).

de Michele, M., Raucoules, D., Idier, D., Smai, F., Foumelis, M. (2021). Shallow bathymetry from Sentinel 2, by joint estimation of waves celerities and wavelengths. *Remote Sensing*, 13(2149).

Nelson, D.L., Garay, M.J., Kahn, R.A., Dunst, B.A. (2013). Stereoscopic height and wind retrievals for aerosol plumes with the MISR interactive eXplorer (MINX). *Remote Sensing*, 5, 4593–4628.

Pailot, S., Bonnétat, S., Harris, A., Calvari, S., de Michele, M., Gurioli, L. (2020). Plume height time-series retrieval using shadow in single spatial resolution satellite images. *Remote Sensing*, 12(3951).

Poupardin, A., Idier, D., de Michele, M., Raucoules, D. (2016). Water depth inversion from a single SPOT-5 dataset. *IEEE Transactions on Geoscience and Remote Sensing*, 54, 2329–2342.

Prata, A.J. and Grant, I.F. (2001). Retrieval of microphysical and morphological properties of volcanic ash plumes from satellite data: Application to Mt Ruapehu, New Zealand. *Quarterly Journal of the Royal Meteorological Society*, 127, 2153–2179.

Prata, A.J. and Turner, P.J. (1997). Cloud top height determination from the ATSR. *Remote Sensing of Environment*, 59(1), 1–13.

Puymbroeck, N.V., Michel, R., Binet, R., Avouac, J.-P., Taboury, J. (2000). Measuring earthquakes from optical satellite images. *Applied Optics*, 39, 3486–3494.

Scollo, S., Folch, A., Coltelli, M., Realmuto, V.J. (2010). Three-dimensional volcanic aerosol dispersal: A comparison between multiangle imaging spectroradiometer (MISR) data and numerical simulations. *Journal of Geophysical Research*, 115(D24210).

Spinetti, C., Barsotti, S., Neri, A., Buongiorno, M., Doumaz, F., Nannipieri, L. (2013). Investigation of the complex dynamics and structure of the 2010 Eyjafjallajökull volcanic ash cloud using multispectral images and numerical simulations. *Journal of Geophysical Research*, 118(10), 4729–4747.

Vadon, H. (2003). 3D navigation over merged panchromatic-multispectral high resolution SPOT5 images. *Proceedings of the 2003 ISPRS Symposium.*

Williams, W.W. (1947). The determination of gradients on enemy-held beaches. *The Geographical Journal*, 109(1–3), 76–90.

Wilson, L., Sparks, R.S., Huang, T.C., Watkins, N.D. (1978). The control of volcanic column heights by eruption energetics and dynamics. *Journal of Geophysical Research*, 83(B4), 1829–1836.

List of Authors

Fabien ALBINO
ISTerre
Grenoble
France

Bas ALTENA
IMAU
Utrecht University
The Netherlands

Grace BATO
NASA Jet Propulsion Laboratory
California Institute of Technology
Pasadena
USA

Olivier CAVALIÉ
CEREGE
Aix-Marseille University
Aix-en-Provence
France

Johann CHAMPENOIS
CEA, DAM, DIF
Arpajon
France

Aya CHEAIB
ISTerre
Grenoble
France

Francesco DE ZAN
Remote Sensing Technology Institute
German Aerospace Center (DLR)
Oberpfaffenhofen
Germany

Amaury DEHECQ
Laboratory of Hydraulics, Hydrology
and Glaciology (VAW)
ETH Zurich
Switzerland
and
IGE
CNRS, IRD, INP
Grenoble Alpes University
France

Benedetta DINI
ISTerre
Grenoble
France

Marie-Pierre DOIN
ISTerre
Grenoble
France

Philippe DURAND
CNES
Toulouse
France

Bénédicte FRUNEAU
LaSTIG
Gustave Eiffel University
Champs-sur-Marne
France

Alex S. GARDNER
NASA Jet Propulsion Laboratory
California Institute of Technology
Pasadena
USA

Giorgio GOMBA
German Aerospace Center (DLR)
Oberpfaffenhofen
Germany

Ramon HANSSEN
Delft University of Technology
The Netherlands

Dinh HO TONG MINH
UMR-TETIS
INRAE
France

James HOLLINGSWORTH
ISTerre
Grenoble
France

Pascal LACROIX
ISTerre
Grenoble
France

Silvan LEINSS
Institute of Environmental Engineering
ETH Zurich
Switzerland
and
LISTIC
Savoie Mont Blanc University
Annecy
France

Shiyi LI
Institute of Environmental Engineering
ETH Zurich
Switzerland

Paul LUNDGREN
NASA Jet Propulsion Laboratory
California Institute of Technology
Pasadena
USA

Mathilde MARCHANDON
ISTerre
Grenoble
France

Stephane MAY
CNES
Toulouse
France

Marcello DE MICHELE
BRGM
Orléans
France

Erwan PATHIER
ISTerre
Grenoble
France

Marc PIERROT DESEILLIGNY
LaSTIG
ENSG-IGN
Gustave Eiffel University
Saint-Mande
France

Virginie PINEL
ISTerre
Grenoble
France

Béatrice PINEL-PUYSSÉGUR
CEA, DAM, DIF
Arpajon
France

Nadine POURTHIÉ
CNES
Toulouse
France

Daniel RAUCOULES
BRGM
Orléans
France

Ewelina RUPNIK
LaSTIG
ENSG-IGN
Gustave Eiffel University
Saint-Mande
France

Céline TISON
CNES
Toulouse
France

Emmanuel TROUVÉ
LISTIC
Savoie Mont Blanc University
Annecy
France

Tim J. WRIGHT
COMET
University of Leeds
UK

Yajing YAN
LISTIC
Savoie Mont Blanc University
Annecy
France

Index

T, V, Y, Z

Printed and bound by CPI Group (UK) Ltd, Croydon, CR0 4YY
16/09/2026
14963172-0003